Springer Series on Naval Architecture, Marine Engineering, Shipbuilding and Shipping

Volume 5

Series editor

Nikolas I. Xiros, University of New Orleans, New Orleans, LA, USA

The Naval Architecture, Marine Engineering, Shipbuilding and Shipping (NAMESS) series publishes state-of-art research and applications in the fields of design, construction, maintenance and operation of marine vessels and structures. The series publishes monographs, edited books, as well as selected Ph.D. theses and conference proceedings focusing on all theoretical and technical aspects of naval architecture (including naval hydrodynamics, ship design, shipbuilding, shipyards, traditional and non-motorized vessels), marine engineering (including ship propulsion, electric power shipboard, ancillary machinery, marine engines and gas turbines, control systems, unmanned surface and underwater marine vehicles) and shipping (including transport logistics, route-planning as well as legislative and economical aspects).

More information about this series at http://www.springer.com/series/10523

Víctor Hugo Chacón

The Due Diligence in Maritime Transportation in the Technological Era

Springer

Víctor Hugo Chacón
Panama
Panama

ISSN 2194-8445 ISSN 2194-8453 (electronic)
Springer Series on Naval Architecture, Marine Engineering, Shipbuilding and Shipping
ISBN 978-3-319-66001-1 ISBN 978-3-319-66002-8 (eBook)
DOI 10.1007/978-3-319-66002-8

Library of Congress Control Number: 2017949489

Printed on acid-free paper

This Springer imprint is published by Springer Nature
The registered company is Springer International Publishing AG
The registered company address is: Gewerbestrasse 11, 6330 Cham, Switzerland

Preface

The idea of developing a research on this topic arose from my practice in maritime litigation before the Maritime Courts of Panama. Most of the cases related to cargo claims I handled were subject to the application of the Hague/Hague-Visby rules. In some of them, the loss or damages could have been prevented by the application of new technologies. Under the duties and liability standard set in the rules, questions often arise as to what new technologies should be required as part of the obligation to exercise due diligence in making the ship seaworthy and to care cargo? Or, when can carriers be held liable for loss of or damage to cargo that could have been prevented by the application of new technological instruments, methods, or devices? In the absence of a response, there is abundant litigation in the courts around the world. The idea to approximate an answer to these questions became then a thesis proposal, which was accepted as a doctoral dissertation by the University of Hamburg, Germany.

I must express my highest respect and gratitude to my supervisor Prof. Dr. Rüdiger Wolfrum. Without his admission and guidance, as his doctoral student, this work would not have been possible. I conducted this project as an associated researcher of the International Max Planck Research School for Maritime Affairs at the University of Hamburg. That allowed me to make the research at the library of the Max Planck Institute for Comparative and International Private Law in Hamburg. To the board of directors and the staff of the School as well as of the Institute, I express my deepest gratitude. This project was possible thanks to a scholarship granted by the National Secretary for Science and Technology (*Secretaría Nacional de Ciencia y Tecnología—SENACYT)*, the Institute for the Promotion of the Human Resources (*Instituto para la Formación y Aprovechamiento de los Recursos Humanos—IFARHU),* both Panamanian institution, and the German Academic Exchange Service (DAAD). To these three institutions, I am deeply indebted.

I must thank also to my colleagues of the IMPRS for maritime affairs, specially to Sarah Gahlen, for her constant and unconditional help, and to Lina Wiedenbach and Young-Kyung Yoon. My high regards go also to the very talented and committed professors, researchers, and doctoral candidates from all around the world who I had to honor and pleasure to meet in the three years and half I spent at the Max Planck Institute in Hamburg. Their advices and friendship made of this study time in Germany a very productive and enjoyable experience. Among them, I cannot avoid mentioning to Dairo Pascual, his wife Natasha Hadsimanovic, and Dorothea Magnus for being a great source of motivation; to Mr. Roland Jörs, who proved to be an excellent coach; and Mrs. Deirdre Meyer for her proofreading services. Also in Hamburg, I'm very thankful to the wonderful people of the parish communities of St. Theresien and St. Elizabeth. Their kindness made my stay in Europe much warmer. Thanks to my relatives and friends in Panama for their support. I must express my special thanks to Olga and Fula Bourdett for their unique and incomparable love, whose support made possible to finish this work.

Finally, but the most important, I must thank my beloved mother, whose early departure shortly before starting this project made this a challenge, but her exemplary life and efforts to grant me an education are reasons enough to honor her in all my endeavors. Because of her and for her are all my achievements.

Panama, Panama Víctor Hugo Chacón
June 2017

Contents

Abbreviations

A.C.	Law Reports, Appeal Cases, England and Wales
A.M.C.	American Maritime Cases
All ER	All England Law Reports
Am. Dec.	American Decisions
Am. J. Comp. L.	American Journal of Comparative Law
Am. L. Rev.	American Law Review
Am. Law Reg.	American Law Register United States
App. Cas.	Law Reports, Appeal Cases (Second Series) England and Wales
B. & Ald.	Barnewall and Alderson's King's Bench Reports
B.C.S.C.	Supreme Court of British Columbia, Canada
Burr.	Burrow's King's Bench Reports tempore Mansfield
C.A.N.Y.	United States Court of Appeal, Second Circuit
C.C.S.N.Y.	Circuit Court, Southern District of New York
C.F.R.	United States Code of Federal Regulations
C.P.D.	Common Pleas Division
Camb.L.J.	Cambridge Law Journal
Can. B. R.	Canadian Bar Review
Ch.D.	Law Reports, Chancery Division, England and Wales
CMI	Comite Maritime International
Co.Rep.	Coke's King's Bench Reports
COGSA	Carriage of Good by Sea Act
Cranch	Cranch's Supreme Court Reports
CSC	Convention on Safe Containers
CSS	Code of Safe Practice for Cargo Stowage and Securing
D.A.R.	Daily Appellate Reporter United States, California
D.C. Cal.	District Court of California
D.C. Mass.	District Court of Massachusetts
D.C.E.D.Wis.	District Court for the Eastern District of Wisconsin
D.C.La.	District Court of Louisiana

D.C.N.Y.	District Court of New York
D.C.Or	District Court of Oregon
D.Md.	District of Maryland
Dig.	Digest of Justinian
DOC	Document of Compliance
Duer	Duer's New York Superior Court Reports
E.D.La.	Eastern District of Louisiana
E.D.N.Y.	Eastern District of New York
E.D.Pa.	Eastern District of Pennsylvania
E.R.	English Report
El. & Bl.	Ellis and Blackburn's Queen's Bench Reports
EWHC	England and Wales High Court (Administrative Court)
Ex. C.R.	Canada Exchequer Court Reports Canada
F. Supp.	Federal Supplement
F.	Federal Reporter, USA
F.Cas.	Federal Cases, USA
Ga.	Georgia Reports
H/H-VR	Hague/Hague-Visby Rules
HGB	Handelsgesetzbuch
How.	Howard's Supreme Court Reports
HR	Hague Rules
Hun	Hun's New York Supreme Court Reports
IMDG	International Maritime Dangerous Goods Code
IMO	International Maritime Organization
ISM	International Safety Management Code
ISPS	International Ship and Port Facility Security Code
ISSC	International Ship Security Certificate
J. Mar. L. & Com.	Journal of Maritime Law and Commerce
J.B.L.	Journal of Business Law
John.	Johnson's New York Reports
K.B.	King Bench
L. Ed	United States Supreme Court Reports, Lawyers' Edition
L.J. Ex.	Law Journal Reports, Exchequer New Series
L.Q.R.	Law Quarterly Review
L.R.A.	Lawyers' Reports Annotated
LASH	Light Aboard Ships:
Ll.L.R./Ll. Rep.	Lloyd's List Law Reports
Lloyd's Rep.	Lloyd's Law Reports
M.L.J.	Mississippi Law Journal
M.P.R.	Maritime Provinces Report, Canada
N.D.Fla.	Northern District of Florida
O.R.L.R.	Ontario Law Reports
Ont. S.C. App. Div.	Supreme Court of Ontario, Appellate Division
Price Exch.	Price's Exchequer Reports
Pro.	Law Reports, Probate England and Wales

QB	Queen Bench
QBD	Law Reports, Queen Bench Division
RSO	Recognized Security Organization
RvdW	Rechtspraak van de Week Netherlands
S. Ct.	Supreme Court of the United States
S.C.R.	Supreme Court Reports, Canada
S.D.Cal.	Southern District of California
S.D.N.Y	Southern District of New York
S.D.R.	Special Drawing Rights
S.E.	South Eastern Reporter
Sandford	Sandford's New York Superior Court Reports
SMC	Safety Management Certificate
SMS	Safety Management System
SOLAS	Safety of Life at Sea Convention
SSA	Ship Security Assessment
Str.	Strange's King's Bench Reports
STWC	International Convention on Standards of Training, Certification and Watchkeeping for Seafarers
Sup.Ct.Rep.	Supreme Court Reporter United States
Taunt.	Taunton's Common Pleas Reports
U. Pa. L. Rev.	University of Pennsylvania Law Review
U.S.App.	United States Appeals Reports
U.S.C.	United States Code
UNCITRAL	United Nations Commission on International Trade Law
UNCLOS	United Nations Convention of Law of the Sea
UNCTAD	United Nations Conference on Trade and Development
Va.J.Int'l L.	Virginia Journal of International Law
W.D.Ky.	Western District of Kentucky
W.D.Mich.	Western District of Michigan
W.D.N.Y.	Western District of New York
W.L.R.	Weekly Law Reports England and Wales
Wall. JR.	J.W. Wallace's United States Circuit Court Reports
Wall.	Wallace's Supreme Court Reports
Wend.	Wendell's New York Reports
WL	Westlaw Transcripts United Kingdom
WLR	Weekly Law Reports England and Wales

Chapter 1
Introduction

1.1 Purpose of the Study

The carriage of goods by sea has been the most useful and effective means of transportation for the international trading of goods. Its paramount importance for the world's economy increased dramatically over the last century when more countries went beyond their borders to compete in the international market creating in large part today's global economy. Boundaries fell away and free trade was opened between all nations. Today, the success of countries' economies is measured, among other factors, on the balance of payment of goods annually imported and exported. This international exchange of goods is possible thanks to vessels that currently move approximately 80% of the world's trade.[1] Because of this, the shipping industry is not only of special importance to the private sector, but to the public sector as well. As a fundamental component of the economic growth and development policy, it has been the subject of public interest and States have endeavored to regulate it.

The maritime transportation is regularly performed under two contractual modalities: charter parties and bills of lading. The research focuses on the latter modality. At the present, the carriage of goods by sea under bills of lading is mostly governed by the International Convention for the Unification of Certain Rules of Law Relating to Bills of Lading, adopted in Brussels in 1924 and commonly known as the Hague Rules. This regulation orders carriers to perform two duties: to practice due diligence to make the ship seaworthy and to care for the cargo. The prevention of cargo damage or losses is a goal of the liability rules set in the governing regime. It meets a social end too, as the economy and in consequence, the social welfare of most nations, depend on the safe arrival of the products at destination.

Both duties have been constantly impacted by the development of technology. Shipping technology has provided new resources, methods and tools to make the

---

[1]UNCTAD, 'Review of maritime transport 2011', 26.

V.H. Chacón, *The Due Diligence in Maritime Transportation in the Technological Era*,
Springer Series on Naval Architecture, Marine Engineering,
Shipbuilding and Shipping 5, DOI 10.1007/978-3-319-66002-8_1

carriage of goods safer than ever. But which of these technologies can be expected by carriers to be applied as part of their obligation to practice due diligence? It is a question that has required lengthy, complicated and expensive litigation. Considering the increasing costs of submitting cargo claims to the judiciary, for both the parties and the states, at least some general parameters that can shed light on the issue should be provided. This is the aim of this work.

This research is presented in three parts. The first part addresses the evolution of the sea common carriers' liability taken from the historical sources of the maritime law and the current international regulations governing the contract. An overview of these historical developments, from the oldest code containing regulations for the maritime commerce to the latest international conventions on the subject, will point out the carrier's duties and the standard of liability set therein. Special emphasis is given to the historical circumstances and causes that drove the adoption of the Hague Rules. The second part presents the interpretation and construction of the two carriers' duties set out in the Hague Rules by the English and American Jurisdictions. Finally, the third part addresses the analyses on new technologies in the case law of the aforementioned jurisdictions. From the case law on this specific subject, I attempt to extract the parameters or situations in which the courts have required the application of new technologies as part of the duties set in the current governing liability regime.

1.2 The Contract of Carriage of Goods by Sea

For the proper understanding of this study, it is necessary to provide first an introductory and general overview of the contract of carriage of goods by sea.

1.2.1 Concept

This contract is commonly defined as the contract of a ship for the carriage of goods by water, usually performed, as mentioned, under the modalities of charter parties or bills of lading, or a combination of both.[2] It is a contractual obligation subscribed between a shipper and a carrier, where the latter undertakes the task of transporting the shipper's goods by sea, from a specific place to a specific destination, against the payment of freight. The carrier assumes some risks of the maritime adventure, which makes the obligation to be classified as one of result.[3]

---

[2]B.A. Garner and H.C. Black, *Black's law dictionary*, 9th edn. (St. Paul, Minn: West, 2009), 69, 375.

[3]F. Sánchez Calero, *El contrato de transporte marítimo de mercancías:* (*Reglas de La Haya-Visby, Hamburgo y Rotterdam*), Colección Grandes Tratados Aranzadi, 2nd edn. (Cizur Menor (Navarra): Thomson Reuters-Aranzadi, 2010), vol. 170, 131.

1.2.2 The Parties Involved

The contract has a bilateral character. As mentioned above, the shipper and the carrier are the main parties. The international conventions on the subject have underlined some definition of both parties, attending to each parties' participation in the contract and its engagement with the other party. The Conventions provide the following definitions.

1.2.2.1 The Shipper

Shipper is defined as the person who ships goods to another or the party who contracts the services of a carrier for the transportation of cargo.[4] The Hague Rules does not present any definition of shipper. The Hamburg Rules include the following definition:

> [A]ny person by whom or in whose name or on whose behalf a contract of carriage of goods by sea has been concluded with a carrier, or any person by whom or in whose name or on whose behalf the goods are actually delivered to the carrier in relation to the contract of carriage by sea.[5]

Though the Rotterdam Rules are the newest regulation enacted for this contract, the definition provided is more ambiguous: "shipper is the party who enters into a contract of carriage with a carrier."[6]

1.2.2.2 The Carrier

Carrier is the person, individual or organization, engaged in transporting passenger or goods for hire.[7] The same shipowner is the carrier when he personally operates his vessel.[8] At the present, vessels often operate under a charter party. In this case, and depending on the type of charter party, the charterer who then resells space on the chartered ship to other particular shippers becomes the carrier for those particulars who ship goods with him.[9] The Hague Rules do not provide a definition based on the function or the activity of the carrier. That is of general knowledge in the industry. The Rules identify the carrier regarding its status or legal relation to the ship and to the

---

[4]Garner and Black, 1503.

[5]The Hamburg Rules Article 1.3.

[6]The Rotterdam Rules Article 1.8.

[7]Garner and Black, 214.

[8]T.J. Schoenbaum, *Admiralty and maritime law*, Practitioner treatise series, 5th edn. ([St. Paul, MN]: Thomson/West, 2011), 795.

[9]S. Girvin, *Carriage of Goods by Sea*, 2nd edn. (Oxford, New York: Oxford University Press, 2011), 18.

shipper. It says then, that the "carriers include the owner or charter who enters into a contract of carriage with a shipper".[10] The Hamburg and Rotterdam rules present similar descriptions.[11] A more detailed definition of the carrier's activity and their respective obligations has been developed according to the distinction made by the courts of two types of carriers: Private carrier and Common carrier.[12]

1.2.2.2.1 Private Carrier

The private carrier is usually the shipowner who lets the whole ship or its whole cargo capacity for one or more voyages or for a specific period of time.[13] This carrier offers his ship for carriage services to a specific shipper and under special agreement. Such agreement commonly appears in a formal letting of a ship known as a charter party,[14] in difference with the common carrier who usually acts under a bill of lading.[15] As the law of charter parties is still governed by the principle of freedom of contract, the parties in a private carriage enjoy some more liberties.[16] The private carrier reserves the right to bargain and the right to decide whether or not he carries a specific cargo as he chooses.[17] He is not necessarily available, nor obliged, to carry goods for the general public. It must be noticed, however, that the determination of the carrier's character based on the modality of the contract subscribed is not always exact. A shipper who takes up the full space on a ship for a specific voyage is contracting with a private carrier, even if the contract of carriage is documented in a bill of lading.[18] The opposite may happen when a shipper lets a full deep tank or some other compartment of a vessel under a charter party, but other tanks or hatches of the same vessel are let to other shippers. In this case, beside the charter party, the carrier is to be held as a common carrier.[19] The difference lies in the exclusivity of the contractual relation with the specific shipper for the use of the whole cargo capacity of the chartered ship. Regarding the standard of liability, private carriers are liable if the damage or loss is caused by their negligence in the performance of their duties.[20]

---

[10]The HR Article 1(a).

[11]Hamburg Rules article 1.1; Rotterdam Rules article 1.5.

[12]This distinction was evidenced from the time of the decision issued by Lord Holt in the case *Coggs v. Bernard*, 2 Lord Raymond 909, 92 E.R. 107, 90 E.R. 971 (1703).

[13]H.N. Longley, *Common Carriage of Cargo* (Albany, N. Y.: Mathew Bender, 1967), 7.

[14]Schoenbaum, 777–78.

[15]Longley, 7.

[16]Schoenbaum, 779.

[17]N.J.J. Gaskell, R.S.T.C. Chorley, C. Debattista and R.J. Swatton, *Chorley and Giles' shipping law*, 8th edn. (London: Pitman Publishing, 1987), 166.

[18]Longley, 8.

[19]Ibid., 9.

[20]Ibid., 8; Gaskell, Chorley, Debattista and Swatton, 167.

1.2.2.2.2 Common Carrier

Common carriers are those who hold themselves, as a regular business, available to carry any goods, from any shipper who pays their charges, as long as there is enough space on the vessel, following a regularly fixed route.[21] The contract under this modality is usually performed under the form of a bill of lading, but for the reason explained above, such a document is not conclusive to determine the character of the carrier. The common carriers were subject in the past to very strict regulations. Unlike the private carriers who were liable for negligence, the common carriers were subject to statutory regulation including strict liability for damages or loss of the goods.[22] Under common law, common carriers were subject to three main obligations:

- They had a duty to serve all who applied for their services,
- Unreasonableness in their rates of charge and operation was prohibited; and,
- They were held to far stricter liability standards than those applied in general business law.[23]

However, the strictness of the liability standard changed radically with the introduction of the Brussels Convention of 1924, commonly known as the Hague Rules. The liability of common carriers under the Hague Rules is measured attending to their negligent activity or omission resulting in damages or losses, similar to the liability for private carriers. Currently, a large number of companies operate as common carriers that own or hire several numbers of ships and offer carriage of goods throughout the world. They are known as cargo liners, whose development since the last century has enabled the rapid growth of the international exchange of goods. Their importance in the current maritime commerce has grown so much, that the recent Rotterdam Rules included a definition for this type of carrier.[24] Most of the cargo liners perform their services under the Hague Rules, as they carry goods from or to States members of the Convention, or they commonly incorporate these rules in their bills of lading as the applicable law to the contract.

[21]R. Colinvaux, *Carver'scarriage by sea*, British shipping laws, 13th edn., 2 vols. (London: Stevens, 1982), 6; L. Gorton, *The concept of the common carrier in Anglo-American law*, Scandinavian university books (Gothenburg: Läromedelaférl. (Akad.-förl.), 1971), vol. 43, 20; Schoenbaum, 777. *See* also *Propeller Niagara v. Cordes*, 62 U.S. 7 (1858).

[22]Longley, 8.

[23]K. Hall, J. W. Ely and J. B. Grossman, *The Oxford companion to the Supreme Court of the United States*, 2nd edn. (Oxford, New York: Oxford University Press, 2005), 197.

[24]Rotterdam Rules, Article 1.3: "'Liner Transportation' means a transportation service that is offered to the public through publication or similar means and includes transportation by ships operation on a regular schedule between specified ports in accordance with publicly available timetables of sailing dates.".

1.2.3 *Respective Obligations*

The determination of the rights and duties of each party depends on the modality of the contract they submit and the applicable law stated. In a Charter Party, the parties enjoy additional freedom; while under a bill of lading, parties are subject to statutory regulations.[25] The doctrine and jurisprudence have distinguished two general groups of obligations in this contract. First, the express obligations refer to the specific characteristics of the contracted carriage. They include the obligation to provide the ship with the special features offered; the time of departure; the port or place of loading and delivery; the amount of freight; etc. On charter parties, the time period of the charter or the specific voyage hired are also particularly important. These obligations or special conditions are stated in the charter party or the bill of lading.

Second, apart from the terms expressly stated in the contract, or set by statue or implied by custom, the courts have recognized and enforced some general duties that must be accomplished by the parties to give business efficacy to the contract.[26] They are automatically incorporated in the contract, unless the parties have agreed to special clauses establishing the contrary.[27] Although the existence of such general implied obligations is the result of jurisprudential enforcement, courts are not allowed to make new dispositions in the contract.[28] This restriction applies even when they consider it reasonable for the correctness and efficacy of the contract.[29]

1.2.3.1 On the Carrier

For carriers, in the common law, the general arising obligation from this contract is summarized as: "[the carrier] is to carry and deliver in safety, answering for all loss or damage which may happen to them (the goods) while they are in his hand as carrier."[30] They are relieved of responsibility for loss of or damages to the cargo, when it is caused by an act of God; King's enemies; defect or infirmity of the goods or their package; or, through a voluntary sacrifice for the safety of the adventure.[31] The access of these exceptions of responsibility is conditional upon the absence of any negligent act by the part of the carrier that exposed the cargo to the cause of loss or damage.[32] To accomplish the main objective of the contract, the common

[25]T. J. Schoenbaum and A. N. Yiannopoulos, *Admiralty and maritime law: Cases and materials*, Contemporary legal education series (Charlottesville, Va: Michie Co., 1984), 364.

[26]T. E. Scrutton and B. Eder, *Scrutton on charterparties and bills of lading*, 22nd edn. (London: Sweet & Maxwell, 2011), 111.

[27]J. F. Wilson, *Carriage of goods by sea*, 7th edn. (New York: Pearson/Longman, 2010), 9.

[28]Gaskell, Chorley, Debattista and Swatton, 182.

[29]Ibid., 182.

[30]Colinvaux, 20.

[31]Ibid., 20.

[32]Ibid., 20.

law jurisprudence has developed three implied undertakings that must be observed by the carriers.

1.2.3.1.1 To Provide a Seaworthy Ship

The seaworthiness of the vessel is generally defined as the condition of being sufficiently staunch, strong and tight, properly equipped and in every way fitted to resist the perils and incidents that a seagoing vessel might ordinarily encounter.[33] Such a condition covers the physical estate of the ship, the crew's competence, the equipment, and its capability to carry the specific cargo. In common law this obligation is absolute and its breach makes the carrier liable regardless of fault.[34] Only when the contract is governed by the Hague/Hague-Visby Rules, is the standard reduced to the exercise of due diligence in making the ship seaworthy.[35]

1.2.3.1.2 Reasonable Dispatch

In absence of a specific date for departure, the vessels must proceed and perform the carriage in a reasonable time.[36] The reasonability of the time of departure will vary case by case, and must take into account, firstly, the different elements of the voyage, such as the conditions of the ship; the state of the cargo handling technology; and, the special circumstances known to the parties.[37] Certainly the condition of the ports, the distance to be sailed and the usual weather condition of the route play an important role. The second element to determine the reasonability of the time of dispatch is, if under the actual circumstances, the carrier was diligence in the execution of the voyage.[38] The carrier, however, is not expected to take more than reasonable measures to accomplish this obligation.[39] To assign carrier's liability for unreasonable delay, the courts must assess whether the circumstances that impeded the carrier to perform the

---

[33]Garner and Black, 1470, 1699.

[34]Wilson, 9.

[35]HR article 3.1.

[36]Wilson, 15.

[37]M. a. K. H. M. Ganado, *Marine Cargo Delays*, The law of delay in the carriage of general cargoes by sea (London, 1990), 36.

[38]*See Dallas W. Dietrich, AS Atlantic Seaboard Flour Mill v. United States Shipping Board Emergency Fleet Corporation*. (*The Panola*) 1925 A.M.C. 1173 (2nd. Cir.). "And the courts have laid down the rule that 'reasonable time' for the performance of acts under a contract is such a period of time as suffices for their performance if the one whose duty it is to perform uses such diligence in the performance as a person of ordinary diligence and prudence would use under like circumstances".

[39]In the case *Briddon v. Great Northern Ry. Co.* (1958) 28 L.J. Ex. 51, the court said at 52: "The contract entered into was to carry… without delay, and in a reasonable time under ordinary circumstances. If a snow –storm occurs which makes it impossible to carry the cattle, except by extraordinary effort, involving additional expense, the company are not bound to use such means and to incur such expense.", as quoted by Ganado at 35.

carriage in a reasonable time, were normal or not. As the conditions in navigation may vary, and may be in some circumstances beyond the carrier's control, provisions in bills of lading and charter parties are generally stated releasing the carrier from liability for delays. Some others go further refusing any liability for delay of any type. These clauses are enforceable by American courts, releasing carriers for delays of 2 weeks,[40] or even longer, and denying compensation for damages resulting, for example, from loss of market.[41] Notwithstanding, other courts have been more cautious in recognizing such clauses and assign liability if the unreasonable delay was caused by the carrier's negligence.[42]

1.2.3.1.3 Not to Deviate from the Agreed Route

The carriage must be performed not only in a reasonable time, but also, following the agreed route.[43] In absence of any agreement regarding the geographical route, the carrier must proceed along the shortest, safest and usual direction to the destination port. A voluntary change without necessity or reasonable cause from the regular course or route shall be considered a deviation.[44] Deviation means "an intentional and unreasonable change in the geographic route of the voyage as contracted."[45] This concept was introduced into the law of carriage from the marine insurance law.[46] Later, it was extended to other causes other than a geographic deviation, to include: "any variation in the conduct of a ship in the carriage of goods whereby the risks incident to the shipment will be increased."[47] Under this extension, deviation was expanded to include: carrying in a different vessel than the agreed one; carrying partly by rail; dry docking with the cargo on board; unreasonable delay; and, carrying on deck.[48] Other situations that under the same

---

[40]*See Parnass International Trade & Oil Corp. v. Sea-Land Service, Inc.* 595 F. Supp. 153, 1985 A.M.C. 485 (S.D.N.Y. 1984) (A short delay of 18 days is not in and of itself unreasonable. It is just such a potential for strikes and port congestion which prohibits carriers from making hard and fast promises such as the one plaintiff contends existed here.); *Pioko Fashions, Inc. v. American President Lines, Ltd.*, 1993 A.M.C. 2615 (W.D. Wash. 1993) (2-week delay in 10,000 mile cargo delivery was not unreasonable deviation as required under limitation in bill of lading); *Quesoro v. Lykes Bros. Steamship Co., Inc.*, 1995 A.M.C. 2054, (S.D.N.Y. 1995).

[41]*See The Panola, see supra* note 38.

[42]Ganado, 38.

[43]W. Tetley, *Marine Cargo Claims*, 4th edn., 2 vols. (Cowansville, [Toronto]: Blais; Thomson Carswell, 2008), Vol. 1, 1811.

[44]*Hostetter v. Park*, 137 U.S. 30, 40; 11, S. Ct. 1, 34 L.Ed. 568 (1890).

[45]Tetley, *Marine Cargo Claims*, Vol. 1, 1812; Wilson, 16.

[46]G. Gilmore and C. L. Black, JR., *The Law of Admiralty*, 2nd. Ed. (Mineola N.Y: The Foundation Press, Inc., 1975), 176.

[47]*Spartus Corp. v. S/S Yafo*, 1979 A.M.C. 2294, 2297–98, 590 F.2d 1310, 1313 (5th Cir. 1979).

[48]Longley, 110–12.

construction were considered deviation or quasi-deviation include: Carrier's corrupt or criminal miss delivery[49]; negligence, gross negligence or willful misconduct;[50] negligent stowage[51]; stowage which greatly exceeds weight limitation of flat rack container; and, safe capacity of cargo cranes and related equipment.[52]

However, English courts have restricted the concept to geographical deviation.[53] American jurisprudence still extends the concept to carrying on deck without authorization or in absence of a custom of carrying such goods in above deck stowage[54]; over carriage, miss delivery and change in the route by truck inland.[55] These cases are referred to as "quasi-deviation".[56] The Hague/Hague-Visby rules do not define deviation nor establishes its consequences, but seems to keep the concept strictly for geographical deviation. The reasonableness of a deviation will depend on the law governing the contract; the surrounding circumstances; and the facts attending to the interest of all the parties.[57] In principle, a deviation is reasonable exclusively in case of rescue of life at sea,[58] which is now also a statutory duty established in national and international regulations.[59] The norm introduced by the HR extended it to the rescue of property at sea as well, and to "any reasonable deviation".[60] Consequently, American and English law also recognizes this same

[49] *See B.M.A. Indus., Ltd. v. Nigerian Star Line, Ltd.*, 1986 A.M.C. 1662, 786 F.2d 90 (2nd Cir. 1986).

[50] *See Iligan Integrated Steel Mills, Inc. v. S.S. John Weyerhaeuser*, 1975 A.M.C. 33, 507 F.2d 68 (2nd Cir. 1974), cert. denied, 421 U.S. 965, 1975 A.M.C. 2158.

[51] *The Chester Valley*, 1940 A.M.C. 555, 110, F.2d 592 (5th Cir. 1940).

[52] *Rockwell International Corp. v. M.V. Incotrans Spirit*, 1989 A.M.C. 887 (S.D.Tex. 1989).

[53] Wilson, 16, note 53; Tetley, *Marine Cargo Claims*, Vol. 1, 1812. *See The Hill Harmony* [1999] 2 Lloyd Rep. 209, 217.

[54] *See St. Johns Corp. v. Companhia Geral Commercial Do Rio de Janeiro*, 263 U.S. 119, 124, 44 S.Ct. 30, 31, 68 L.Ed. 201 (1923); *Sedco, Inc. v. S.S. Strathewe*, 1986 A.M.C. 2801, 800 F.2d 27 (2nd Cir. 1986).

[55] Tetley, *Marine Cargo Claims*, Vol. 1, 1812.

[56] Ibid., 1812.

[57] Ibid., 1814–15. *See Stag Line v. Foscolo, Mango & Co.* [1932] A.C. 328.

[58] *See Scaramanga & Co. v. Stamp* (1880) 5 CPD 295 (CA), 304.

[59] Girvin, 403. *See* UK Merchant Shipping Act of 1995 section 93(1), and UNCLOS Article 98.

[60] HR Article 4(4): "Any deviation in saving or attempting to save life or property at sea or any reasonable deviation shall not be deemed to be an infringement or breach of this convention or of the contract of carriage, and the carrier shall not be liable for any loss or damage resulting therefrom.".

cause as valid.[61] In addition, a deviation may be allowed to avoid danger of the ship or to the cargo;[62] or when it is made necessary by some default on the part of the charterer.[63] Carriage of containers on deck has been also considered a reasonable deviation. Containerships are specially designed to transport containers safely on deck, so the cargo is not exposed to greater risks.[64] With the pass of time, it became a well-established trade custom in a world-wide basis.[65]

Unreasonable deviation is regarded as fundamental breach in the common law,[66] and the carrier is held as an insurer of all damages caused to the cargo by the deviation.[67] Both English and American legislations hold the carrier liable for damages or losses resulted from unreasonable deviation and deprive the carrier from relying on any legal or contractual stipulation in his favor.[68] Under the HR an unreasonable deviation will produce similar consequences, depriving the carrier to rely on the exceptions and limitations of liability.[69] In some cases, however, such as the excluded situations listed above, American courts have allowed the limitation per package.[70] Under English law, the contractual freight is also reduced to a

---

[61]Section 4.4 of the US COGSA 1936 and UK COGSA 1971.

[62]Girvin, at 403–04: "The danger to the vessel may result from natural causes, such as storms, ice, or fog, or political factors, such as the outbreak of war or the fear of capture by hostile forces." *Citing Duncan Köster* (*The Teutonia*) (1982) LR 4 PC 171, 179. Also deviation for urgent repairs might be considered reasonable. *See Kish v. Taylor* [1912] AC 604.

[63]Wilson, 17–19.

[64]*See Du Pont Nemour International, S.A. v. S.S. The Mormacvega*, 493 F2d. 97, 101–02 (C.A.N. Y. 1974); *Electro-Tec Corp. v. S/S Dart Atlantica* (1984) 598 F. Supp. 929 (D.Md. 1984).

[65]*See Konica Bus. Machines v. The Vessel Sea-Land Consumer*, 47 F.3d 314, 315 (9th Cir. 1995), 1995 A.M.C. 1065; *Konica Business Machines, Inc. v. Vessel "Sea-Land Consumer"* 153 F3d. 1076. 1998 A.M.C. 2705, 98 Cal. Daily Op. Serv. 6955, 98 Daily Journal D.A.R. 9595. However, cargo stored in an open "flat rack" container stowed on deck was held as an unreasonable deviation in *Constructores Tecnicos S. de R.L. v. Sea-Land Service Inc.*, 956 F.2d 841, 1992 A.M.C. 1284 (5th Cir. 1991).

[66]Wilson, 20–21.

[67]*See Insurance Co. of North America v. Blue Star (North America), Ltd.* (S.D.N.Y., 1997). Not Reported in F. Supp., 1997 WL 345235 (S.D.N.Y.), 1997 A.M.C. 2434. District Judge Sweet cited: *Du Pont de Nemours Int'l S.A. v. S.S. Mormacvega, supra* note 64; *Hearne v. Marine Ins. Co.*, 87 U.S. (20 Wall.) 488, 22 L.Ed. 395 (1874); *Oliver v. Maryland Ins. Co.*, 11 U.S. (7 Cranch) 487, 3 L.Ed. 414 (1813).

[68]Longley, 118; Tetley, *Marine Cargo Claims*, Vol. 1, 1839; *see Encyclopaedia Britannica v. S.S. Hong Kong Producer*, 422 F.2d 7, 16–17 (2nd Cir. 1969), cert. denied, 397 U.S. 964 (1970).

[69]Tetley, *Marine Cargo Claims*, Vol. 1, 1830: "The carrier may lose one or all of the following rights: The package limitations; the 1 year delay for suit; the defense of due diligence at art. 3(1); the exculpatory defenses at art. 4(2)(a) to (q); other limitation and exclusion of the contract such as jurisdiction and arbitration clauses.".

[70]Gilmore and Black, JR., at 181: "The Seventh Circuit has held that an unreasonable deviation did not oust the $500.00 per package limitation in Cogsa 4(5)".

reasonable sum if the goods are delivered in the destination port.[71] General average cannot be claimed unless the breach of contract by deviation is waived.[72]

1.2.3.2 On the Shipper or Charterer

1.2.3.2.1 Not to Ship Dangerous Goods

Shippers shall not ship dangerous goods without previous notification to the carrier regarding its dangerous condition. The objective of this obligation is to provide the carrier with sufficient knowledge about the potential danger.[73] Such notification enables them to refuse to carry the goods or to take reasonable measures to assure the safety of the vessel, crew and other cargo on board.[74] When the carrier is already aware of the condition, there is no obligation to give notice.[75] Under English law the concept of dangerous goods is not merely restricted to inflammable or explosive goods. It also includes everything that could put the vessel or other cargo in danger, including for example, infected cargo or cargo prohibited or subject to quarantine, or cargo which may not be allowed in the destination country.[76] Notwithstanding, as the determination of what a dangerous cargo is presents frequent problems, the International Maritime Organization adopted in 1965 the International Maritime Dangerous Goods Code (IMDG).[77] This code establishes and regularly updates a list of goods considered as dangerous.[78]

Under the Hague/Hague-Visby Rules, the carrier is allowed to discharge, destroy or render innocuous any inflammable, explosive and dangerous goods loaded without the shipper's knowledge or consent.[79] These actions release them from paying compensation and hold the shipper liable for all damages or expenses directly or indirectly arising from these measures.[80] The situation varies when the carrier has knowledge of the potential danger and the goods become, after loaded, a

[71]W. a. I.E. R. H. Payne, *Carriage of Good by Sea*, 13. ed. E.R. Hardy Ivamy (London: Butterworths, 1989), 23.

[72]Ibid., 23.

[73]Girvin, 312.

[74]Ibid. 312; Wilson, 35. *See Brass v. Maitland* (1856) 6 El. & Bl. 470; 119 E.R. 940.

[75]Girvin, 312.

[76]Wilson, 34. *See Mitchell, Cotts v. Steel* [1916] 2 K.B. 610.

[77]Available at www.5.imo.org/SharePoint/mainframe.asp?topic_id=158. (Last visited: May 21st, 2012).

[78]International Maritime Organization, *International Maritime Dangerous Goods Code: IMDG-Code; incorporating amendment 34-08* (London: International Maritime Organization, 2008). The code makes a classification of goods in explosives, gases, flammable liquids, flammable solids, oxidizing substances and organic peroxides, toxic and infectious substances, radioactive materials, corrosive substances, miscellaneous dangerous substances and articles.

[79]The HR, article 4.6.

[80]Wilson, 36. The HR, article 4.6.

real danger to the ship or to other cargo. In this case, the carrier shall proceed in the same way, without any liability for the shipper and releasing him from any expenses.[81]

1.2.3.2.2 To Nominate a Safe Port

This obligation falls mostly upon the shipper under a charter party, where the charterer determines the ports the ship must visit. Shippers, under a common carriage governed by a bill of lading, must accept the ports the carrier specifies for loading and discharge. A shipper contracts with a common carrier depending on whether the ports offered suit his need of carriage or not. The shipper here has no power to decide or to order the carrier into which port he must call. It is the charterer, contracting with a private carrier in a charter party, who must nominate safe ports to protect the safety and integrity of the vessel. A safe port means that the loading and destination ports are adequate:

> […]in the relevant period of time, the particular ship can reach it, use it and return from it without, in the absence of some abnormal occurrence, being exposed to danger which cannot be avoided by good navigation and seamanship.[82]

The relevant period of time means the entire period from the moment of the vessel's arrival till the time of her departure.[83] To be qualified as a safe port, there must be sufficient depth of water; absence of ice or rocks limiting the free access to the port; adequate administrative systems such as safety equipment buoys, warning lights and radars; absence of political hostilities[84]; adequate weather forecasting systems; availability of trained pilots and tugs; enough and always available sea room and room to maneuver.[85] Natural hazards and political danger to the physical safety of the vessel must be factored in as well as, the special characteristics of the vessel and the season when the port is nominated.[86]

The effect of this warranty allows the carrier or shipowner to refuse to proceed to the nominated port without breaching the contract.[87] If the ship proceeds complying the charterer's order, he is liable for any damages to the vessel and for extra charges incurred in loading or unloading, if the unsafe conditions require such additional expenses.[88] The shipper is released of liability in two situations. First, when the port becomes unsafe after it was nominated. In this case, the charterer must substitute

[81] Ibid., 36: The shipper, however, is required to pay its contribution in general average.

[82] Ibid., 26. *See The Easter City*, [1958] 2 Lloyd's Rep. 127, 131.

[83] Ibid., 26.

[84] Ibid., 26.

[85] Ibid., 27. *See Islander Shipping Enterprises, S.A. v. Empresa Maritima del Estado, S.A.* (*The Khian Sea*), [1979] 1 Lloyd's Rep. 545, 547.

[86] Schoenbaum. Vol. 2, 31–33.

[87] Ibid., Vol. 2, 32.

[88] Ibid., Vol. 2, 32–33.

another port, attending the circumstances.[89] The second is when the master proceeds to the unsafe port with knowledge of its bad condition.[90] In this case, the damages must be divided proportionately if the fault is shared between the charterer and the master.[91]

[89]Girvin, 319, *citing Vardinoyannis v. The Egyptian General Petroleum Corp.* (*The Evaggelos Th*) [1971] 2 Lloyd's Rep. 200; Schoenbaum. Vol. 2, 33.

[90]Girvin, 334; Schoenbaum, Vol. 2, 33.

[91]Schoenbaum, Vol. 2, 33.

Chapter 2
The Origin of the Obligation of Practicing Due Diligence in Maritime Transportation

The first part of this work presents an overview of the sea carrier's duties and liabilities during the centuries before the enactment of the Hague/Hague Visby rules, and addresses the historical origin of the current sea carrier duties to practice due diligence in making the ship seaworthy and to care the cargo. Maritime transportation, as a cornerstone of the commerce, development and civilization, has been subject of regulations since ancient times. Rules governing the carrier's duties are found back in history. A review from the ancient sources of regulation for carriage of goods by sea law, including those during the middle ages and in the modern era, reveals a constant application of an absolute liability standard on sea carriers. Such standard changed during the XIX century under English law, when the global trade increased, along with the advent of new philosophical ideas that prompted some carrier's practices that required later urgent attention from the international community. Such efforts ended on the existing standards contained in the Hague/Hague-Visby Rules. After that, attempts for more modern regulation have been tried, but without success. From the standards of liability demanded to maritime carriers for centuries before the current era, this part provides and aid to a better understanding of the duties required to them, in what we now call the technological era.

2.1 The Evolution of the Standard of Carrier's Liability in Maritime Transportation

2.1.1 Introduction

2.1.1.1 The Shift of the Standard of Liability

Long before the adoption of the International Convention for the Unification of Certain Rules of Law Relating to Bills of Lading of 1924 (HR), the common carriers liability was absolute. They were obliged to restore the carried goods in

V.H. Chacón, *The Due Diligence in Maritime Transportation in the Technological Era*,
Springer Series on Naval Architecture, Marine Engineering,
Shipbuilding and Shipping 5, DOI 10.1007/978-3-319-66002-8_2

every case, with few exceptions made such as damages or losses resulted from acts of God or Enemies of the King. Accordingly, the implied obligation to provide a seaworthy vessel was absolute as well. This condition held the common carriers as insurers of the cargo. Liability was enforced regardless of his diligent activity or careful actions in performing the carriage. Hence, it was irrelevant whether or not the carriers applied new technologies to prevent and avoid risks and to provide a better care of the cargo. With such a standard, there was no place for any discussion on this subject, and that, certainly, would have made this work completely meaningless. But at a certain point in history, shipowners absolutely refused to continue working under such a heavy burden, and introduced some contractual provisions to release themselves from such a strict standard. These provisions became in time so extreme that prompted international statutory regulation.

As a result, after long years of discussions, the international community adopted the Hague Rules, which shifted dramatically the absolute obligation to a standard based on negligence in the performance of some minimum duties. The convention has been ratified by some 80 sovereign States and is included as applicable law in the bills of lading of the largest cargo liners of the world. Nowadays, most movement of goods by common sea carriers across the planet is governed by these rules. The liability of common sea carriers for damages or losses is decided according to their performance of these duties. What due diligence is and how it is practiced, has become the core of courts discussions when deciding cargo claims. And here is where the application of new technologies, as a measure of due diligence in making the ship seaworthy and for the care of the cargo, plays doubtless an important role at the present.

But before going into the analysis on the implementation of new technologies as part of the due diligence obligation, an understanding of the origin of the standard of liability and content of these duties in itself is needed. To provide a more holistic approach to the problem, we find it necessary to understand, first, where this duty of practicing due diligence in maritime transportation is coming from; and secondly, how and why it became the current standard to assess liability. For this purpose, a review of historical sources of maritime law is required. There we find the previous standard applied to sea common carriers and the reasons supporting it. That will lead us to an understanding of the causes for its evolution to the current liability system.

2.1.1.2 The Historical Maritime Law

Looking back in history, maritime navigation is perhaps the oldest means of transportation developed by humankind. Since ancient times, the necessity of employing vessels to carry passengers and goods resulted in early maritime law.[1] Therefore, it is possible to find historical maritime law going back hundreds of

[1]Schoenbaum, 3.

years ago. One example of that is the Hammurabi's Code, the oldest codification containing rules relating to the carriage of goods by sea. Maritime law was the consequence of the international exchange of goods. International trade is not a new practice at all. It is possible to find some signs of international trade after the First Punic War (268–241 B.C.), when the Carthaginian Hamilca Barca founded commercial colonies in Spain.[2] The development of the trade around the Mediterranean Sea, especially during the late Roman Empire and the Middle Ages, gave rise to some international systems of maritime law.[3] Rhodian and Roman law are usually referred to as the first sources of maritime law and contractual law, respectively. They are still subject of attention in legal studies and jurisdictional decisions that confirm the value of their historical review.[4]

Much later, in the eleventh century, the Crusades stimulated again international trade. The Christian control of ports in coastal countries such as Syria, Egypt, Lebanon and Turkey, opened a stable trade connection with the Italian cities of Venice, Genoa and Pisa, all of which highly profited from this exchange.[5] During this same time, there was also an outstanding commercial growth of the Hansa towns of Lübeck, Hamburg, Bremen, Visby, etc. All these maritime commercial activities in different points of history demanded regulations. In the middle ages, cities involved in international trade started writing down the customs and statutes to be observed by merchants which were then enforced by court systems created by the local governments.[6] The states at a certain point came to supervise and partially control the commercial activity because of its contribution to the national economy.[7] By the beginning of the thirteenth century, the trade grew so much in all

---

[2]S. S. Jados, *Consulate of the sea, and related documents* (University, Ala: University of Alabama Press 1975), vii.

[3]Schoenbaum, 3.

[4]H. E. Anderson, III, 'Risk, Shipping, and Roman Law' (2009) 34, Tulane Maritime Law Journal, 183–210, at 183: "Of course, since there are modern legal equivalents to the well established procedure for general average, some courts have argued they need not actually rely on Roman law. In reality, many U.S. courts have mentioned the Rhodian maritime laws in their opinions. Still, in the English case of Taylor et al. v. Curtis (The Hibernia), Chief Justice Gibbs remarked: 'It seems totally unnecessary to go to the Rhodian or Roman law for what common sense and common justice must suggest to everyone; and though it be pleasing to learned curiosity to perceive the custom of our own times confirmed by such ancient precedents, we should be satisfied with finding the analogy, without grounding ourselves upon it as the reason.' With deference to the learned judge, the real relevance of the Roman law, or any historical legal device for that matter, is not so much the actual rules established, but that such law informs us of the historical legal and customary practices and guides us with respect to the prevailing commercial or political situations that were the origins of such rule… Furthermore, it is important to understand and realize the context of the origins of these maritime legal principles in order to assess the relevance of modern practices. Only through a careful review of the historical context can future improvements be made to the current law, which originated from older law that contemplated similar physical risks through different commercial and legal arrangements."

[5]Jados, vii.

[6]Ibid., viii.

[7]Ibid., x. It seems to have happened especially after the 15th century.

commercial cities that their national pride was associated with the volume of commerce they handled.[8] Along with this salient commercial growth, unethical practices arose among merchants.[9] This, among other reasons, demanded the enactment of legal codes with uniform regulations to be strictly observed by the people involved in the commerce.[10] This gave rise to some other bodies of maritime law such as the Consulate of the Sea, the Laws of Visby and the Rules of Oleron. Despite the larger number of historical codes containing maritime legislation,[11] these three are considered the leading historical sources of European codification of maritime laws.[12] They contributed consequently to the formation of an internationally accepted maritime law.[13]

This commercial maritime activity and the regulations built upon it, was developed during times when maritime navigation was still very rudimentary. Carriers were exposed to the catastrophic risks; pirates and privateers, unpredictable rough weather conditions; lack of proper knowledge and navigational techniques; and absence of the technological improvement available nowadays.[14]

As human society has experienced an evolution in all aspects, maritime law, in the same way, has faced changes through the centuries, along with the development and improvement of international trade and navigation practices. Our contemporary

[8]Ibid., at x: "By the beginning of the thirteenth century in all large commercial cities there appeared a new class of people eager to share in the profits of trade. The concentration of money in commercial cities provided new opportunities for investment of surplus capital in new industries producing a variety of merchandise for new and expanding markets. New trade routes were opened, better merchandising methods were devised, and the first sign of modern capitalism appeared. The rise of nationalistic feeling became associated with commercialism."

[9]Ibid., at xi: "Religious convictions appeared to have no restraining influence in the merchant class…When commerce developed to the level of controlling national economies, whatever position the Church might have taken on the moral and ethical aspects of commercialism was completely ignored by the merchant princes, and the papacy no longer had the military to enforce its dicta."

[10]Ibid., xii.

[11]E.C. Benedict and A.W. Knauth, *Benedict on admiralty*, 7th edn. (New York, NY: Matthew Bender, 1969, i.e. 1958]; v. 1, 1974), vol. 1 Chapter 1. E.g. The Maritime Laws of the Kingdom of Jerusalem; Le Guidon de la Mer; The Laws of Hansa Towns; Ordinnance de la Marine of 1681; etc.; Jados., at xiii: "Customs of the city of Amalfi, Ordinances of Trani…"

[12]F. R. Sanborn, *Origins of the early English maritime and commercial law* (New York: The Century Co., 1930), 63. *See* Holdsworth; G.J. Mangone, *United States admiralty law* (The Hague; Boston: Kluwer Law International, 1997), 7

[13]Jados, xiii.

[14]Ibid., at vii: "On the seas the merchants were exposed to pirates, and privateers, to shipwreck and death. Meager knowledge of navigation, lack of proper navigational instruments, the nonexistence of lighthouses and beacon lights, dependence upon favorable winds, and the inability of vessels to withstand severe storms often ended in aimless sailing until the depletions of water and food resulted in the death of all aboard."

maritime law is the result of the evolution of principles contained in those ancient maritime laws and customs of the antiquity.[15] The legal and commercial reasoning of jurist and merchants in past times, in the solution of juridical problems emerged from the maritime trade, are worthy to be confronted with the interpretation and application of our contemporary maritime law.[16] It is more interesting when it seems they were more concerned in observing the values of justice, fairness and equity in creating and applying law. Hence, legal history deserves attention for a better understanding of the reasoning and elements that prompted its evolution, and how they were motivated by their time and the external social forces that worked on them.[17]

The study of old legal regimes governing this contract has the potential to produce risky results. It may be subject to speculation regarding their exact content, its interpretation, and the real commercial practices that surrounded them. The attempt is more problematic considering the access to the data as well as dealing with different legal cultures.[18] However, a detailed historical analysis is not the main objective of this work. We try to provide at least an introductory and general overview of the main historic bodies of maritime law, addressing concretely the sea carrier's duties and liability rules. Some scholars in the maritime law field have already written about the historical sources of maritime law,[19] but with a few exceptions,[20] in a general way and not with specific emphasis on the liability rules set out in those legal bodies. This specific issue has been more commonly addressed by authors writing on the origin of the Hague Rules, but mostly departing from the

[15]Schoenbaum, 3.

[16]Benedict and Knauth, Vol. 1, at § 1: "In adverting to the precedents and rules of practice of a bygone age, it should be our task to take account of the times and circumstances in which they were set and to use them not as shackles to bind but as guides to lead us in our attempts to find a solution for our legal problems. History will also teach us a certain humility; for, in this field of maritime law, we shall find that the forces of nature which our ancestor had to contend with and to provide legal solution for their ravages, have scant respect for the sophisticated handiwork of our times, that we rely much on the systems devised in the past to alleviate the hardships wrought by maritime casualties and disasters, that to this day we have not been able to make a reasonable codification of our laws and private litigants have to bear a enormous legal cost of establishing legal principles which ought to be the responsibility of the lawgivers to establish."

[17]D. Ibbetson, 'The Challenges of Comparative Legal History' (2013) 1, Comparative Legal History, 1–11., 11; O. W. Holmes, *The common law* (London: MacMillan, 1968), Ed. by Mark De Wolfe Howe, at 5: "The Law embodies the story of a nation's development through many centuries, and it cannot be dealt with as if it contained only the axioms and corollaries of a book of mathematics. In order to know what it is, we must know what it has been, and what it tends to become. We must alternately consult history and existing theories of legislation."

[18]Ibbetson, 11.

[19]J. Kent, *Commentaries on American Law*, 14. ed. by John M. Gould (Boston: Little, Brown, and Company, 1896); Sanborn; Mangone; Schoenbaum, among others.

[20]E.G.M. Fletcher, *The Carrier's Liability* (London: Stevens, 1932); K.F. Krieger, *Ursprung und Wurzeln der Rôles d'Oléron*, Quellen und Darstellungen zur Hansischen Geschichte, n. F. (Köln: Böhlau, 1970), vol. 15; A. J. M. Meyer-Termeer, 'Die Haftung der Schiffer im griechischen und römischen Recht'.

absolute standard applied by the British judges of the common law[21]; with short references to previous systems.[22]

Why was the common carrier's liability strict under Anglo-American law prior to the Hague Rules? Why did it change to a system based on negligence? The answer must be found in legal history. A research on the topic shows that the enforcement of strict liability can be traced far back to the time when the English judges applied it to common carriers. That standard governed the business even in times when the technologies used by the shipping industry nowadays, and that substantially improve it, were simple unimaginable. The revision of the ocean carrier liability in these historical legal bodies provides an important source of reference and comparison to assess the duties assigned now to them by the Hague Rules. To understand the radical change in the liability standard for common sea carriers posited in the Hague Rules, the legislative history and the events that prompted its adoption must be addressed as well.[23] The understanding of the reasons for change and their surrounding historical facts enable us to have a more critic view and to nourish the further analysis of the reasonableness and convenience of maintaining, replacing, amend or abolish the regulations resulting from this evolution.

There have been some more recent attempts to update the liability rules set in the HR, with the adoption of two protocols amending part of the original Convention. As these protocols were still insufficient, the international Community organized through the UNCITRAL and UCTAD, the redaction and adoption of two new conventions, proposing the Hamburg Rules and the Rotterdam Rules as new regulations to govern this contract. They have similar, but not identical, liability regimes, meant to replace the HR. Although none of these conventions are significantly applied, they are worthy to be at least briefly revised. They show the legal thinking post-Hague Rules, as well as, the awareness and concern of the international community about the need for an update in the liability regime.

---

[21]A.W. Knauth, *The American Law of ocean bills of lading*, 4th edn. (Baltimore: American Maritime Cases, 1953); M.A. Clarke, *Aspects of the Hague Rules: A comparative study in English and French law* (The Hague: Nijhoff, 1976); M. F. Sturley, *The Legislative History of the Carriage of Good by Sea Act and the Travaux préparatoires of the Hague Rules*, 3 vols. (Littleton, Colorado: Rothman, 1990).

[22]A. v. Ziegler, *Haftungsgrundlage im internationalen Seefrachtrecht: Evolution der Haftungsgrundlage von den Haager Regeln zu den Hamburger Regeln unter Berücksichtigung des nationalen und internationalen Seerechts von Grossbritannien, den Vereinigten Staaten von Amerika, Kanada, Frankreich, Deutschland und der Schweiz* (Zürich, Baden-Baden: Schulthess; Nomos, 2002); H. Karan, *The carrier's liability under international maritime conventions: The Hague, Hague-Visby, and Hamburg rules* (Lewiston, NY [u.a.]: E. Mellen Press, 2004).

[23]W. Tetley, 'Interpretation and Construction of the Hague, Hague/Visby and Hamburg Rules' (2004) 10. The Journal of International Maritime Law, 30–70, 58. Viscount Simonds on this regard stated in *The Muncaster Castle* [1961] AC 807 at 836, [1961] 1 Lloyd's Rep. 57, at 67: "To ascertain [the] meaning [of the Hague Rules] it is, in my opinion, necessary to pay particular regard to their history, origin and context, and, as I think the courts below have not paid sufficient regard to this aspect of the case, I must deal with it at some length."

2.1.2 *Historical Sources of Maritime Law*

2.1.2.1 Hammurabi's Code of Laws

The oldest records of maritime law are found in the code of Hammurabi, written in the ancient Babylon between 2000 B.C. and 1600 B.C.[24] The code is a compilation of much older Sumerian customary law containing, among others, rules for marine collisions and ship leasing.[25] Concrete regulations for maritime navigation, including shipbuilding, the payment of shipbuilders and shipmen, hire of ships and responsibility for the cargo shipped are found in rules 234 to 240.[26] The rule 237 establishes an obligation of care and liability for the loss of a vessel and her cargo:

> 237. If a man has hired a boatman and boat, and laden her with corn, wool, oil, dates, or any other kind of freight, and if the boatman is careless and sinks the boat, and her cargo is lost; then the boatman shall replace the boat he has sink [sic] and all her cargo that he has lost.[27]

The expression of a loss caused by the "careless" conduct of the boatman suggests his liability for negligence. However, Driver and Miles affirm that there is no case in this code where liability is undoubtedly based on negligence.[28] The Babylonian law, as expected, did not establish general principles or general theories on liability, or negligence, etc.[29] The Babylonians legislated on specific and concrete acts; and, in some cases where negligence or "carelessness" was stated, proof was required.[30] For this case, it was not required to prove negligence, or that the shipman had taken reasonable care, to escape liability.[31] The shipman simple had to replace everything (cargo and boat), independently of what his careless act or omission was.[32] He could not rebut his liability before the court.[33] Same authors point out that the liability of the shipman was contractual and comparable to that of the "common calling", found later in English Law, which obliges the carriers to restate the lost goods, without any concerns for negligence or due care.[34] Though not totally clear, the standard then seems to have been absolute. Because of its

---

[24]Benedict and Knauth, Vol. 1. §2.; Jados, xii.

[25]Schoenbaum, 3.

[26]C. Edwards, *The Hammurabi Code: and the Sinaitic Legislation* (London: Watts & Co., 1904), 66–67. Also on boatman liability: "Rule 236. If a man has given his boat on hire to a boatman, and the boatman is careless, and the boat is sunk and lost; then the boatman shall replace the boat to the boat owner."

[27]Ibid., 66.

[28]G. R. Driver and J. C. Miles, *The Babylonian Laws*, 2 vols. (Oxford: Clarendon Press, 1952), Vol. 1, 465.

[29]Ibid., 462.

[30]Ibid., 462.

[31]Ibid, 465.

[32]Ibid, 464.

[33]Ibid, 466.

[34]Ibid, 464.

antiquity and imprecision, the influence of the Hammurabi code in later bodies of laws cannot be easily, or clearly traced. But it works as a reference of the oldest standard applied to sea carriers.

2.1.2.2 The Roman Law

2.1.2.2.1 Historical Context

Given the power and expansion of the Roman Empire and the outstanding development of their legal system; many, if not most of the modern legal theories ground their principles on the legal reasoning of Roman jurists. Their main laws are found in the Digest of Justinian; a compilation of analyzes, opinions and decisions of the Roman jurists, completed around 533 A.D.[35] Despite the impressive development of the Roman law, maritime law was not a subject of great attention by Roman jurists. Rather, it is said, it was borrowed from the Rhodian Law, where the Romans considered the maritime law to have originated.[36] Such a conclusion comes from the frequently cited answer of Emperor Antonius Pius to a Nicomedian merchant who claimed jurisdiction on a maritime dispute. The emperor Antonius Pius answered with this famous sentence: "I am the lord of the world, it is true, but the law is the lord of the sea. Let the maritime law of Rhodes be applied to any question where no law of ours is inconsistent with it."[37]

The absence of their own maritime laws is attributed to the fact that Rome was more a military than a commercial or maritime power.[38] Agriculture had a bigger importance for their economy[39] and therefore, there were initially more regulations on the use of land.[40] Trade was even conceived as an indecorous matter and, in some cases, discouraged.[41] However, after the Punic wars, (ca. 246–146 BC), the Romans started to develop their navy.[42] During the last two centuries of the Roman Republic (ca. 509 B.C.–26 B.C.), the carriage of goods by sea increased due to the trading of slaves, the need to secure the supply of food, and the importation of

---

[35]Schoenbaum, 5.

[36]Anderson, III, 195.

[37]T. E. Dönges, *The Liablility for Safe Carriage of Good in Roman-Dutch Law* (Cape Town: Juta & Co. Ltd., 1928), 4.

[38]A. C. Schomberg, *A treatise on the maritime laws of Rhodes: By Alexander C. Schomberg* (Oxford: sold by D. Prince and J. Cooke. J. F. and C. Rivington, P. Elmsly, and T. Payne and Son, London, 1786), 7; Kent, Vol. III, 5.

[39]Anderson, III, 195.

[40]Meyer-Termeer, 147.

[41]Schomberg, 15; Kent, Vol. III, 5–6; Dönges, 2. An example of this is a law introduced by the Emperor Claudius forbidding men of noble or illustrious families and members of the Senate to own ships larger than necessary to carry their own corn and fruits.

[42]Sanborn, 7; Anderson, III, 184.

luxury goods.[43] Some engagement in commerce is reported with vessels carrying up to 400 tons of cargo calling on ports in the Mediterranean, Black Sea and some Atlantic ports.[44]

Notwithstanding, their maritime legislation was not in an independent body, but included in their civil law, and had no separate maritime tribunals.[45] Paulus and Ulpian are two of the most cited jurists in the provisions of the Digest of Justinian, relating to maritime law.[46] Despite the lack of an extensively developed maritime legislation, there were, among others, rules regarding the ownership of a vessel, the danger of pirates and collisions, general average, the charter of a vessel, salvage, maritime loans and liability for freight.[47]

2.1.2.2.2 The Concept of Diligence

As mentioned, the carrier activity was originally within the scope of the civil law. There was not an exclusive jurisdiction for shipping law in a separate branch. The carrier's liability was judged according to the degree of diligence demanded by the type of contract they subscribed. Carriage was practiced as a *depositum* or a *locatio operis*. Both contracts imposed on the debtor a special duty of diligence or care of the goods entrusted on him. There we find the first sources of the concept of diligence as used today in the Hague Rules.

(1) **Diligence and the Relation with *Culpa***

Roman jurists analyzed the concept of *diligentia* in conjunction to the concept of *culpa*. According to the classical jurisprudence, *culpa* was defined as the lack of due diligence in the compliance of an obligation.[48] The concept was applied to obligations *ex delicto* as well.[49] To differentiate it from the *aquilian culpa*, it was referred to as contractual *culpa*.[50] Diligence was primarily documented as a positive duty of Roman officials and guardianship of minors, and then used to extend

---

[43]Meyer-Termeer, 147; R. Zimmermann, *The law of obligations: Roman foundations of the civilian tradition*, 1st edn. (Cape Town: Juta & Co. Ltd., 1990), 406.

[44]Anderson, III, 184–185. *See* Lionel Casson, "*Ships and Seamanship in the Ancient World 369*" (1971).

[45]Schoenbaum, 4.

[46]Anderson, III, 196.

[47]Schoenbaum, 5.

[48]A. Torrent, *Manual de derecho privado romano* (Madrid: Edisofer, 2002), 360; M. J. García Garrido, *Derecho privado romano: Casos, acciones, instituciones*, Derecho/ ediciones académicas, 13th edn. (Madrid: Ediciones académicas, DL 2004), 419.

[49]W. L. Burdick, *The Principles of Roman law and their relation to modern law*, (Repr.) (Holmes Beach, Florida: Gaunt & Sons, Inc., 1989), 415–416.

[50]García Garrido, 419.

liability not only for contractual *dolus*, but also for contractual *culpa.*[51] *Culpa* or negligence was differentiated from *dolus* through the absence of a conscious intention to produce damage to the creditor.[52] Diligence was especially required in contract of *negotiorum gestio*, but later was extended to all *bona fides* contracts, including partnership, sale, hire, loan, mandate and deposit.[53] In these contracts, the debtor was liable not only for *dolus,* but if he did not manage the creditor's goods or business with the expected due diligence, he was liable also for *culpa* for any damage caused.[54]

Though it is not certain that the Romans developed a classification for three degrees of *culpa*, later commentators distinguished in the Roman texts some nuances or different degrees of this concept.[55] The first is the *culpa lata* or gross negligence: "not realizing what everyone realizes", understood as excessive negligence, even comparable to *dolus.*[56] The second is *culpa levis* or ordinary negligence, which depending on the standard of diligence to be confronted, is subdivided in *culpa levis in abstracto* and *culpa levis in concreto.*[57] The third is *culpa levissima,* or the lack of fulfillment of a special attention or diligence for a specific obligation.[58] The problem of determining the type of *culpa* to be attributed to a specific debtor was then solved by the determination of different standards or degrees of diligence.[59] In consequence, they distinguished three degrees: *diligentia bonus pater familias*, *diligentia quam suam in rebus* and *custodia.* The liability of a wrongdoer was determined precisely attending to the degree of diligence he was required to practice in the performance of the obligation contracted.

---

[51]H. Hausmaninger, 'Diligentia Quam In Suis: A Standard of Contractual Liability from Ancient Roman to Modern Soviet Law' (1985) 18, Cornell Int'l L.J., 179–202, 185.

[52]García Garrido, 418.

[53]Ibid., 418–419.

[54]Ibid., 419.

[55]Note, 'Three Degrees of Negligence, The' (1873) 8, Am. L. Rev., 649–68, at 665: "Since the publication of Sir William Jones essay, and within the present century, it has been demonstrated, almost beyond cavil, that there does not exist in the Roman law a uniform division of negligence into degrees. It is generally admitted that the Florentine manuscript of the Digest is the oldest and the most valuable of the manuscripts; that its reading of the text of the law Contractus is the correct reading; … that Pothier was wrong; and that the symmetrical scheme of a threefold division of negligence is not supported by the texts of the Roman law, but was the work of the commentators". Zimmermann at, 192 suggest also that it was the medieval lawyer who came to distinguish different degrees of negligence. Torrent, 361.

[56]García Garrido, 420; Dig. 50.16.213; Dig. 16.3.32.

[57]Torrent, 361; García Garrido, 420.

[58]Torrent, 361; García Garrido, 420.

[59]Notwithstanding, it has been argued that the Romans described these different types of negligence, but did not clearly define the associated degrees of diligence needed. *See* W. W. Buckland, *A Manual of Roman private law* (Cambrige: University Press, 1925), 337.

(2) **Degrees of Diligence in Roman Law**

aa) Diligentia Bonus Paterfamilias

This degree of diligence is also called *exacta diligentia or exactissima diligentia.* The debtor was compelled to exercise the most rigorous level of care, taking all the measures that a *bonus paterfamilias* would take. The *paterfamilias*, as known, placed such a fundamental importance on Roman private law that his actions were taken as a standard of conduct. Under this degree, liability was attributed according to an objective criterion, regardless of negligence.[60] Failing to accomplish this degree of diligence made the debtor responsible for *culpa levis* in *abstracto*,[61] but also for culpa levissima.[62] According to Gaius, this degree was required in contracts of consumption such as *mutuum*, where the debtor was bound even if the object loaned became lost or damaged by accident.[63] The same degree was required to debtors in the quasi contract of *negotiorum gestio* (management of other business); *commodatum,* (gratuitous loan to a person for his personal use); *deponens* (gratuitous deposit); and some others.[64]

bb) Diligentia quam suis rebus

This is a lower standard of care than the *diligentia exactissima*. The debtor was not obliged to practice such an extreme level of care. His liability was not measured attending to the abstract standard for the diligent *paterfamilias.*[65] *Diligentia quam in suis rebus* was the diligence that an ordinary person would show in his own interest, when regularly managing his own affairs or business.[66] Damages caused by the failure in the fulfillment of this standard, made the debtor liable for *culpa levis in concreto.*[67] The first mention of this degree of diligence was made by Celsus,[68] referring to the contract of deposit.[69] It was then developed later by Gaius, as a special and independent standard of *culpa*-liability separate from the

---

[60]Zimmermann, 192, at 193, on the reason for this high standard Zimermann says that: "[T]he answer to this question lies in Justinian's tendency, originating in Greek philosophy and reinforced by Christian religion, to make fault the central element of the law relating to liability."

[61]Torrent, 361; García Garrido, 420.

[62]Zimmermann, 192.

[63]Dig. 44.7.1.4: "[The debtor] is obliged to observe the most exact diligence in guarding the property, and it is no sufficient for him to practice the diligence which applies to his own affairs, if someone could have guarded the property in a more diligent manner."

[64]Note, 'Three Degrees of Negligence', at 667: "...the pledgor and the pledgee; the vendor and the vendee."

[65]Zimmermann, 210.

[66]Torrent, 361.

[67]Ibid., 361.

[68]Hausmaninger, 183.

[69]Dig. 16.3.32: "The statement made by Nerva that gross fault is equivalent to fraud was not accepted by Proculus but seems to me to be very true. For even if a person is not careful in the degree required by the nature of man, still, unless he shows in the deposit the care customary with him, he is not free from fraud; for good faith is not maintained if he shows less care than in relation to his own affairs."

exactissima diligentia rule.[70] It was required to be exercised in contracts of partnership, joint ownership, tutors, curators, and by the husband in relation to his wife's properties.[71]

cc) Custodia

Finally, we have the special type of diligence called *custodia*, applied to specific contracts and related to the care and preservation of tangible goods.[72] In the early Roman law, the carriers were liable only for *custodia*,[73] meaning the act of guarding, watching, caring or keeping goods safely. Under Justinian law, responsibility for *custodia* appears within the standard of *diligentia exactissima*.[74] *Custodian prestare* was the behavior required to debtors to keep the object safe during the time it was under their tenancy; and to give it back in the same condition they had received it.[75] This standard implies an absolute obligation of the debtor to restore the goods that are under his *custodia*, even when the goods are damaged or lost for any cause, and regardless of his diligent or negligent activity.[76]

2.1.2.2.3 Standard of Care for Sea Common Carriers

(1) **The Original Standard**

In Roman times, the carriage of goods was originally performed under two categories of contracts: *Depositum*, executed for free; and, *Locatio operis*, in the special form of *Locatio conductio*, which required payment.[77] In the first case, the depositary was compelled to practice *diligentia cujusvis hominis*[78] or *diligentia quam in sua,* holding him liable only for *dolus* (fraud and malice). The owner of goods could file an action *in depositum* for damages or loss of his property. In the second, the *Location operis*, the debtor was required to practice *omnis diligentia* or *bonus pater familias*.[79]

The contract of carriage of goods was mostly performed under the contract of *locatio conductio*, entered into with an *exercitor* or *magister navis*.[80] The *exercitor* was the person who received the profits of the ship, be it the shipowner or bareboat

---

[70]Ibid., 200.

[71]Note, 'Three Degrees of Negligence', 667–668.

[72]Ibid., 665.

[73]Karan, 8.

[74]Torrent, 363.

[75]Zimmermann, 194.

[76]Torrent, 362.

[77]Dönges, 20.

[78]Ibid, 29.

[79]Ibid, 30.

[80]Meyer-Termeer, 177.

charter.[81] The contract was for the hire of the whole ship or part of the ship.[82] It was in the second case, the hire of space on the ship, were an obligation with the *excercitor* to perform the carriage came about. Thus, the carrier was responsible for damages or losses caused by *dolus* or for *culpa* of himself or his servants.[83] The liability was even for "*culpa levis*", and the affected creditor had an *actio ex locatio* or *actio ex conductio* to claim his losses or damages.[84] Under this system, the *exercitor* was exonerated from liability when damages of losses were caused by theft, or caused by third parties, or by *force majeur* such as robbery, trap of pirates, shipwreck, fire, and unusual water levels.[85]

(2) **The Praetor's Edict**

Besides the regulation and the actions granted to the cargo owner under the two aforementioned contracts, another more specific and stricter regulation for the common carriers was later introduced by a *Praetorian* Edict. The exact date of its introduction is unknown, but for sure, it was after the enactment of the *locatio conductio*, which already existed by the second half of the 2nd century B.C.[86] Labe commented on this about the year 10 or 11 A.D.[87] Some authors have tried to figure out the date of its enactment, determining that it was probably around or after the 1st century B.C.[88] Interestingly, this concurred with the expansion of the maritime commerce in Rome.

The common carriers activity came to be regulated by a *Praetorian* Edict. The same granted an *actio de recepto*, dogmatically considered as one of the most interesting and historically significant of the *pacta praetoria*.[89] *Receptum*, in this case, was the undertaking by a sea carrier, who warrants that the customer's goods are to be kept safe while on the ship or in his premises.[90] That made that the carriers liability was not based on *culpa* but on *receptum*.[91]

---

[81]Sanborn, 10; Meyer-Termeer, 150.

[82]Meyer-Termeer, 177.

[83]Ibid., 177.

[84]Dig. 4.9.3.1.

[85]Meyer-Termeer, 177.

[86]Ibid., 185, at 177: „Daraus ist zu schliessen, dass es die *location conductio* schon in der ersten Hälfte des 2. Jahrhunderts v. Chr. gab."

[87]Ibid., 185.

[88]Ibid., at 185–186: "Sehr viele Autoren sind der Ansich, das *receptum nautarum* sei vor oder spätestens im 1. Jahrhundert v. Chr. enstanden; als argumentführen sie vor allen an, dass in dem Edikt das Wort "*nautae*" und nicht "*exercitores*" verwendet wird, woraus sie schliessen, dass das Edikt über das receptum nautarum in einer Zeit erlassen worden sei, in der das Wort *exercitor* noch nicht bestanden habe, so dass es älteren Ursprung sei als die *actio exercitoria*".

[89]Zimmermann, 515.

[90]Ibid., 515.

[91]Dönges, 28.

aa) Content of the Edict

The *Praetor Edict* establishes that "*Nautae, caupones, stabularii, quod cujusque salvum fore receperint, nisi restituent, in eos judicium dabo*."[92] Ulpian explains that shipmasters (*nautae*), innkeepers (*caoupones*) and stablers (*stabularii*) must restore the property, luggage or traveler's effects entrusted on them. Although this edict makes references to cases not related to shipping (innkeepers and stablers), Ulpian makes it clear that the edict has primarily a maritime character.[93] It relates mainly to the *nautae* activity; the other two cases being extensions to this edict.[94] This is the main instrument of liability in Roman law for those who regularly practiced the carriage of goods by sea, establishing an absolute obligation to restore the goods, regardless of whether the carriage was gratuitously or under payment.[95] The carrier warrants the goods will be *res salva fore*.[96] This *receptum* makes public carriers liable for *custodia*.[97] It demands a more active duty on the vigilance of the cargo that falls also into the category of *exactissima diligentia o diligentia diligentissimi paterfamilias*.[98] The obligation was not limited to protecting the cargo against theft, but according to Gaius and Paul, also against damages.[99] That implies liability for *culpa in non-faciendo*, for omission incurred by the carrier or his servants in preventing theft and damages caused by third parties.[100]

However, even if the carrier practiced such a high level of diligence, in the end it was irrelevant. The edict established an absolute obligation for the *nautae* to restore the goods,[101] regardless of whether the goods were destroyed, lost or damaged without the sea carriers own fault, or the fault of his employees; or by other incidents not considered as *vis maior*.[102] The edict established a special regime of absolute or strict liability[103] that required sea common carriers to act as insurers of the goods.[104]

---

[92]Dig. 4.9.1.

[93]Dönges, 9.

[94]Ibid., 9: "In Dig. IV, 9.1. 1–4, where Ulpian treats of such important matters as the reason for the introduction of the Edict (§1), the person liable under the Edict (§§2 and 3), and the extension, in practice, of this liability to "exercitores ratium et lintrarii," (§ 4), he always speaks of, or at any rate, implies 'nautae.' In §5 'caupones et stabularii' are mentioned for the first time. Just as in §4 I suggest, he is here referring to an extension, beyond nautae, of the person to whom the Edict applied".

[95]Ibid., 9, 22.

[96]Meyer-Termeer, 197.

[97]Dönges, 24; Zimmermann, 515; Dig. 4.9.5.

[98]Dönges, 32.

[99]Dig. 4.9.5.1; 4.9.6.4.

[100]Dönges, 31.

[101]Ibid., 32; Zimmermann, 515.

[102]Zimmermann, 1121.

[103]G. C. J. J. van Den Bergh, 'Custodiam Praestare: Custodia-Liability or Liability for Failing Custodia?' (1975) 43, Tijdschrift voor rechtsgeschiedenis - The Legal History Review, 61–72, 62; Torrent, 362.

[104]Zimmermann, 515.

bb) The Reasons for the Edict

Ulpian comments on the reasons for the enactment of this special and stricter regulation. He argues that the obligation was not too strict because, in contrast to the later conception of common carrier, the carrier was not obliged to take everyone's goods; it was up to the carrier's own discretion which goods he undertook for transportation.[105] He describes it as *maxima utilitas huius edicti*, to preserve good faith, to insure the safety of the goods delivered, and to prevent fraud and robbery.[106] On preventing fraud and robbery he additionally says:

> […] it is necessary to confide largely in the honesty of such men; and if they were not held very strictly to their duty, they might yield to the temptation to commit a breach of contract, and even into secret leagues with thieves.[107]

The strict standard was necessary to prevent collusion between carriers and thieves in stealing the goods; a situation that seemed to occur often during that time.[108] Sea carriers and innkeepers did not enjoy a good reputation and Rome was full of robbers.[109] Pomponios also suggests that this was a measure taken by the *Praetor* to make known that he had an interest in protecting people against dishonesty.[110]

On the other hand, cargo owners already possessed possible actions against the carrier, who under the *locatio operis* were liable for *custodia*.[111] In addition, also available were the *actiones furti* and the *action* and *damni in factum aversus nautae,* for claims of theft and damage respectively.[112] Zimmerman says that the introduction of this action was because it was based in a general provision of care of the goods, a more sophisticated creation than the older restrictive remedies.[113] Another explanation is found in the interest to remove difficulties regarding evidence, as shipper might not have been able to determine how their cargo was damaged or lost and who might have been responsible for said damages.[114]

---

[105]Dig. 4.9.1.1.

[106]Fletcher, 96.

[107]T. M. Mackenzie and J. Kirkpatrick, *Studies in Roman law: With comparative views of the laws of France, England and Scotland*, 7th edn. (Holmes Beach, Fl: Wm. W. Gaunt, 1991), 222–223; Dig. 4.9.1.1.

[108]Dig. 4.9.1, Ulpian further says: "Let no one think that the obligation placed on them is too strict; for it is in their own discretion whether to receive anyone; and unless this provision were laid down, there would be given the means for conspiring with thieves against those whom they receive, since even now they do not refrain from mischief of this kind."

[109]Fletcher, 96.

[110]Zimmermann, 516. Dig. 4. 9. 3. 1.

[111]Ibid., 517.

[112]Ibid., 517.

[113]Ibid., 517.

[114]Fletcher, at 96, Fletcher quotes the explanation provided by J.B.C. Stephen in (1896) L.Q.R. XII. 119: "It was, therefore, better that the carrier should be held liable for all loss or damage (since he could best take precautions against such loss) than that the freighter should be deprived of his remedy".

cc) To Whom it Applies

Specifically referring to the *nautae,* the action could be brought against the "seaman", a term that might include all the crew people on the ship. Ulpian clarifies that it does not mean all the crew members. *Nautae* means the person who managed the ship and must answer for the actions of the master's and those ordinary sailors acting under his orders.[115] The seaman's or carrier's liability was extended to wrongful or negligent acts performed by all his sailors, be free men or slaves. Gaius further states that a man who runs a ship must be responsible for the wrongful acts caused not only by him, but also by the crew members, because he is using the services of such bad men for running his vessel, and in consequence he is liable, not on the basis of contract but in *quasi-delict.*[116]

In cases where slaves caused damages to the property of his own master, while they were employed on the ship, the carrier was also liable. This was explained by Paul: if the cargo owner had his slave working in the vessel where his goods were carried, and this slave stole or caused damaged to his own master's cargo, the cargo owner still had an *actio in factum* against the carrier.[117] The liability is also extended to acts performed by some other persons on the ship holding some level of authority over her, such as ship′s guards or cabin stewards, who received goods on behalf of the shipowner.[118] Ulpian further states that the carrier is liable also for damage or losses resulting from acts of passengers.[119]

Similar liability is for the transport of columns, where the *conductor operis* was liable for his fault and for the fault of his employees.[120] Roman law established carrier's liability for wrong or negligent actions of his crew or servants, in contraposition of what we have today as an exclusion of liability for nautical fault, listed in the Hague Rules. This provision shows the rigorous standard of care they were compelled to apply in the course of their business.

dd) Goods Covered

The rule applies to cargo received by the *exercitor* or by the master,[121] including goods received by agents especially appointed by the carrier; or by other servants in his service with an implied position to receive goods.[122] The rule is even extended to passenger's personal baggage, and provisions shipped on board for the

---

[115]Dig. 4.9.2.

[116]Dig. 44.7.5.6.

[117]Dig. 4.9.6. Ulpian further states in Dig. 4.9.7.4, that the person in charge of the ship is liable for losses caused by another's slaves because when employing them on the ship "he ought to investigate their reliability and integrity."

[118]Dig. 4.9.3.

[119]Dig. 4.9.8.

[120]Zimmermann, at 1121: "Again we are dealing with vicarious liability stricto sensu, albeit in a contractual context" *See* Dig. 19. 2. 25. 7.

[121]Dig. 4.9.1.3.

[122]Dig. 4.9.1.2; Dig. 4.9.1.3.

voyage.[123] It even applied to goods that did not belong to the person who put them on board, but who had a real interest in their safe arrival at the destination.[124] But the carrier was liable only for losses occurred to goods while on board the ship.[125] In any case, it applies to the transport of goods of private persons. There is no reference in the Digest about transport of goods for the State subject to this regulation.[126]

ee) Possible Exceptions

Originally, this guaranty was absolute and comprised all forms of *vis maior.*[127] Later some exceptions were granted in cases of the damages or losses resulted from shipwreck or piracy, and to other forms of *vis major* or *damnum fatale.*[128] Carriers however, could not rely on these exceptions when they were responsible for exposing the goods to those risks.[129] In addition to these exceptions, it is worth noting that carriers were not deprived of the right to contractually negotiate this strict rule. Ulpian listed some causes which relieve the carriers of liability.[130] Among them, the carrier had the possibility to exclude his liability by declaring that each passenger must take care of his own goods and that he will not be responsible for any loss. If the passenger agreed with that, the carrier was relieved from liability.[131] This contractual exclusion, however, applied only for passenger's property, as they could take care of their own goods during the voyage. Nothing is said about property of people not traveling on the ship.

ff) Impact in Future Legislation

During the first centuries of the Christian era, the Romans held control over the, then considered center of the world, including the Mediterranean Sea; from Syria and Egypt to Britain; including the Atlantic coast of Spain, Portugal, France and England.[132] Their maritime law became so uniform and universal that many of its principles remain today.[133] Such uniformity responded to the needs of the sea-born

---

[123]Dig. 4.9.1.6.

[124]Sanborn, 11; Dig. 4. 9.1.7.

[125]Dig. 4.9.7.

[126]Meyer-Termeer, 190.

[127]Zimmermann, 515.

[128]Ibid. 515; Introduced by Labeo in Digest 4.9.3.1.

[129]Dönges, 33.

[130]Ibid., at 102: "…Ulpian, in his eighteenth book on the Edict, where he enumerates the circumstances under which the carrier will not be liable, distinctly affirms this right in the following words: 'Item si praedixerit, ut unusquisque vectorum res suas servet neque damnun se praestaturum, et consenserint vectores praedictione, non convenitur'. …This right of modifying the strictness of the liability under the Edict is affirmed by Lauterbach, Glück, Windscheid and Heineccius."

[131]Dig. 4.9.7.

[132]Mangone, 5.

[133]Sanborn, 19.

trade, which changed little during that time.[134] Though the Roman Empire had mostly disintegrated by the fifth century in much of Western Europe, they kept control of commercial cities such as Naples, Amalfi, Gaeta, and the eastern part of Sicily until the eleventh century.[135] The maritime law of the late Roman Empire was applied in these port cities of Italy,[136] where the growth of medieval commerce started, and subsequently provided some codes of maritime law.[137] Sanborn says that the laws of the late Empire were later collected in separate centuries into two codes: The Rhodian Sea Law and the Basilica.[138] Then, these laws spread to France, Spain, England, the Low Countries and Visby in the Baltic, and were described as a "*Law common to all Nations.*"[139] Indeed, its principles are found in the *Lex Mercatoria*, as well as in concrete bodies of maritime laws, such as the Consolato del Mare; the Laws of Oleron; The Laws of Visby and the Ordinance de la Marine of Luis XIV.[140] The success of its adoption and application in all these European trade centers is attributed, among other reasons, to "its respected history and sophistication in the settlement of shipping disputes".[141]

Today the influence of the Roman law in our contemporary law is indisputable. It exists not only in countries with a civil law system. It has influenced the legal system of the common law countries as well, where many of their legal institutions are based in Roman laws principles.[142] Kent points out that these laws reflected the wisdom of the philosophy of the great Roman Jurists, and especially "the spirit of equity, in all its purity and simplicity seems to have pervaded those ancient institutions".[143] Indeed, the vast impact of the Roman law is attributed also to its philosophical content, also influenced by Greek philosophy.[144] Hence, the standard of liability set in the *Praetors* edict became part of the European *ius commune*.[145] It strongly underlined the maritime policy of modern Europe and North America, as we will see later.[146]

---

[134]Ibid., 19. *See* Desjarding, Arthur, "*Introduction Historique à l'Etude du Droit Commercial Maritime*" (Paris, 1890), 2; Mangone, 5.

[135]Mangone, 5.

[136]Ibid., 5.

[137]Sanborn, 35. E.g. Trani, Amalfi, etc., at 26: "We need not wonder, therefore, nor consider it to be the flowering of some incomprehensible seed, when we see codes of maritime law and commercial courts springing up in Italy in the XI and XII Centuries almost immediately, as historical time goes after the official of the Empire had left the peninsula forever."

[138]Ibid., 35, 39.

[139]Ibid., 27, 41.

[140]Burdick, 76.

[141]Mangone, 6.

[142]C. P. Sherman, *Roman Law in the modern world*, 2nd edn., 3 vols. (New York: Baker, Voorhis & Co., 1924), Vol. 1, 321.

[143]Kent, Vol. III, 7–8, note (a). Papinian, Paul, Julian, Labeo, Ulpian and Scaevola.

[144]J. Gordley, *The philosophical origins of modern contract doctrine*, Clarendon law series (Oxford [England], New York: Clarendon Press; Oxford University Press, 1991), 30.

[145]Zimmermann, 520.

[146]Mackenzie and Kirkpatrick, 222.

2.1.2.3 The Sea Law of Rhodes

2.1.2.3.1 Historical Context

Rhodes is an island in the South Aegean Sea, a few hundred kilometers from Greece and some twenty kilometers from modern Turkey. It is considered the origin of the early maritime law.[147] Rhodes was a maritime center whose citizens are held as the earliest people that created, digested and promulgated a maritime legislation.[148] They had a strong commercial power that gave great authority to their law.[149] It was a sort of *lex mercatoria* common for the states bordering on the Mediterranean Sea.[150] Its principles were accepted and applied by Greeks and Romans.[151] Today it is not possible to set an exact date when these laws started to be compiled. There is only a general presumption that it was probably about nine centuries before the Christian era; or perhaps later, when Rhodes began to have superiority on the seas, some two centuries before the foundation of Rome.[152] Sanborn states its date around the 3rd or 2nd century B.C.[153]; concurring also with the expansion of the Roman maritime commerce.[154] However, there is no convincing and precise historical evidence of the exact content of such law, with the exception of two references made in the Digest of Justinian; and comments made by authors such as Cicero and Strabo.[155] None of them refers to the liability for carriage of goods.

There is a code entitled the Sea Law of Rhodes (*Homos Rodion Nautikos*), compiled hundreds of years after the possible original law.[156] The accuracy of this

---

[147]Anderson, III, 195; B. Farthing, *International shipping: An introduction to the policies politics and institutions of the maritime world*, 2nd edn. (London [etc.]: Lloyd's of London Press, 1993), 2.

[148]Kent, Vol. III, 4. However, it is also said that they have adopted many of the previous Phoenician statutes. *See* Jados, at xii: "The Phoenicians, a great seafaring people, promulgated other laws that governed sea commerce in the Mediterranean, about 2000 B.C.…The island of Rhodes, a commercial center, adopted many of these Phoenician statutes, later referred to erroneously as the Rhodian Laws."

[149]Sanborn, 5.

[150]Zimmermann, 407–408.

[151]Sanborn, 5.

[152]Schomberg, 37–38.

[153]Sanborn, 5.

[154]Ibid., 7; Meyer-Termeer, 147–148; Anderson, III, 184.

[155]R. D. Benedict, 'The Historical Position of the Rhodian Law' (1909) 18, The Yale Law Journal, 223–42, at 230: "The maritime power of Rhodes is stated to have been prominent during the three or four centuries preceding the Christian Era, and this, therefore, would naturally be the period during which the Rhodian Maritime Law, if any existed, would have taken form and substance. But these two references by Cicero and Strabo, both made not far from the time the Christian Era, seem to be the only references to Rhodian Law which either Greek or Roman literature can furnish us."

[156]Jados, at xii: "It is claimed that the Byzantine emperor, Leo the Isaurian (714–741), issued a code of laws called in Greek, *Nomos Rodion Nautikos,* named after the Rhodian Laws. However, upon close examination, the Isaurian statutes have actually no relation to the earlier Rhodian Laws."

compilation is subject to doubt regarding its original content, as changes may have been introduced over the years. Such a code is held as the first collection of maritime laws of the later Roman Empire.[157] Ashburner wrote the most accurate work on the "Nautical Law of Rhodes" in 1909.[158] He affirms that the code was put together probably between 600 A.D. and 800 A.D., from different materials and different epochs.[159] The provisions found there correspond mainly to maritime regulations in the Byzantine period, which governed the maritime industry in the eastern Mediterranean during that time.[160] Sanborn also points out that it was in use in the south of Italy for some six or seven hundred years.[161] Therefore, we cannot consider it as the law governing the maritime commerce before or during the Roman Empire, but as the regulation after the same, during the first part of the Middle Ages.

2.1.2.3.2 The Standard of Carrier's Liability

This code does not provide a specific rule with the obligation of restoring the cargo, or the carrier's liability for not restoring goods. An important point remarked by Ashburner and echoed by Schoenbaum, is that the Part III of this code, in conjunction with the book LIII of the Basilica, constituted a more complete body of maritime laws for the Byzantine period, around the IX century A.D., and even before.[162] The book LIII of the Basilica did not contain new regulations but reproduced the provisions relating to nautical law of the Justinian Digest.[163] The strict rule of liability of the *Praetor's* edict found in the Digest 4.9.1 was included in the Basilica.[164] Hence, this absolute standard was maintained during the Byzantine period. Ashburner also confirms that, pointing out that "some statutes speak as if the shipowner's obligation was absolute provided that the goods have been written by the *scribanus* in his *quaternus* or *cartularium*".[165] Liability could also be incurred in the process of loading and unloading the goods to and from the ship.[166]

---

[157]Sanborn, 35.

[158]Schoenbaum, 7.

[159]W. Ashburner, *Homos Rodion Nautikos*, The Rhodian Sea-Law, Repr. of the ed. Oxford 1909 (Aalen: Scintia, 1976)., 112–113.

[160]Schoenbaum, 7.

[161]Sanborn, 37. *See* Wagner, Dr. Rudolf and Pappenheim, Dr. Max: "*Handbuch des Seerechts*" (Leibzig, 1884, 1906) *Systematischen Handbuch der deutschen Rechtwissencahft*, edige by K. Binding, 60.

[162]Schoenbaum, 7.

[163]Ashburner, civ, cxcviii.

[164]Ibid., cv.

[165]Ibid., at cxcviii. He cites some statutes such as: "St. Ragus. VII, 6; St. Phara, V, 1; St. Pera, V, 14; St. Massil. IV, 26,, 127; C. Tortosaq, IX, 27, 9. The Ordin. Trani, 16, speak as if the scrivano was personally liable."

[166]Ibid, cc.

However, this absolute standard is not limited to a general rule to restore the goods. Unlike the very general provisions contained in the Justinian Digest, this code introduces some more specific regulation on technical issues of the maritime navigation, which had to be regarded by both carriers and shippers. The standard is more evident when dealing with cargo damage, and refers concretely to damages arising from a lack of seaworthiness and improper care of the cargo.

(1) **The Duty to Provide a Seaworthy Ship**

The code did not explicitly establish an obligation on the shipowner to provide a vessel in proper condition for navigation. But, though it is not expressly stated, from this period and these statutes comes the obligation of the carrier to provide a seaworthy ship. Perhaps not in all its aspects but at least in one of the most important: water-tightness. Some statues of the same period state that the vessel must be well *calcata,* meaning watertight.[167] Liability for bad *calcatura* was set statutorily. If goods were damaged by water, because the vessel was not watertight, the shipowner was absolutely liable, and must restore the goods in the same quality and quantity.[168] As in Roman law, exceptions to liability are given if the carrier can prove that the damages were caused by a storm or became wet during the extinguishing of a fire.[169]

In addition, Chapter 11 grants merchants, willing to hire a vessel, the right to inquire and check before loading their cargo that she was well prepared, properly equipped and manned.[170] Chapter 11, it is said, might have been established by a later Emperor, but most of its content comes from customs of the middle Ages.[171] It seems more like an advice for merchants than a proper legal provision.[172] However, it presents three specific aspects of the ship that the merchant is supposed to inspect before loading: (1) the condition of the vessel itself; (2) the tackle; and, (3) the mariners.[173] These three elements correspond to the modern concepts of: seaworthiness, equipment worthiness and human worthiness. Since that time, these elements have been a subject of statutory attention as part of the proper integrity of a ship and for the correct performance of the carriage. Nevertheless, it does not

---

[167]Ibid., clxxxii.

[168]Ibid, cc.

[169]Ibid., cc.

[170]Ibid., at 92. "Chapter 11: The merchant and the passengers are not to load heavy and valuable cargoes on an old ship. If they load them, if while the ship is on its voyage it is damaged or destroyed, he who loaded an old ship has himself to thank for what has happened. When merchants are hiring ships, let them make precise inquiry from the other merchants who sailed before them putting in their cargoes, if the ship is completely prepared, with a strong sail yard, sails, skins, anchors, ropes of hemp of the first quality, boats in perfect order, suitable tillers, sailors fit for their work, good seamen, brisk and smart, the ship's side staunch. In a word let the merchant make inquiry into everything and then proceed to load."

[171]Sanborn, 38; Ashburner, cxiii.

[172]Ibid., lxxvi.

[173]Ibid., clxxxiii.

establish whether the merchant assumes part of the risk when he decides to load his cargo into a vessel that is not in her optimal condition.

On equipment worthiness, the code orders the carrier to furnish proper equipment for the carriage of specific goods commonly carried in those times. The carrier had to foresee possible risks to occur during the voyage, such as storms, which doubtless have always been one of the main and most common risks faced in maritime navigation. The code contains some provisions assigning as a duty of the carrier, to be prepared for the possible occurrence of such risks. Chapter 34, for example, outlines the case of the carriage of linen and silk. For its proper carriage, the captain must "supply good skins, in order that in a storm no harm may be done to the freight by the dashing of the waves."[174]

(2) **The Duty of Care of the Cargo**

Further duties are set regarding the obligation of care of the cargo. Here, the duty of care is partially shared with the merchant who was supposed to travel with his cargo. The liability for water-damaged cargo is allocated depending on who discovers the water in the ship and reports it to the other party. The captain and the crew are liable if the passengers warn them about the water in the ship, and they do nothing about it. Alternately, if the crew informs the passengers and cargo owners of the situation but they do nothing, then the former are freed from liability.[175]

Chapter 38, regarding the carriage of corn, provides additional obligations in the case of water-damaged cargo. Under a gale, the captain must "provide skins and the sailor work the pumps."[176] During such meteorological events, the sailors have to prevent the corn from being spoiled by water from the bilge. If they are negligent in doing so, they have to pay a penalty. If the cargo suffers water damage due to a gale, the captain, sailors and the merchant, share the losses. In this case, even during acts of God, such as a gale, the captain must still bear part of the merchant's loss.[177] As it is understood that the general standard of liability for damages was strict, this rule may have been intended to establish a fine against negligent sailors. The punishment was more severe for sailors who stole, who had "to make it good twofold and lose his whole gain."[178] In any case, according to Chapter 44, there is an express obligation to make up the deficiencies in quality and quantity of the goods, when they are injured by water from the bilge.[179] In this case, the captain had to restitute the goods as he had received them.

In general, the Sea law of Rhodes is mostly remembered for its provisions on jettison and general average. Regarding the contract of carriage of goods, it

[174]Ibid, 109.

[175]Ibid., cxcix.

[176]Ibid, 112.

[177]Ibid, cxcix.

[178]Ibid., 112.

[179]Ibid., cxcix, at 117: "…But if the cargo is hurt more from the bilge and not from the gale, let the captain take the freight and hand the goods dry in quantity as he took them".

provides the first outlines of the concept of seaworthiness and the adoption of measures regarding the proper care of the cargo during voyage.

2.1.2.4 The Rules of Oleron

2.1.2.4.1 Historical Context

Oleron was an island and trading center located in the Atlantic, in the Aquitaine bay of France, not far from Bordeaux, once under the control of the English Crown.[180] In the middle of the 12th century, these rules were introduced in England either by Eleanor, Duchess of Guienne and wife of Henry II, or by their son Richard I (Coeur de Lion).[181] Their exact origin is still obscure, being disputed by France and England.[182] The Rules originally consisted of twenty-four judgments copied later into the Black Book of English Admiralty.[183] These judgments were not exactly ordinances; neither law properly promulgated, but a compilation of judgments showing how the maritime court of Oleron would decide some cases.[184] Pronounced by the "*prud' hommens*" or merchants, it became a part of what was then known as the law merchant, also derived from Roman law tradition and Italian sources.[185] They were applied as the general maritime law in the North Sea and Atlantic Ocean, becoming the main source of the later maritime laws of Visby and the Hansa Towns.[186] They passed into the English law through the adoption into their customs by the English ports of London and Bristol at the beginning of the 13th century.[187] Later, they were applied in the courts of Admiralty in England.[188] Its application during the Middle Ages outlined the basements of the English maritime law.[189] Highly praised by its equity and wisdom,[190] they are unanimously considered to be the direct foundation of modern European maritime law.[191] But not only in Europe, even in the 19th century they were still being cited in some decisions of the Supreme Courts of the United States.[192]

---

[180]Schoenbaum and Yiannopoulos, 7.

[181]Sherman, Vol.1, 353.

[182]Benedict and Knauth, Vol. 1. §6; Schoenbaum, 10.

[183]Mangone, 7.

[184]Schoenbaum, 10, at note 10.

[185]G. Miller, *The legal and economic basis of international trade* (Westport, Conn: Quorum Books, 1996), 100; Schoenbaum, 10.

[186]Schoenbaum, 11.

[187]Fletcher, 43; Schoenbaum, 11.

[188]Miller, 100.

[189]Ibid., 46.

[190]Schomberg, 88, 90.

[191]Benedict and Knauth, Vol. 1. §6; Schoenbaum, 10; Schomberg, 88.

[192]*See New Jersey Steam Navigation Co. v. Merchants' Bank of Boston*, 47 U.S. 6 How. 344, 12 L. Ed. 465 (1848).

2.1.2.4.2 The Standard of Liability

As mentioned, these rules were not an exact legal code,[193] and like the previous commented historic sources, these rules did not explicitly state what the carrier's standard of liability was. Fletcher points out that the Oleron's rules were not made by lawyers and were not derived from any theoretical reasoning.[194] Hence, he affirms, it is not possible to determine a general principle of liability, nor is it possible to affirm that the master was liable for negligence.[195] But if Roman law still influenced the maritime practices of the people of this area at that time, consequently, it is possible to expect that their judgments may have consider the standard set in the *Praetors' edict*. One thing that is quite clear is the possibility for the shipowner to be relieved from liability for damage caused by *damnum fatale*, such as shipwreck and piracy,[196] as it was in Roman law.[197] But the liability standard in itself requires further and more detailed analysis to reach a more accurate conclusion.

The Rules X and XI, regarding breach of contract by damage or loss of cargo can shed some light on finding the standard of liability.[198] These two provisions made the master liable for damages to or loss of the cargo caused by two specific events.[199] The first regards defective gears (cordage, robes and slings); which in

---

[193]Miller, 100.

[194]Fletcher, 45.

[195]Ibid., 45.

[196]Schoenbaum, 10, at note 11.

[197]Introduced by Labe in Digest 4.9.3.1.

[198]E. Cleirac, *The ancient sea-laws of Oleron, Wisby, and the Hanse-towns, still in force. Taken out of a French book, intitled, Les us & coutumes de la mer*, 31686th edn. (Abingdon, Oxon: Professional Books, 1981), 6. "Rule X. When the master fraight [sic] a ship, he ought to shew his merchant the cordage that belong to her. And, if they see any Thing [sic] amiss or wanting, he must rectifie [sic] it. For, if for want of good cordage any pipe, hogshead, or other vessel, should happen to be spoiled or lost, the master and mariners ought to make it good to the merchants. So also, if the ropes or sling break, the master not having shewed to the merchants, he must make satisfaction for the damage. But if the merchant say, that the cordage is good & sufficient, and rest satisfied therewith, and afterwards it happens that they break; in that case each of them shall share the damage, viz. the merchant to whom the goods belongs, and the said master with his mariners.

Rule XI. A vessel being laden with wines, or other goods, and hoysing [sic] sail at Bourdeaux, or any other place, if the master and his marines have not trimmed their sails as they ought to have done, and it happens that ill weather overtakes them at sea, so that the main yard shakes, or breaks one of the pipes or hogsheads; the ship being arrived at her port of discharge, the merchant says to the master, that by reason of his yard his wine was lost. In that case, if the master replies, it was not so, both he and his marines (be it four or six, or such of them as the merchant shall think best) must take their Oath, that the wine was not destroyed by them, nor by the main yard, or thought their default, as the merchant charge them: and then the said master and his mariners shall be acquitted thereof. But, if they refuse to make oath to that effect, they are then obliged to make satisfaction for the same. For they ought to have ordered their sails aright, before they sailed from the port where they took in their lading."

[199]Krieger, Vol. 15, 79.

modern maritime law corresponds to a want of equipment worthiness. The second relates to the want of proper trimming of the main yard and sails on the vessel which relates to the obligation of care of the cargo. Both cases demanded actions of the master and his mariners prior to the beginning of the voyage in order to assure the safe carriage of the goods, even in case of "ill weather", as stated in rule XI. The difference between them is in the possibilities for the master to ameliorate or exclude his liability.

Rule X covers the carrier's obligation to furnish good equipment (cordage). It is possible to reduce his liability if he shows the cordages to the merchant before the voyage. The merchant assumes solidarity in the damage by giving his approval for it, and damage caused by defect in the cordages. However, even when the master shows the merchant the cordage, he remains at least partially liable. It might be the case of what we now call an exception of contributory negligence on the part of the cargo owner.

The case described in Rule XI relates to damages caused by the movement or shake of the main yard during ill weather. The master may avoid liability by an oath made by him and his mariners, denying the cause of damage was the shake of the main yard, alleged by the merchant. Ill weather is an "act of God" that normally releases the carrier from liability, but does not release him if the lack of proper trim exposed the cargo to the harmful effects of the storm on the vessel. Therefore, the occurrence of an exculpatory cause did not free the carrier immediately from liability, because he was expected to have prepared his vessel even to face the consequences in the case of ill weather. The carrier cannot escape liability for damages caused due to lack of proper trim. Here the influence of the Roman's logic is noteworthy. The carrier will not be relieved from liability for a *vis major*, if he exposed the cargo to that risk; or, knowing it, did not take measures to overcome them, nor in the occurrence of an exculpatory cause, where he negligently contributed to the loss or damage. This concept still remains in the HR.[200]

But the mere reading of the liabilities described in these two rules do not allow to clearly identifying what standard was applied to carriers. Krieger is recognized for carrying out the most detailed and accurate research on the Rules of Oleron.[201] He affirms that the criteria exposed in the commented rules is based on the German law principle which states that the person who caused the damage must respond for it, independently of the existence of *culpa* or not, or of any obligation of diligence.[202]

---

[200]HR Article 4(q).

[201]Schoenbaum, 10, note10.

[202]Krieger, Vol. 15, at 80: "Es handelt sich hierbei um nicht anderes als um die Anwendung des germanischen Rechtsgrundsatzes:„Wer Schaden tut, muβ Schaden besern", wonach die Haftung für nicht vorsätzlich zugefüge Schäden allein danach beurteilt wird, wer den Schaden verursacht hat, ohne das Merkaml der„culpa" im Sinne einer moderner Fahrlässigkeit zu berücksichtigen. In diesem Sinne ist auch die komplizierte Regelung des Kapitels X der Rôles d'Oléron zu verstehen. Ausgehend von dem genannten Grundsatz gibt sie dem Kapitän die Möglichkeit, die Kaufleute zur Mitwirkung und damit zur Mitverantwortung für die Auswahlt des Arbeitsgerätes heranzuziehen und somit die Härte dieser Erfolgshaftung zu mildern".

If the elements of negligence or diligence are indifferent to assign liability, then the standard is closer to the strict one. This again, is not surprising keeping in mind the spread of the Roman concepts still in vogue during that time.

An important point is that these rules assume, just like in the Rhodian Sea Law, that the merchants are traveling in the ships with their cargo and that they can personally take care of their property while on board.[203] Thus, the cargo is not completely left under total *custodia* of the carrier. The presence of the merchant on the ship was of such importance, that they had to be consulted in all decisions on cargo handling in case of perils at sea.[204] In this case, carriers' duties are partially shared with the merchant, and apparently, the liability as well, depending on the case.

2.1.2.5 The Consulate of the Sea

2.1.2.5.1 Historical Context

The exact origin of these regulations is again not clear. Azuni affirms the Pisans compiled it[205]; while Grotious and Marquardus say it was compiled from the maritime ordinances of Greek emperors, emperors of Germany and some others.[206] However, it is mostly accepted that its origin is found in Catalonia, Spain.[207] Barcelona became one of the most important and richest trading centers of Europe in the 12th century, trading with the Mediterranean cities of Pisa, Genoa, Sicily, Greece and Egypt.[208] King James I of Aragon in 1258 established 21 regulations for navigation prepared by the "prudent men" of the maritime guild of Barcelona.[209] A public maritime court was created by 1347, chaired by consuls responsible for the administration of the maritime law.[210] Ashburner sets the date of the *Consolato* at the end of the 14th century, although he admits that it contains "large masses of much earlier material."[211] The compilation is estimated to have been completed by 1370.[212] The earliest printed edition dates from the 14th of July

---

[203]Knauth, 115.

[204]Miller, 100.

[205]Jados, xv.

[206]Kent, Vol. III, at 9: "… it was a collection made in the time of the crusades, from the maritime ordinances of the Greek emperors, of the emperors of Germany, the kings of France, Spain, Syria, Cyprus, the Baleares, and from those of the republics of Venice and Genoa."

[207]Jados, v, at xv: "Pardessus, refuting Azuni's contention, ascribed it to a collection made in Barcelona, written in a Romance dialect, under the title *Consulat de la Mar*, and published in the fourteenth century". *See* Pardessus, Collection, Vol. V, 367.

[208]Ibid., v; Mangone, 10.

[209]Mangone, 10.

[210]Ibid.

[211]Dönges, 14.

[212]Ibid., 11.

of 1494, divided into three parts with 334 articles.[213] It is frequently accepted as the earliest general code of maritime laws in modern Europe that is still in preservation.[214] Its importance was such that it was referred to by Azuni as a legislation "whose authority is above all others."[215] The provisions were observed not only in Spanish ports, but also in other Mediterranean cities of Italy and France.[216] Its second part states, among others matters, the responsibilities of the shipowners for cargo carriage.

2.1.2.5.2 The Standard of Liability

The code first presents, in Chapter 61, a general obligation of the patron to protect and care for the life and property of the passengers and merchants. The term "patron" means the person in command of the vessel, and is used indistinguishably for captain, master of the vessel or commander.[217] The obligation to protect the cargo from damages is developed with detailed provisions relating to two main aspects: the vessel's seaworthiness and the care of the cargo.

(1) **The Vessel's Seaworthiness**

On the ship's seaworthiness, the Consulate pays also particular attention to the vessel's water-tightness. The code establishes liability for damages caused by water coming from the deck, ships sides, from the bilge or other part of the vessel, where the cargo has contact with water, or for lack of proper caulk. In these cases, Chapter 63 establishes that the carrier must compensate the merchants for all the damage caused by water seepage.[218] The strictness of the provision was such that if the carrier is not able to pay back the damages to the merchant, the ship must be sold for that purpose. The carrier's obligation to provide a watertight seaworthy ship seems to be absolute, but still allowed for two exceptions. The first relates to cargo damage caused by water entering into the ship due to rough weather or storm.[219] It is the common example of act of God, largely admitted as an exception. The second exception may challenge our understanding of the absoluteness of the obligation. Chapter 66 relieves the carrier for damages caused by water leaking through the vessel's bottom, though being well caulked.[220] However, the same Chapter 66

---

[213]Ibid; Mangone, 11.

[214]Dönges, 14.

[215]Jados, xiii. *See* D. A. Azuni, Droit Maritime de L'Europe (Paris, 1805) Vol. I, 392.

[216]Mangone, 11.

[217]Jados, at xvi: "This often leads into confusion since a patron may have been an able-bodied seaman who merely leased the vessel from its owner in order to engage in trade, or he may have been hired by the owner of the vessel to command it for a predetermined fee or a share on the profit the vessel would earn."

[218]Ibid., 36.

[219]Ibid, Ch. 65.

[220]Ibid, 37, Ch. 66.

clarifies that the exception is based in the assumption that the merchant has checked the vessel previously to hire her, and confirmed for himself that she is watertight. When he notices a want of proper caulk, but does not report it to the master, he therefore assumes part of the risk. Notwithstanding, if the merchant reports the leak to the "patron", he is responsible to accomplish any promise he made to the merchant in this regard.[221] This condition to assign liability demands that the merchant be familiar with the structural or technical condition of the vessel. Probably most of the merchants during that time were familiar with the condition of the ship, or perhaps, it applied more to charterers. But, for those merchants who did not have proper skills to determine whether the vessel was well caulked or not, the code provides an explanation of what was considered to be well caulked:

> Chapter 66: …In order to avoid all disputes between the patron and the merchants over these matters, our ancestors, in order to explain what they meant by proper pitching and tarring of the vessel, stated the following: If the deck of the vessel was tarred up to the deck rail or above it, and also up to or above the openings for anchor chains, the patron of the vessel cannot be held responsible for the damage or waterlogging of the cargo, even if the water seeped through the deck.[222]

It was a technical description based in the custom for making the ship watertight. This aspect of the seaworthiness condition was statutorily stated. Its mere compliance relieved the carrier for damages caused by water seepage.

Another interesting and quaint provision of these rules regards the equipment worthiness. Damages caused to cargo by rats on the ships were common in that time. To prevent them, the code ordered shipowners to have cats on the ships. The code provides that if the cargo reported damages by rats and there was not a cat on the vessel, the shipowner was liable for not having such "equipment" on board.[223] The matter was serious. The code clarifies that if the shipowner had cats before the voyage began, and they died after the departure, he was not liable for damages caused by rats before the arrival to a port where he could buy a new cat. In this case, he was not liable to compensate because it was not his fault.[224] This last reference suggests that the liability here is based on *culpa* from the part of the shipowner when the expression so stated. It must be noticed however, that the cat's death might fall into the concept of act of God or of *vis mayor*, when the shipowner could not have total control over the cat's life. One wonders what might have happened in case of damages caused by those same cats. The code is silent on this possible issue.

(2) **The Care of the Cargo**

The Consulate of the Sea presents a special concern on cargo stowage. It provides detailed rules indicating where to place the goods and the order of storage into the ship. There was an express prohibition for a shipowner or helmsman to place the

---

[221]Ibid.

[222]Ibid., 37.

[223]Ibid., Ch. 67.

[224]Ibid., 38, Ch. 68.

goods on the ships in damp places, near to the mast, or the steering wheel, or in the bottom of the vessel, or in the prow, or in any other place where it can suffer damage.[225] In addition, the carrier was not allowed to place cargo at the bottom without flooring it; neither to use cargo to make basement or flooring for other cargo, unless they are all of the same weight.[226] If the cargo at the bottom suffers damages because of this improper stowage, the shipowner was responsible and must make good the damage.[227] It even goes further into detailed explanations on cargo stowage: the heavier cargo should be placed underneath those that are lighter as a matter of preventing damage to the lighter cargo.[228] The carrier was also liable for damages to or losses of cargo carried on upper open deck without consent of the merchant.[229] The processes of loading and unloading had to be undertaken by the "patron", if so was agreed with the merchant,[230] but without liability if the cargo gets waterlogged during these operations.[231] An absolute obligation to restore the carried goods is made clear not for damages but for losses. Chapter 67 provides that:

> Any goods or possessions loaded aboard the vessel and entered in the ship's register, which are subsequently lost, will be the responsibility of the patron of the vessel and its owners must be compensated by him for their loss.[232]

'No mention to the cause of the loss or any specific exclusion or exception for this obligation is stated. The code follows the ancient tradition of holding the shipowner strictly liable for the loss of the cargo. The "patron", whom we understand to be the shipowner or carrier, as well as any other shareholder, must pay them but according to the participation or share they have in the vessel. The ship itself was responsible for all damages.[233] This is perhaps one of the first mentions of the limitation of liability to the value of the vessel. When the selling of the ship was necessary to pay off merchants for their cargo losses, it deprived even other creditors and shareholders, with the sole exception of sailors' wages who had a privilege over all other debts.[234] In some cases, such as the carriage of cargo on deck without consent of the merchant, Chapter 185 establishes that if the result of the sale of the vessel is not enough to compensate the loss, and the patron still has additional goods, these goods shall also be sold to pay the merchant.

---

[225]De Capmany y de Monpalau, D. Antonio. Translator of "*Código de las Costumbres Marítimas de Barcelona, hasta aquí vulgarmente llamado Libro del Consulado.*" Vo. 1. Madrid, 1791, 130, Ch. 62.

[226]Ibid., 134–135, Chapters 68–70.

[227]Ibid., 134, Ch. 69.

[228]Ibid., 135, Ch. 70.

[229]Ibid., 138, Ch. 185.

[230]Ibid., 136, Ch. 72. The merchant could make such an arrangement directly with the sailors.

[231]Ibid., 136, Ch. 71.

[232]Ibid., 133, Ch. 66.

[233]Ibid., 136, Ch. 71.

[234]Ibid., 130 and 138, Ch. 62 and 185.

In general, the Consulate shows a more protectionist regulation for the cargo interest. For its time, the Consulate is perhaps the most complete set of rules for the maritime commerce and navigation. As seen, it is not restricted to rules establishing compensation for cargo damages or losses. It provides the perhaps first technical descriptions, or objective rules to be followed in making the ship water-tight and how to perform the stowage of the cargo. There is not a unique and general rule establishing carrier's liability for cargo damages or losses. They are dispersed along the code depending on the cause of the damage. First, damages caused by water-logged cargo for want of water tightness; secondly, those damages caused for want of proper stowage or due care. The standard of liability might offer some doubts regarding the cargo damage caused by other reasons. But it is clear, however, that it was strict in cases of cargo loss.

2.1.3 *Anglo-American Maritime Law*

2.1.3.1 The Maritime Law of England

English Law has been the most influential and direct reference of the modern maritime law. It took special preponderance since the industrial revolution when England became the major ship owning nation of the world. There have been discussions about the original liability standard of the common carrier in English Law.[235] Most of the available decisions issued by English judges on the matter show that the common carriers were subject to strict liability. It is evident since the end of the 17th century and beginning of the 18th. Previous to that period, this standard is not so clear. However, it was the enforcement of this strict standard during the 18th and 19th centuries that catalyzed the change for the current regulation contained in the Hague Rules. We try here to present the sources and reasons of this absolute regime applied in England to common carriers and the reasons that motivated it's evolution through the adoption of international regulation.

2.1.3.1.1 The Influence of Roman Law on the Admiralty Courts

By the 13th Century, the courts of the Mediterranean cities of Pisa, Trani and Amalfi, were popular forums of maritime justice.[236] There, civil law was applied for the resolution of conflicts relating to freight, damage to goods shipped, marine wages, and maritime contracts in general.[237] The model was followed by England,

[235]The doctoral thesis of E.G.M. Fletcher, *see supra note* 20, challenges the historical continuity of an absolute liability for common carrier under English law.

[236]G. a. J. C. F. Bruce, *A treatise on the jurisdiction and practice of the English Courts in Admiralty actions and appeals*, 3rd edn. (London: Maxwell, 1902), 4.

[237]Ibid., 4.

through the Admiral, a state officer entrusted to solve maritime conflicts.[238] Though the origin of the English Court of Admiralty is uncertain,[239] according to the records, it was formally established by Edward III.[240] It gained greater importance from the 14th century, following the Battle of Sluys in 1340 when England took command of the sea.[241] This court was highly influenced by Roman law, introduced, as mentioned before, through the adoption of the Rules of Oleron.[242]

But the influence of Roman law on English law was not only due to the application of the Rules of Oleron.[243] Before that, following the Norman invasion in 1066, there were other concrete events that prompted the study and application of Roman legal institutions in England.[244] The Roman influence on English Law during the 12th and 13th centuries was such, that it is remembered as the "Roman epoch of English legal history".[245] Thus, the maritime law of England grew from the customs of merchants of the Mediterranean and Atlantic sea which was originally based on the principles of equity and justice derived, in much of its completeness, from the Roman law.[246] Therefore, the primary body of English maritime law was a collection of prior laws from Rhodes and the Rolls of Oleron, among

[238]Ibid., at 3: "Whether judicial functions were originally conferred upon him or not may be matter of doubt, but as soon as maritime affairs began to assume importance, matters happening at sea, and not within any county from whence a jury could be summoned, requiring judicial investigation, were referred to him for adjudication."

[239]Ibid., 3.

[240]Sherman, Vol. 1, 3.

[241]R. C. Fitzgerald, 'Admiralty and prize jurisdiction in the British commonwealth of nations' (1948) 60, The Juridical Review, 106–22, 106.

[242]Sherman, Vol. 1, at 364–365: "The Court of Admiralty, established in the 14th century during the reign of Edward III, owing to its necessary relations with foreign countries gradually adopted procedure and rules based on the roman civil law, the Court of Admiralty came to observe the partially Romanized rules of Oléron. The Court of Admiralty retained its importance, name jurisdiction, and roman law tendencies until very modern times, when in the 19th century it became a part of he consolidated English High Court of Judicature".

[243]Burdick, 76.

[244]Ibid., at 65–67, 70–71. Burdick lists some of the other sources: 1) the designation by King William of the Italian scholar very well versed in Roman Law, Lanfranc as prime minister, chief adviser and in 1070, Archbishop of Canterbury; 2) The existence of the ecclesiastical courts applying the Canon Law of the Continent, highly influenced by Roman law; and, 3) The designation of Vacarius, circa 1143, a learned teacher of Roman Law at Bologna, as first professor and founder of the school of law at the University of Oxford. Furthermore, two books on the law and customs of England, one of the jurists Ranulph de Glanville between 1180 and 1190, and the other of Bracton, between 1250 and 1258, show the influence of Roman law in the legal reasoning of English jurists.

[245]Ibid., 74.

[246]Ibid., 76.

others, called the *Black Book of Admiralty*, compiled in the early fourteenth century in the reign of Edward II.[247] With Roman law as an important source, the rule set in the *Praetors Edict* was observed, as indeed was later confirmed.

2.1.3.1.2 The Sea Carrier as a Bailee

Although, the Rules of Oleron were applied in the admiralty court of England, in the development of English maritime law, the first analyses of the carrier's liability were not made exactly on transport law, but under the law of bailment.[248] The carrier was held as a bailee and his liabilities were firstly analyzed in such a context. Since the origin of the law of bailment in England, the bailee's liability was subject to the general principle of strict liability.[249] Holmes says that the English law of bailment was of German origin, which in turn, was developed by German philosophers who were also professors of Roman law.[250] Fletcher acknowledges further that the source of the absolute liability, in force since that period, could have been a consequence of Germanic introduction from the time of the Conquest, in addition to the influence of Roman law.[251] The Germanic common law of the Norman Conquest made bailees of all sources absolutely responsible for goods delivered, even when lost by theft and regardless of negligence.[252] The strict liability was based on the possessory remedies granted to the bailees, which allowed them to claim against third parties; and in consequence, remained strictly liable before the bailor for the lost goods.[253] This standard, according to Holmes and Holdsworth, remained throughout the middle ages.[254] Fletcher disagrees with this conclusion arguing the lack of proper evidence to support the carrier's absolute liability had been in fact continuous and steady.[255]

During the 13th century, this strict rule was subject of doctrinal analysis in the works of Bracton (ca. 1250), and Britton (1287). Bracton tried to introduce the

---

[247]E. Gold, *Maritime transport: The evolution of international marine policy and shipping law* (Lexington, Mass: Lexington Books, 1981), 39, *citing* C. John Colombos, *Internatinal Law of the Sea,* 6th ed. (London: Longmans, 1960) pp. 34–35.

[248]Fletcher, 1.

[249]T. A. Street, *The Foundation of legal liability*, A presentation of the theory and development of the common law, 3 vols. (Northport, Long Island, N.Y.: Thompson, 1906), Vol. 2, 255.

[250]Holmes, Ed. by Mark De Wolfe Howe, 133, 138; Gorton, vol. 43, at 62 says that Holdsworth shares Holme´s opinion on the German origin.

[251]Fletcher, xi. He also suggests that the absolute liability of the carrier could have been an Elizabethan innovation applicable to carrier by land or due to application of the Praetor Edict.

[252]Ibid., 2.

[253]Gorton, vol. 43, 62. *See* Holmes, Ed. by Mark De Wolfe Howe, 138; Holdsworth, *A History of*, vol. 3, 337; J. Basedow, 'Common Carriers Continuity and Disintegration in U.S transportation law' (1983) 13, Transportation Law Journal, 1–42, 5.

[254]Ibid., 63.

[255]Fletcher, xiii, 11; Gorton, Vol. 43, 63.

Roman concepts of different degrees of diligence, but he did not succeed.[256] Following Bracton, Britton suggested releasing the bailee from liability when the item borrowed was lost or damaged by fire, water, robbery or larceny, unless these cases occurred through their own fault or negligence.[257] His suggestion however, does not seem to have been largely adopted by the courts, but by the 14th century, the strict rule was moderated in certain cases.[258] In this period is where Fletcher points out that the "liability of a bailee was something substantially less that an absolute liability".[259] However, by the 15th century, a case against the *Marshal of the Marshalsea* (1455) about the escape of a prisoner, demonstrates the application of the original strict standard.[260] The court held the Marshal absolutely liable because he had the possibility, as a bailee, to recover the goods or compensation from the wrongdoer who caused the damage or loss.[261]

In the 16th century, Story reports that during the time of Henry VIII (1509–1547), the carriers where subject to liability for robbery, only if the carrier had exposed the cargo to risks such as traveling by dangerous roads, inconvenient hours, or at night.[262] But later, under the Elizabeth commercial reign (1558–1603), the common carrier was held again responsible for all losses, excluding only acts of God or enemies of the King.[263] Thus, in the 16th century we find the first reported cases analyzing the carrier's liability.[264] In *Woodlife's Case* (1596),[265] a factor was sued to restore goods entrusted to him. The defendant alleged that such goods, along with his own goods, were the subjects of robbery. Popham C. J. stated that:

> […] It is a good plea before auditors, and there is a difference between carriers and other servants and factors, for carriers are paid for their carriage and take upon them safely to carry and deliver the things received.[266]

---

[256]Street, Vol. 2, 254.

[257]Ibid., 256.

[258]Ibid., 259. He refers to the *Bonion's Case* (1315).

[259]Fletcher, 18.

[260]Holmes, Ed. by Mark De Wolfe Howe, 140. Jailers were hold as bailees in charge of cattle (the prisoners). A prisoner escaped by the action of subjects of the king who broke the prison and released the prisoner. The court stated that if the prisoner were released by king's enemies, then it would be a cause of exclusion of liability as there were not any to claim, but as the Marshal had an action against the subjects of the king, then he was answerable to the bailor.

[261]Ibid., 140.

[262]J. Story, *Commentaries on the law of Bailments: with illustrations*, 9th edn. (Boston: Little, Brown, 1878), 460. *See Jones on Bailm.* 103; Saint Germain *Doctor and Student*, Dial. 2 ch. 38.

[263]Ibid.; Fletcher, at xi also points that the strict standard may have been an innovation in the time of Elizabeth applicable to land carriers, then extended to sea carriers.

[264]Holmes, Ed. by Mark De Wolfe Howe, 143.

[265](1596) Moore 462.

[266]As cited by Street, Vol. 2, 263.

The strict liability seems to be based on the payment they receive for the service of carriage.[267] This reasoning is said to be characteristic of the 16th and 17th centuries when the doctrine of consideration appeared for the first time.[268] Still, the rights and duties of carriers were set in the law of bailment, in which bailment may or may not have been for payment.[269] This is evident in *Southcott v. Bennett* (1601).[270] This was an action on detinue against an ordinary bailee who pleaded robbery of the goods without his negligence.[271] The court held strictly liable all bailees, even those who accepted the possession of goods from others as a favor.[272] Sir Edward Coke in his report on this case stated that "[a]n obligation to keep is the same as an obligation to keep safely; in either case the bailee is liable if the goods are stolen."[273] The case confirmed the absolute liability of the bailee, and was held up as the leading case for the next hundred years.[274]

Notwithstanding, Street reports two cases where a plea of robbery was accepted as an exception of liability.[275] After them, in *Morse v. Slew* (1671),[276] another sea carrier alleged to have been robbed. Sir Mathew Hale held the defendant liable, despite having had enough number of watchmen for the protection of the ship. The argument of the decision was based again on the reward the carrier receives for his work.[277] What appears here is that during these centuries, the standard was not totally clear to have been completely absolute, or at least, robbery was admitted as an exculpatory cause in some cases.

2.1.3.1.3 The Distinction of Common Carrier

Since the 14th century and prior to the appearance of the modern bill of lading and the charter party, the English law took into consideration some general obligations

[267]Street, Vol. 2, 299.

[268]Gorton, Vol. 43, 53.

[269]Ibid.

[270](1601) 4 Co. Rep. 83b; 76 E.R. 1061.

[271]Street, Vol. 2, 263; Gorton, vol. 43, 65.

[272]Holmes, Ed. by Mark De Wolfe Howe, 142. But in the gratuitous bailment, the bailee could evade liability if prior to taking possession of the goods, he had expressly rejected responsibility for them; Fletcher, 27.

[273]Fletcher, 32.

[274]Holmes, Ed. by Mark De Wolfe Howe, 142.

[275]Street, Vol. 2, 265–266. In two cases in 17th century robbery was accepted as a good plea: *Williams v. Hide* (1628) and *Vere v. Smith* (1661).

[276]86 E.R. 129.

[277]Street, Vol. 2, 269.

of the carriers with independence of the contract.[278] However, proper distinction of the character and duties of common carriers came centuries after in *Coggs v. Bernard* (1703).[279] The reasoning expressed therein made this case one of the most important for determining the liability of the common carrier.[280] Lord Holt C. J. overturned previous decisions on the strict liability for all kind of carriers, and established a differentiation between general bailees and common callings; and specifically, between private and common carriers.[281] Different liability standards were assigned to each one. The first were liable for negligence; while all bailees for reward, practicing a public employment, expressly mentioning common hoymen, master of ships and common carriers, were strictly liable.[282] Lord Holt C. J. explained the reasons supporting the assignation of this demanding standard for common carriers as follow:

> A common carrier by custom or usage may lawfully claim a reward: and where a man carrying goods is of a public employment, as a carrier, hoyman, & c. he must answer for all events, excepting the acts of God, and the enemies of the King; and this is a political establishment, for the safety of all persons concerned, and whose affairs necessitate them to intrust [sic] such carriers. For by this means all private combinations between them and highwaymen and other robbers, are prevented, which cannot easily be discovered.[283]

Those persons practicing business that fell into the category of common callings (innkeepers, farriers, tailors, ferrymen, gaelors, etc.) were expected to avoid losses by unskillfulness or through improper preparation of the business.[284] Among them, common carriers were those who exercised carriage as a public employment, open to undertake carriage of goods for every person in general, holding themselves out as ready to perform transportation from place to place for a reward and as a regular business, not just a casual occupation.[285] This special category was coupled with special duties to serve every shipper without discrimination and with care, in an implied assumption on their part.[286] Their liability arose not only for damages or losses caused by their unskillfulness or the improper preparation of the business, but for every damage to or loss of the cargo regardless of their negligence, with the

[278]A. D. Hughes, *Casebook on carriage of goods by sea*, 2nd edn. (London: Blackstone, 1999), at 4: "In the *Bukton v Tounesende (the Humber Ferryman)* YB 22 Liber Assisarum No. 41, f.94 (1348), there were some rules protecting the customers and would-be customers of those who performed the 'common callings', including innkeepers and common carriers."

[279](1703) 2 Lord Raymond 909, 92 E.R. 107, 90 E.R. 971 (1703).

[280]Gorton, vol. 43, 59.

[281]Ibid., 66.

[282]Holmes, Ed. by Mark De Wolfe Howe, 149.

[283]90 E.R. 971 (1703).

[284]J. H. Beale, JR., 'The Carrier's Liability: Its History' (1897) 11, Harvard Law Review, 158–68, 163.

[285]Story, 465.

[286]Basedow, at 5: "Perhaps the better view is that the special obligations of the common callings were worked out during and not before the development of assumpsit, subsequent to the introduction of the action sur le case in 1285."

exception of those caused by acts of God or King's enemies. The reasons for assigning strict liability to common carriers were not only based on receiving a reward for the employment.[287] Lord Holt cited Roman law to ground this decision. Holmes noted that, but added that the mention to the exceptions of acts of God and enemies of the King were characteristic of English Law.[288]

Indeed, Lord Holt was well learned in Roman law and introduced some of its concepts in some of his judicial decisions.[289] He cited as authority the work of Bracton, who in turn, adopted the divisions, language and reasoning of the Roman texts.[290] It is evident in the arguments in support of this decision. He invokes practically the same reasons given by Ulpian in Digest 4.9.1, for the *Praetor's edict*; regarding the prevention of possible collusion between carriers and thieves.[291] In a later case, *Lane v. Cotton* (1706),[292] he expressly acknowledged that the principles of the English law are based on those of the Roman law, and added a more detailed explanation upholding the standard:

> […] for what is the reason that a carrier or innkeeper is bound to keep such goods as he receives at his peril? It is grounded upon great equity and justice; for if they were not chargeable for loss of goods, without assigning any particular default in them, they having such opportunity as they have by the trust reposed in them to cheat all people, they would be so apt to play the rogue and cheat people, without almost a possibility of redress, by reason of the difficulty of proving a default particularly in them, that the inconveniency would be very great. And though one may think it a hard case that a poor carrier who is robbed on the road, without any manner of default in him, should be answerable for all the goods he takes; yet the inconveniency would be far more intolerable if it were not so, for it would be in his power to combine with robbers, or to pretend a robbery or some other accident, without a possibility of remedy to the party; and the law will not expose him to so great temptation, but he must be honest at his peril. And this is the reason of the civil law in this case, which though I am loth to quote, yet inasmuch as the laws of all nations are doubtless raised out of the ruins of the civil law, as all Governments are sprung out of the ruins of the Roman Empire, it must be owned that the principles of our law are borrowed from the civil law, therefore grounded upon the same reason in many things.[293]

Concern for equity and justice, and the observation of the reality of the transport industry in that time, made to Lord Holt to apply this standard as a measure of public policy. His decisions undoubtedly, settled with more clarity the common carriers liability since the beginning of the 18th century.

---

[287]Story, 461 *See* Jones on Bailment at 103–104.

[288]Holmes, Ed. by Mark De Wolfe Howe, 155, 157.

[289]R. Zimmermann, 'Europa und das römische Recht' (2002) 202, Archiv für die civilistische Praxis, 243–316., 291–292; Gorton, Vol. 43, 60.

[290]*Nugent v. Smith* (1875) 1 C.P.D. 19, Brett J., at 28–29.

[291]Dig. 4.9.1.1, *See supra* bb) The reason for the Edict.

[292]12 Modern 472(1706); 88 ER 1458.

[293]Ibid., 481–481.

2.1.3.1.4 The Common Sea Carrier as an Insurer of the Cargo

Although the term "insurer" was not expressly mentioned in *Coggs v. Bernard* or in *Lane v. Cotton*, in these cases (as well in some others),[294] the common sea carrier was already treated as such.[295] It was in *Forward v. Pittard* (1785),[296] where Lord Mansfield used for the first time this term to describe the common carrier's obligation.[297]

Lord Mansfield, as well as Lord Holt, was well learned in Roman civil law.[298] He regarded the law merchant as convenient for the regulation of maritime matters, especially for its principles of equity and usages of trade on which it was based.[299] He had previously introduced into the English bar in the case *Luke v. Lyde*—(1759),[300] concepts of the Rhodian Laws, the Consolato del Mare, the laws of Oleron, among others.[301] It is in this case where he called the maritime law "not a law of a particular country, but the general law of nations", quoting Cicero to support this statement.[302] With this background, the arguments supporting his decision in *Forward v. Pittard* on the common carrier's liability are not surprising:

> By the nature of his contract, he [the common carrier] is liable for all due care and diligence, and for any negligence he is suable on his contract. But there is a further degree of responsibility by the custom of the realm, that is, by the common law; a carrier is in the nature of an insurer. It is laid down that he is liable for every accident, except by the act of God, or the King's enemies. ...But to prevent litigation, collusion, and the necessity of going into circumstances impossible to be unraveled, the law presumes against the carrier, unless he shows it was done by the King's enemies or by such act as could not happen by the intervention of man, as storms, lightning, and tempests.[303]

---

[294]*See Gibbon v. Paynton* 98 E.R. 199; (1769) 4 Burrow 2298. "...a common carrier, in respect of the premium he is to receive, runs the risk of them, and must make good the loss, though it happen without any fault in him; the reward making him answerable for their safe delivery."

[295]Street, Vol. 2, 301.

[296]99 E.R. 953, (1785) 1 Term Reports 27.

[297]Street, Vol. 2, 302.

[298]J. C. B. Campbell, *The lives of the chief justices of England: from the Norman conquest till the death of Mansfield*, 2 vols. (Philadelphia, 1851), vol. 2, at 258: "While at Oxford he attended lectures on the Pandects of Justinian, and during his working life he maintained that Roman civil law to be the foundation of jurisprudence."; Street, Vol. 2., 141; Burdick., at 76–77: "Mansfield was accused by his enemies of introducing into English Law principles unknown to its courts. 'The Roman code, the law of nations, and the opinions of foreign civilians are your perpetual theme,' said one of the letters of 'Junius'."

[299]Kent, II, 1–2.

[300](1759) 2 Burr. 882, 889–890.

[301]Kent, Vol. III, at 17: "...the treatises of Roccus, the laws of Wisby the maritime ordinances of Louis XIV, and the commentary of Valin."; Campbell, Vol. 2., at 258: "He thoroughly grounded himself in ancient and modern history by a perusal of the most eminent original historians."

[302](1759) 2 Burr 887; Kent, II. Vol. III, 1; Street, Vol. 2, 331; Campbell, Vol. 2, at 249: "Cicero indeed was his favorite, whose work Lord Mansfield translated into English while studying at Oxford."

[303]*See supra note* 296, at 956 or 33.

When referring to custom of the realm, he clarifies it means the common law.[304] His decision echoes the same reasons given by Lord Holt, which in turn, as seen, were the same as Ulpian on the necessity to prevent collusion, but adding the need to prevent litigation and the impossibility in most cases to prove the cause of damage or loss. Further explanations for holding common carriers as insurers were later largely presented in *Riley v. Horne* (1828).[305] In addition to fraud prevention, Judge Best also pointed in more detail to the difficulties for the cargo owner of proving the cause of damage or loss.

This character of an "insurer"[306] held the carrier liable for his own actions, as well as for those of the master, his agents and servants.[307] This liability covered the entire voyage until delivery of the goods. In *Hyde v. Trent and Mersey Navigation Company* (1793), the opinion of the majority of the court was that the carrier holds the risk for the goods until a personal delivery at the house or place of deposit of the consignee.[308] The shipowner could exclude his liability as aforementioned, when the damages or losses occurred through an act of God or enemies of the Queen. Lightning fell within the exception of an act of God.[309] Kent also explained that if the location of a rock or a sand bar was generally known, and the ship crashes or strands upon it without action of adverse winds, the loss or damages is imputed to the master negligence and not to an act of God or peril of the sea.[310] Another example presented by Story regarded the barge master who, under tempestuous weather, shoots a bridge; he is liable for his temerity and imprudence.[311] The general rule was that the carrier was responsible for every loss not occasioned by

[304]Custom of the realm was used as a synonym of common law. *See Nugent v. Smith* (1875) 1 C.P. D. 19, at 23; Gorton, vol. 43, 101.

[305]130 E.R. 1044; (1828) 5 Bingham 217. Best C.J. said at 220: "When goods are delivered to a carrier, they are usually no longer under the eye of the owner; he seldom follows, or sends any servants with them, to the place of their destinations. If they should be lost or injured by the grossest negligence of the carrier or his servants, or stolen by them, or by thieves in collusion with them, the owner would be unable to prove either of these causes of loss. His witnesses must be the carrier's servants; and they, knowing that they could not be contradicted, would excuse their masters and themselves. To give due security to property, the law has added to that responsibility of a carrier which immediately arises out of his contract to carry for a reward, namely, that of taking all reasonable care of it, the responsibility of an insurer. From his liability as an insurer, the carrier is only to be relieved by two things both so well known to all the country, when they happen, that no person would be so rash as to attempt to prove that they had happened when they had not, namely, the act of God, and the King′s enemies."

[306]Beale, JR., 168.

[307]Kent, Vol. II, 600. *Boucher v. Lawson* (1815) Cas T H 194; Story, 482.

[308](1793) 5 T.R. 389, at 399 judge Grose said: "The law, which makes carriers answerable as insurers, is indeed a hard law: but it is founded on wisdom, and was established to prevent fraud. But it seems to me, that it would be of little importance to determine that carriers were liable as insurers, unless they were also bound to see that the goods were carried home to their place of destination; since as many frauds may be practised [sic] in the delivery as in the carriage of them."

[309]Kent, Vol. III, 350.

[310]Ibid. Vol. III, 351–353; also in Story, 496.

[311]Story, 462. *See Jones on Bailm.*, at 107; *Amies v. Stevens*, 1 Str. 128.

the act of God, or of the King's enemies.[312] But of course it did not include damages caused by the inherent deterioration of quality or quantity of the goods, or for the ordinary wear, tear and chafing of the goods in their course of transportation, or for acts attributable to the shipper.[313] Being judged under an absolute standard, considerations on the duties of providing a seaworthy vessel and the care of the cargo were not so relevant. They became more important later. This rule governed the activity of common carrier not only in England but also in all nations of Europe under civil law.[314] Notwithstanding, the standard was much more rigid in English law than in the continental countries. In France for example, there was exclusion of liability for losses coming from superior force during robbery, considered within the scope of *dannun fatale.*[315]

2.1.3.1.5 Additional Reasons for the Absolute Standards

Under the absolute standard, the common sea carriers were held liable in a stricter manner than other businesses also exercising public callings. Why was this standard more rigorous in England? Primarily, it was, as we have stated above, the consequence of the historical sources of the English maritime law, clearly acknowledged in the reasons given by Lord Holt and Lord Mansfield in the aforementioned cases. Decades later, in *Nugent v. Smith* (1875), it was expressly recognized by Brett J., that this standard was directly taken from the *Praetor's* edict of the Roman law.[316] Public policy was pointed out as the main reason for its assignation on common carrier. He said that this exceptional liability rule was adopted in the common law as measure of public policy because the conditions of this trade in that time in England were similar to those in the ancient Rome that led to the Roman *Praetor* to adopt this special regulation for sea carriers.[317] But beyond the historical antecedents of this standard, there were some considerations of morality and political and commercial reasons inherent to that time as well, that supported its applications.

---

[312]Ibid., 463.

[313]Ibid., 463.

[314]Kent, Vol. II, 598, 604.

[315]Ibid.

[316](1875) 1 C.P.D. 19, at 24: "The reason of the implied promise, given by Lord Holt in Coggs v. Bernard, and by Best, C. J., in Riley v. Horne, founded on the reason on which the Prætor allowed the exceptional liability of ship-masters, inn-keepers, &c., applies at least quite as strongly to the part of the carriage by sea beyond the realm as to the part within it."; *See also* Story, 461; Gorton, vol. 43., 64.

[317]Ibid., at 29–30: "…but the exception [of common carriers liability], both in the Roman Empire and in England, was no natural exception, but one depending entirely on public policy, arising from the manner in which some particular kinds of business were carried on in both places… The two trades, therefore, carried on in England under the same conditions as the three enumerated in the edict, were, the ship-masters and innkeepers. The conditions which had induced the Prætor as matter of policy to hold them to a strict liability in Rome were the same conditions as existed in the mode of carrying on the same business in England."

(1) **Consideration of Morality**

It is said that early historical English law contained no contemporary philosophy but was rather elastic and practical.[318] This seems to have later changed. Reflections on morality, policy and concern for the social interest reflected the philosophical thinking of the epoch. During the middle ages the medieval state pursued a moral ideal where commerce and industry were regarded as relations between persons, not only as an exchange of goods or services.[319] Economic relations were conceived "as part of an eternal order inspired by God", and pricing and service were moral issues and not mere methods of profit maximization.[320] Honesty and fairness were highly regarded, and to take advantage of special situations in detriment of the community was frowned upon.[321]

Graveson points out that English law observed during its formation the general principles of natural law or natural justice, whether in their medieval conception of eternal law, or in the secularized version of the 17th and 18th century.[322] Judges and legislator accepted them consciously as they were considered a higher law that must be complied with.[323] The principles based in such theories became afterward part of the substantive law, remaining as fundamental postulates also reflected in the constitution and the tradition of English law.[324]

On this regard, James Gordley points out that the natural laws created by the late scholastics became the basis for the modern contract doctrine. During the sixteenth and early seventeenth centuries, they developed a contractual theory based on a synthesis of Roman law with the Aristotelian and Thomistic moral philosophy.[325] Later, in the seventeenth century, Grotius, the founder of the northern school of natural law, took over and propagated almost in the same terms such contractual doctrines, but without its Aristotelian and Thomistic philosophical content.[326] His successors followed the same path.[327] However, though these principles were not

---

[318]Fletcher, 3.

[319]W. Holdsworth, *A History of English Law*, 4th. reprint. (London: Mathuen & Co. Ltd., 1936), vol. 2, 468.

[320]Basedow, 6, note 20, *See*. Holdsworth, Vol. 2, 468–69.

[321]Holdsworth, Vol. 2, 469.

[322]R. H. Graveson, 'The Spirit of English Law' (1948) Vol. 60, The Juridical Review, 83–105, 100.

[323]Ibid., 89.

[324]Ibid., 89–90.

[325]Gordley, 69–70. The works of the scholastics in unifying the Greek philosophy, Thomistic morality and Roman law, started at the beginning of the sixteenth century with the founder of the Spanish natural law school, the Dominican Francisco de Vitoria and his pupils Diego Covarruvias (1512–1577) and Domingo de Soto (1494–1560), and were completed by the end of the same century with the work of the Jesuits Francisco Suarez (1548–1617), Luis de Molina (1535–1600) and Leonard Lessius (1554–1623).

[326]Ibid., 71.

[327]Ibid., at 71: "Indeed, the doctrines remained much the same in the work of his successors, Samuel Pufendorf (1632–1694) and Jean Barbeyrac (1674–1744), and in that of the French jurists Jean Domat (1625–1695) and Robert Pothier (1699–1772), who were to have a great influence both on the drafters of the French Civil Code and on the nineteenth-century common lawyers."

explained in philosophical or moral terms, its content was still based in the same grounds. This is what Graveson refers to as the secular version of the natural laws. If the English judges followed such principles, as Graveson notes, it is quite evident in Lord Mansfield, who came to confirm more forcefully the character of insurer of the common carriers.

For Lord Mansfield, recognized as the "most accomplished judge who ever presided over the courts of King's Bench," and whose decisions shaped the commercial law of England, the observation of ethic and morality seems to have been particularly important.[328] With this in mind, when Lord Mansfield decided maritime claims, it would be expected that his decisions were not only based on the judicial precedents of English law, but also on his direct study of the Roman sources, and the principles of natural laws, which intrinsically contained philosophical and moral considerations.

Therefore, the maritime law "created" by the judicial decisions during this period is recognized, in words of Kent, to be "reasoned at large and practically applied."[329] He further says, that the arguments of the bar as well as the opinions of the bench were "intermingled with the gravest reflections, the most scrupulous morality, the soundest policy, and a thorough acquaintance with all the various topics that concern the great social interest of mankind."[330]

Concretely with regards to the strict liability assigned to common carriers, it was applied not merely because the *Praetor's* edict said so, but also because, under further analysis, it was considered to be a "great principle of public policy, which has proved to be of eminent value to the morals and commerce of the nation in succeeding generations."[331] This conclusion was during a time when the common good was, perhaps, more valued as a main purpose of law, preferring the protection of the public by requiring more efforts on individuals offering a service on a commercial basis. The observation of morality demanded the prevention of collusions. It was also more reasonable to assign more responsibilities to the party who had more control on the shipping operation. Similarly, it was seen as fair to avoid having the weakest party bear the burden of proof for the causes of damage or loss of the cargo. It was clear that the shipper was at real disadvantage in gathering the evidence of such events. Considering these arguments, the application of the strict liability rule on common carriers sounds quite logical.

---

[328]Campbell, Vol. 2, 235, 253, 254, 258. He is remembered for applying ethics diligently and recommending the philosophical works of Cicero. On international law, he gave full recognition to the ideas of Hugo Grotius. Besides knowing his legal theories, he had a special interest in reading the juridical writers of France, probably Domat and Pothier, successors of Grotius in the northern school of law; Street, Vol. 2. 143–44. He even observed moral obligation as a source of consideration, giving origin to the theory of moral consideration, which he held as enough in itself to support a promise. Though this theory did not last longer, it was recognized for many years after his death, but apparently not exactly in maritime cases.

[329]Kent, Vol. III, 17.

[330]Ibid.

[331]Ibid., Vol. II, 602.

(2) **Political and Commercial Reasons**

At the end of the 18th century the strict standard of liability was only assigned to common carriers and innkeepers, excluding other types of common callings.[332] The mere public policy concern showed in the courts' decisions might be insufficient in explaining the reasons for applying a heavier burden exclusively on common carriers and not to all types of similar professions. Basedow noted that and presented three additional hypotheses. The first was the possibility for monopolistic practices in the transport sector. But this is immediately discarded due to the lack of sufficient evidence of any monopolistic power on the part of the carriers.[333]

The second refers to two legal reasons. On one side, Basedow points that with the advent of the concept of general assumpsit, it became the legal instrument to require common carrier to serve to every customer and with care.[334] With the evolution of contract theory, assumpsit adopted later a more specific meaning in contractual law and was not applicable as the basis of common carriers' obligation. Their obligations became based on the "custom of the realm", that was explained in the aforementioned cases, but without regards to its historical reasons.[335] The other legal reason was the confusion between accidents and act of God as an exception for carrier′s liability.[336] The cases provided a vague meaning of acts of God that created some ambiguities on the concept.[337] That made more difficult to distinguish them from mere accidents.

A third political reason, and perhaps more convincing, was constructed by Oliver Wendell Holmes.[338] The English feudal society depended strongly in the carriage of commodities and goods, a task that could not be entrusted to dishonest people that might enter into collusion to rob the goods.[339] Since the 17th century, and particularly after second half of the 18th century, the industrial production prompted the commercial expansion of England. The creation of wealth for the

---

[332]Basedow, 6.

[333]Ibid., 7. *See* Asterburn, *The Origin and First Test of Public Callings*, 75 U. Pa. L. Rev. 411, 420 (1926).

[334]Ibid., at 5: "The special duties of the common callings were based on an implied assumpsit on their part. The 'holding out' to the general public was regarded as a general or universal assumpsit of both serving the public without discrimination and carrying out this service carefully."

[335]Ibid., 7.

[336]Ibid., 7.

[337]Ibid., 8.

[338]Ibid., 8. *See* Holmes, *Common Carriers and the Common Law*, 13 Am. L. Rev. 611 (1879).

[339]Ibid., at 8: "The English feudal society, during the 17th and 18th centuries, spent only part of the year on the land from which it derived its income. For much of the year the nobility lived in towns supported by income from the surrounding estates. Hence, the aristocracy depended heavily upon both the availability and the safety of the carriage for passenger and goods. The movement of commodities could not be entrusted to the arbitrary, profit-oriented decisions of those engaged in the industry. The liability of the carrier had to be tightened to forestall collusion with thieves. Although the same danger existed with respect to other bailees, they were less important to the nobility. The professions which survived as common callings into the 19th century can be easily linked to the infrastructure of transportation."

country depended also on a reliable public transportation for such industrial production, and that demanded a higher standard of liability on common carriers. This thesis, however, reflects again the concern of the courts to prevent dishonest acts on the part of the carriers, in a commercial activity that was fundamental for the economic growth of England. This situation is probably, what Brett J. identifies as the similar conditions to those that motivated the Roman regulation contained in the *Praetor's* edict.[340] Indeed, and as aforementioned, it is estimated that the *Praetor's* edict appeared around the 1st century B.C., during the expansion of the maritime commerce of Rome.[341] The carriage of goods in the Roman society of that time, and in England of the 17 and 18th century, were under similar circumstances. In both cases, demanded safe carriage. English judges confirmed the value, fairness and convenience of the application of the strict standard created by the Romans. And, though the rule was very demanding, its application during those centuries does not seem to have discouraged the shipping industry.

2.1.3.1.6 The Decline of the Absolute Standard

The English courts seem to be consistent, at least since the 18th century, in the application of the rule holding common carriers strictly liable.[342] However, in the 19th century, a new ideological current and new commercial events, made this strict rule to lose application. Through the introduction of contractual clauses, carriers found the way to release themselves from this heavy standard.

(1) **The Liberalism Theory**

This absolute standard began to lose application at the beginning of the 19th century as a result of the new philosophical ideas in vogue in England. The philosophical thinking of Bentham and Austin eclipsed the general principles of natural justice that had oriented the English law during the previous centuries.[343] It gave rise to the will theories, supported in the preaching of freedom by metaphysical and political philosophers.[344] Austin's ideas of liberalism, which enjoyed extended acceptance in the English society of the 19th century, were based in an analysis restricted to concrete rules and principles of law, but did not take into account the large legal tradition and history of the English law.[345] Therefore, as the

---

[340]*See supra* note 317.

[341]*See supra* note 88 and accompanying text.

[342]Kent, Vol. II, 602.

[343]Graveson, 100.

[344]Gordley, 214.

[345]Graveson, at 84: "It was that Austin's analytical approach was in itself restricted to concrete rules an principles or law rather than the deeper cohesive and synthetic force lying equally behind the legislative command and its judicial execution. Austin's strong, beneficial and lasting influence on the development of jurisprudential thought in England lay within the bounds of his self-imposed limitations."

common law legal system responded to the national character of the English society, it was strongly influenced by the ideas of individual liberty, which had become more than just a principle of law but a way of life.[346] Prominent economists of the epoch such as Smith, Ricardo and Mills remarked on the "freedom of bargaining as a fundamental and indispensable requisite for progress."[347] The philosophy of freedom and individualism had to be supported by the law which must recognize the will of the parties and allow them to agree to whatever terms they desire by contract.[348] Such theories on individualism allowed some excesses in the use of the contractual liberty.[349]

This was also the time of the *Pax Victoriana*, when "free enterprise was at its highest, little tolerance for regulations, fewer rules and even less interference" were part of the Protestant ethic.[350] In addition, by the second half of the 19th century, the governments of the maritime states had more concern with "promoting the commercial interest of shipowners than with enforcing their obligation to the general community."[351] Hence, the concern for public policy showed by the courts in the previous centuries lost its strength.

(2) **The Second Industrial Revolution**

The second industrial revolution took place in the second half of the 19th century and brought one of the largest economic booms in history, boosting international trade. The opening of the Suez Canal in 1869 caused a commercial revolution.[352] Although the development of the steamship occurred in the beginning of the 19th century, creating a safer and faster means of transport, its impact was not felt until the second part of the century.[353] The introduction of the iron hull and screw propeller as technological advances for the shipping industry also favored its expansion.[354] British carriers gained a prominent position in intercontinental trade. At the same time, others challenges arose as a result of this massive shipping expansion. Among others, shipbuilding became more important, with ever increasing costs to build ships, as they became larger and more sophisticated.[355] It was then necessary to encourage the investment in shipbuilding, but the absolute

[346]Ibid., 85–86.

[347]Gordley, 214. *See* Williston, '*Freedom of Contract*', *Cornell Law Quarterly*, 6 (1921), 365, at 366–369.

[348]Ibid., 214.

[349]Graveson, 92.

[350]Gold, 119.

[351]Ibid., 119.

[352]Ibid., 95.

[353]Ibid., 91–92.

[354]Ibid., 91–92.

[355]E. P. Wheeler, *The modern law of carriers: The limitation of the common-law liability of common carriers under the law merchant, statutes and special contracts.* (New York: Baker, Voorhis & Co., 1890), 2.

standard was seen as an obstacle for investors to buy ships.[356] In addition, with more ships calling to ports, the danger of collision increased, and the long-standing strict liability was considered an unbearable burden.[357] But this increase in the sophistication of the new vessels resulted also in less safety. Shipowners did not have to sail themselves their vessels, like in the past when both, owner and merchant, usually sailed together. Vessel and cargo were insured, even in some cases, over insured, pursuing to have the possibility to make profits from an eventual total loss.[358] As competition in shipping was very high, shipowners cut also some expenses, being the safety standards the easiest one.[359]

(3) **The Excesses in the Principle of Freedom of Contract**

With liberalism in mind and new business opportunities provided by the second industrial revolution, the strict liability became a kind of shackle for shipowner's business growth projections. Although the conception of the common carrier as an insurer of cargo remained for the rest of the 18th and 19th century, carriers found a method to elude the burden of this standard by the inclusion of clauses excluding liability for certain risks in the contract.

The principle of freedom of contract allowed parties to bargain their own conditions in the bills of lading, including clauses excepting some perils of the sea.[360] The existence of such clauses in the bills of lading is reported as early as the 16th century in English law.[361] The practice was increasingly adopted and the validity of such clauses was not subject of dispute.[362] Notwithstanding, Kent says that it was not properly known until the same case of *Forward v. Pittard* (1785).[363] At the beginning of the 19th century, courts were gradually accepting the economic pressures for a more flexible legal framework for the development of an international trade allowing self-determination and freedom of contract.[364] In 1804 the exclusion of liability by notice was largely recognized and settled by judicial decision in the case *Nicholson v. William*.[365] However, the exclusion of liability

---

356Ibid., 2.

357D. Rabe, *Seehandelsrecht: Fünftes Buch des Handelsgesetzbuches: mit Nebenvorschriften und Internationalen Übereinkommen*, 4th edn. (München: C. H. Beck, 2000), 299.

358Gold, 118.

359Ibid., 118–119.

360Story, 489; Fletcher, 52.

361Fletcher, 52, at 86: "The earliest reference to an exception expressly exempting the master from liability occurs in 1545 in the record of *Holderness v. Elderness*."

362Ibid. Concretely about carriers, Fletcher mentions two cases from the 17th Century where these clauses were accepted, at 179: "…such a right was recognized by Lord Coke as available to any bailee; it was judicially confirmed in *Kenrig v. Eggleston*, (1648) and later by Sir Matthew Hale in *Morse v. Slue*, (1671)"; Colinvaux., at 627 also cites *Paradine v. Jane* (1647).

363Kent, Vol. II, 607.

364Basedow, 18; Also Wheeler, 75, *See Covington v. Willan*, Gow 115 (1819) and *Peek v. North Staffordshire R. Co.*, 10 House of Lords Ca. 473 (1863).

365Kent, Vol. II, 607.

was limited to protect the carrier from extraordinary events, but neither for the consequences resulting from the lack of ordinary personal care and diligence,[366] nor for gross negligence.[367]

Problems arose around 1880 when British shipowners went so far as to contract out liability in a larger extension, excluding the liability for their own negligence.[368] By that time, the principle of freedom of contract became almost sacred.[369] Under this principle, any carrier accepted goods to be carried "when he liked, as he liked and whenever he liked".[370] They included in their general form of bills of lading, among other conditions: (1) clauses excluding liability even for their own negligence, known as "negligence clauses"; (2) a lien on the cargo for indebtedness of the cargo owner; and, (3) the appointment of the British courts and British law as the exclusive forum and applicable law for cases in all trades and in all places.[371] A report issued by the Imperial Shipping Committee in 1921 summarized the situation as follows:

> There is nothing in English law to stop [shipowners] from contracting out of the whole or any part of his liability, and, by a practice which has gradually extended since about 1880, British shipowners do habitually in their bills of lading contract themselves out their common law liability to a large extent.[372]

The original strict allocation of risk on the common carrier was viewed by the British courts as a default principle applicable only in the absence of stipulations to the contrary.[373] The same conception was accepted by other European courts.[374] Some English judges considered that the doctrine of exempting carriers from liability by notice had been carried too far, even lamenting its introduction into the Westminster Hall.[375] However, the English courts continued accepting as valid such practices. The stipulations introduced by the British shipowners in their contracts, lead to a state of practically no liability at all for any damage to or loss of the cargo carried, resulting in the obvious discontent of the cargo owners. The general principle of strict liability for common carriers, historically applied under

---

[366]Ibid. It is concluded from the language of the courts in the cases *Bodenham v. Bennet* 4 Price Exch. 31, and *Garnett v. Willan,* 5 B. & Ald. 53.

[367]*Riley v. Horne* 130 E.R. 1044; (1828) 5 Bingham 217, 225.

[368]Report of the Imperial Shipping Committee on the Limitation of Shipowner's Liability by Clauses in Bills of Lading and on Certain Other Matters Relating to Bill of Ladings (1921), 7, reprinted by Sturley in *The Legislative History…,* Vol. 2, 135.

[369]R. P. Colinvaux, *The Carriage of Goods by Sea Act, 1924* (London: Stevens & Sons Limited, 1954), 1.

[370]Knauth, 116.

[371]Ibid., 120.

[372]As cited by Colinvaux, 1.

[373]M. F. Sturley, 'The History of COGSA and the Hague Rules"' (1991) 22, Journal of Maritime Law and Commerce, 1–57, 5.

[374]Ibid., 5; Knauth, 119.

[375]Kent, Vol. II, 607–608. *Smith v. Horne*, 8 Taunt. 144.

English law, became in reality a dead letter. And with no liability at all, carriers showed less interest in ensuring the seaworthiness of the vessels and in providing a proper care of the cargo. Hence, by the middle of the 1880's, "badly built, ill-found, grossly overload and often over insured vessels" were put on the sea, becoming frequently "coffin ships" that took the cargo and the crew's life's to the depths of the seas.[376]

2.1.3.2 International Attempts for Regulation and Unification

Doubtless, freedom of contract has been one of the cornerstones of the fast commercial growth reported since the 19th century. But when such a freedom is extended to the point of breaking with basic ethical principles, an important legal institution, as contractual freedom is, can be used to support and legitimize practices that are, in reality, unfair and abusive. Such practices, rather than promoting commerce, hamper and discourage it. Given this situation, the international community undertook the discussion, redaction and adoption of special regulations for this contract, as an attempt to stop the practices developed in that time by the shipowners.

2.1.3.2.1 The Liverpool Conference of 1882

Attempting to find a solution to the practices carried out by British shipowners, the International Law Association[377] held a meeting in Liverpool in August 1882, where a model of bill of lading known as the "Conference form" was proposed.[378] The main idea behind this "form" was to change the long applied strict liability rule, into a liability based on negligence in "all matters relating to the ordinary course of the voyage."[379] The discussions revolved around the duties of the carriers in the performance of the carriage, and the assignation of liability for breach of those duties. This proposal, introduced for the first time the concept of practicing "due diligence" to make the ship seaworthy, as the basis for liability, instead of the warranty of seaworthiness.[380] The ship was not held responsible for losses caused

---

[376]Gold, 119, at 121: "For some reason, the shipping industry has always taken an almost perverse pride in maintaining that cruel discipline, inhuman conditions, and inadequate compensation provided the only environment in which sailors could adequately perform their work. It was in a time when there was not international safety regulation, and few national rules, where only some tacit rules stated mostly for the benefit of by underwriters that seek to reduce their risks." *See* Fayle, *History of World's Shipping Industry*, pp. 286–287.

[377]The International Law Association was founded in 1873 with the original name of Association for the Reform and Codification of the Law of Nations. In 1895, it changed to the present name.

[378]Knauth, 119.

[379]Sturley, *The History of COGSA…*, 7.

[380]Ibid., 7.

by some events, unless they result from want of due diligence of the shipowner or her manager.[381] This provision implied the practice of due diligence to prevent occurrence of such events.

The model included liability for negligent acts of the servants in the stowage or delivery of cargo and similar operations. The necessity of a fast, accurate and careful stowage of cargo and its proper delivery were taken for granted. Shipowners had assumed such responsibility for centuries, all over the world, and the problems surrounding adaptation to early steam navigation were estimated to have been already overcome.[382] On the other hand, it upheld the exoneration of liability for accidents of navigation, even those caused by "negligence, default or errors in judgments of the Pilot, Master, Mariners, or other Servants of the Shipowner."[383] By that time, steam navigation had increased substantially and rapidly, but was still a novelty for many masters and seamen who now had to deal with this new technology and with larger numbers of vessels calling at traditional ports. Therefore, errors in navigation were the main cause of maritime casualties that ended in cargo loss or damage with subsequent cargo claims.[384] This was the reason for the shipowner's rejection to assume liability for nautical faults.

In the end, the "Conference form" was not generally accepted.[385] The New York Produce Exchange adopted similar conceptions contained in that model in 1883 and 1884, but both of them were not subject to major application.[386]

---

[381]Int'l Law Association, Report of the Tenth Annual Conference held at Liverpool August 8th–11th, 1882, at 104, reprinted in *The Legislative History...*, Vol. 2, 62. The Conference form stated: "…Ship not answerable for losses through *Explosion, bursting of Boilers, breakage of Shafts, or* any latent defect in the *Machinery or Hull*, not resulting from want of due diligence by the Owner of the Ship, or any of them, or by the Ship's Husband or Manager;…".

[382]Ibid., 77 reprinted in *The Legislative History...*, Vol. 2, 35.

[383]Ibid., 104, reprinted in *The Legislative History...*, Vol. 2, 62.

[384]Ibid, 77 reprinted in *The Legislative History...*, Vol. 2, 35. Mr. Richard Lownders, President of the Liverpool Chamber of Commerce and Deputy-Chairman of the Committee and Chairman of the Liverpool Committee, presented an interesting exposition on this issue: "…Every year harbours and roadsteads and particular ocean-tracks are more and more crowded, the pace of steamship is more rapid, and, perhaps not least formidable, Board of Trade inquiries grow more penetrating and ingenious. If an unfortunate captain or officer, however well certificated, in the course of a long voyage makes one mistake, the chances of its doing damage, and the chances of its being found out, are vastly more that they used to be; and, what a shipowner naturally does not like, the chances of his being called upon to pay for a valuable cargo are in the same proportion more that they ever used to be. Nor can it even be said that a shipowner has had time in the course of centuries to get to this infliction, and hardened against it, for, in truth, it is only of late years – only since two decisions of our Courts not more that twenty years old –that this liability for sea perils, occasioned by the faults of seaman, has been brought home to the shipowners."

[385]Sturley, *The History of COGSA...*, 7.

[386]Ibid., 7–8.

2.1.3.2.2 The Hamburg Conference of 1885

In the 12th Conference of the International Law Association held in Hamburg in 1885, further discussions on the subject were reassumed with the final adoption of the "*Hamburg Rules of Affreightment*".[387] The Hamburg Chamber of Commerce proposed a set of rules that in the first provision declared unlawful any clause or covenant lessening or avoiding the carrier's liability, but this proposal did not prosper. A second proposal of modification of the former Liverpool bill of lading was also submitted. The discussion revolved mostly around setting the responsibility for negligence. The rule I stated a general obligation to make the ship in all respects seaworthy and fit for the stowage and proper delivery at destination:

> The shipowner shall be responsible, that his vessel is properly equipped, manned, provisioned, and fitted out, and in all respects seaworthy and capable of performing her intended voyage, and for the stowage and right delivery of the goods. He shall also be responsible for the barratry, faults, and negligence, but not for errors in judgment, of the master, officers and crew.[388]

The same rule kept the exclusion liability for nautical fault but restricted to errors of judgment, not for fault or negligence of the master and crewmembers, as the previous "Conference form" did. The reason for this differentiation respond to the argument that most of the casualties resulting in cargo damage or loss, were caused by acts of negligence or default by the master, officers or crewmembers, that were avoidable by practicing due diligence.[389] In addition, it included liability for unreasonable delay.[390] Another difference is that this proposal was not a model bill of lading, but a set of rules to be inserted in the bills of lading.[391] These rules at the end suffered the same fate as the former Liverpool form.[392]

---

[387]International Law Association, Report of the Tenth Annual Conference held at Hamburg August 18th–21th, 1885, reprinted in *The Legislative History*..., Vol. 2, 73.

[388]Ibid., 165, reprinted in *The Legislative History*..., Vol. 2, 122.

[389]Ibid., 76, reprinted in *The Legislative History*...,Vol. 2, at 90. In this regard, the proposal of the chairman of the committee gives some reasoning for such an amendment: "Dr. Wendt proposed that the words 'the act, neglect, or default', in the phrase "act, neglect, or default of the master or crew in navigating the vessels," be omitted, and replaced by the words "error of judgment." Although he did not think it necessary to give all his reasons for doing so, he might state that the experience he had gained by his investigations into maritime casualties for many years past, enabled him to assure the Conference that at least nine-tenth of all such disasters as stranding, collisions at sea, and so forth, were due to some neglect or other of the crew which might be prevented if due diligence were exercised. He therefore thought that such 'neglect or default' the shipowner ought to be made answerable, but not for mere 'error of judgment'."

[390]Ibid., 165, Rule II, reprinted in *The Legislative History*..., Vol. 2, 122.

[391]Sturley, *The History of COGSA*..., 8.

[392]Ibid., at 8: "A few German companies adopted the rules but elsewhere they had little immediate impact. In 1887, the Association "rescinded" them and reaffirmed the principles of the Conference form. The format of the Hamburg Rules nevertheless remained persuasive, and future efforts at achieving uniformity did so through uniform rules rather that a model bill of lading."

2.1.3.2.3 Results of These Conferences

Although these attempts did not enjoy major success, the surrounding discussion presented some reasoning that laid the foundation of the future statutory regulations. For example, they presented a fragmentation of the carrier's duties and the allocation of liability in accordance to those failures in the specific duty that caused the damage to or loss of the cargo. If the cause was errors of judgment in navigation as well as other *vis major*, there was no liability at all. If the cause was lack of seaworthiness, the carrier was liable if he did not practice due diligence in making the ship seaworthy. Finally, if the cause was negligence in stowage and proper delivery, total liability was assigned.

Despite the efforts of the International Law Association, the situation remained practically the same. Strongly criticized by cargo interests, the Glasgow Corn Trade Association issued a series of resolutions in 1890 condemning the practices of the sea carriers, which were described as having "surpassed all bounds of reasons and fairness."[393] However, the British shipowners had a strong influence in the English Parliament and the complaints of the cargo owners were not attended.[394] The "negligence clauses", certainly considered unethical and an unfair practice, became the catalyst for statutory regulations, not in England but in the United States, both countries representing at that time, the main interests of both sides of the industry.

2.1.3.3 The Maritime Law of the United States of America

2.1.3.3.1 The Standard of Liability for Sea Carriers

The maritime law of the United States of America was, at least since its independence in 1776, the same as the maritime law of Europe.[395] The Article III of the American Constitution conferred upon the Supreme Court and the Federal Courts of the United States jurisdiction on admiralty and maritime matters.[396] The power vested in the Supreme Court enables her to underline some substantive rules that have binding effect upon all the lower federal, district, and circuit courts.[397] Kent affirms that the decisions of the Federal Courts on commercial cases reflected the moral and intellectual character of the United States, especially in the admiralty courts, which particularly showed great research and familiarity with the maritime law principles of Europe.[398] Therefore, the absolute liability of the sea common

[393]Knauth, 120.

[394]Ibid., 120.

[395]Kent, Vol. III, 1; Wheeler, 1.

[396]Wheeler, 1; Tetley, *Interpretation and Construction…*, 42; *See* 28 US Code s. 1333.

[397]Tetley, *Interpretation and Construction…*, 42, 46. They are bound under the principle of *stare decisis.*

[398]Kent, Vol. III, 17.

carrier in England was generally understood to be the same in the United States.[399] Whether the carrier was at fault or not, it was immaterial for the setting of his liability for cargo damage or loss.[400] The common sea carrier was a warrantor of safe arrival, unless one of the common law exceptions was invoked.[401] The cases *Schiefflen v. Harvey* (1810)[402] and *Elliot v. Russell* (1813)[403] were the first actions against common carriers reported in New York.[404] In the second case, the strict principle of liability was applied. It confirmed the common carrier as warrantor of the safe delivery of goods in all, but the excepted cases of an act of God and public enemies, with no distinction between carriers neither by land nor by water.[405] Chief Justice Kent affirmed that this principle appeared to be sound and wise and generally accepted among nations.[406] By the middle of the 19th century, the standard was still in force. In 1848, in the case *New Jersey Steam Nav. Co. v. Merchant's Bank of Boston*,[407] the U.S. Supreme Court said:

> The general liability of the carrier, independently of any special agreement, is familiar. He is chargeable as an insurer of the goods and accountable for any damage or loss that may happen to them in the course of the conveyance, unless arising from inevitable accident – in other words, the act of God or the public enemy.[408]

Ten years later, such absoluteness was again confirmed in the *Propeller Niagara v. Cordes* (1858).[409] In this case, the U.S. Supreme Court provided a more detailed description of the concept of common carrier,[410] with additional remarks on the extension of its obligation relating not only to the seaworthiness but also to the duty of care. The shipowner was held liable even for the damage to cargo resulting from a stranded ship. The court made some statements that reaffirmed explicitly the carrier's implied obligation of providing a seaworthy ship, which include a proper and skillful master and crew. The master in this case, after the vessel's stranding did not do anything to protect the cargo, when there were many possibilities to reduce the losses. Such inactivity from the part of the carrier, (among other circumstances

---

[399] Story, 470; Wheeler, 1.

[400] Gilmore and Black, JR., 140.

[401] Wheeler, 1; Gilmore and Black, JR., 140.

[402] 6 John. 1709.

[403] 10 John 1.

[404] C. Warren, *History of the Harvard Law School and of early legal conditions in America* (Union, N.J: Lawbook Exchange, 1999), 248.

[405] Kent, Vol. II, 608–609. Contractual exclusion of peril for the sea was allowed, but not for negligence.

[406] Ibid., 609: "It was further shown that the marine law of Europe went to the same extent, as did also the civil law, and the law of those nations in Europe which have made the civil law the basis of their municipal jurisprudence.

[407] 47 U.S. (6 How.) 344, 12 L.Ed. 465 (1848).

[408] Ibid, 381.

[409] 62 U.S. (21 How.) 7.

[410] Gorton, vol. 43, 20.

evidenced in the trial), made the court to conclude that the master was not properly qualified for the type of vessel he was sailing. Regarding the obligation of care of the goods, the remarks were even more significant. It stated first, that even if the carrier had not received the payment in advance for the carriage, he is subject to the same strict liability; second, he is obliged to make a proper stowage; and third, that a stranded vessel did not release the master of his duty to care for the cargo, which remained throughout the time in his charge and accordingly, all possible measures must be taken to save it.[411] Anything shorter than this—affirmed the judge—would be inconsistent with to the nature of the undertaking and the contract, as it is "universally understood in courts of justice".[412] Mr. Justice Clifford also pointed out that this rule was of centurial existence and as time and experience had showed its convenience and fairness, he esteemed it not reasonable to change it by judicial power, but only by statutory regulation.[413] In the decision, it was accepted that the right and obligations of the parties are set and regulated by the contract and some references were made to the possibility of exclusion of liability in the bill of lading for some other events that may have occurred without the fault or negligence of the carrier. It made clear, however, that this right did not include any exoneration of liability for fault or negligence. The absolute warranty of providing a seaworthy vessel at the beginning of the voyage was well established and recognized by the American Courts.[414] The enforcement of this standard continued until the increasingly insertion in the bills of ladings of the clauses excepting liability, forced the enactment of special legislation.

2.1.3.3.2 Restrictions to the Principle of Freedom of Contract

The principle of freedom of contract was also accepted in the United States but under some restrictions. The first decision analyzing exclusion of common carrier's

---

[411]*See supra* note 409, at 26: "[The] Duties remain to be performed by the owner, or the master as the agent of the owner, after the vessel is wrecked or disabled and after he has ascertained that he can neither procure another vessel nor repair his own, and those too of a very important character, arising immediately out of his original undertaking to carry the goods safely to their place of destination. His obligation to take all possible care of the goods still continues, and is by no means discharged or lessened, while it appears that the goods have not perished with the wreck, and certainly not where, as in this case, the vessel is only stranded on the beach."

[412]Ibid, 27.

[413]Ibid., at 25: "Most of the rules of law prescribing the duties of a carrier for hire, and regulating the manner of their exercise, have existed for centuries, and they cannot be modified or relaxed except by the interposition of the legislative power of the Constitution. Time and experience have shown their value and demonstrated their utility and justice, and they ought not and cannot be changed by the judiciary."

[414]*See Richelieu & O. Nav. Co. v. Boston Marine Ins. Co.,* 136 U. S. 408, 10 Sup. Ct. 934 (1890); *The Edwin I. Morrison,* 153 U. S. 199, 14 Sup. Ct. 823 (1894); *The Caledonia*, 157 U. S. 124, 15 Sup. Ct. 537. (1895); *The Irrawaddy*, 171 U. S. 192, 193, *sub nom. Flint* v. *Christall*, 43 L.Ed. 132, 18 Sup. Ct. Rep. 833 (1898).

liability was the *Cole v. Goodwin and Story* (1838),[415] decided by the New York State Court. A stage coach gave a general notice that the carriage of "the baggage of passengers is at the risk of the owners." The court denied the exoneration of liability in the argument that the defendant was a common carrier in respect to the passenger's baggage, and relieved for loss only if caused through acts of God or public enemies. The Federal Courts allowed the exoneration of liability for certain circumstances, but, differing from the Common law and European courts, not for the effects of negligence or for not providing a seaworthy ship.[416] In *Dorr v. The New Jersey Steam Navigation Company* (1850),[417] Judge Campbell made a distinction of the two different liabilities arising for common carriers from the contract of carriage. The first is for losses by accident or mistakes that make him liable as an insurer. The second is for losses caused by negligence or default of himself or his servants that make him liable as an ordinary bailee. He found reasonable the exclusion of liability as an insurer by express agreement in the contract, to protect himself for misfortune, but not for losses caused by his negligence.

Citing this case, the negligence clauses were largely analyzed and rejected by the Supreme Court of the United States in *Railroad Co. v. Lockwood* (1873).[418] The case was about the carriage of goods by train, but the arguments presented apply for every type of common carriers. Justice Bradley made a thorough revision of the case law existing on the subject, finding that in most of the cases, the judges denied the recognition of such clauses. He explained that the policy holding common carriers as insurer of the cargo, responded to the need of secure the "utmost care and diligence" in the compliance of those duties, as they were fundamental for the "welfare of every civilized society".[419] The carriers, being few powerful corporations, were in a clear commercial superiority that impeded shippers to negotiate any of the contractual terms of the bills of lading. It was also acknowledged that the improvement of society reduced the possibilities of collusion and bad faith from the part of the carriers and that make reasonable to ameliorate the strict rule. Hence, it was acceptable to exclude liability for damages caused by certain inevitable accidents or any superior force, as it was allowed in the civil law. But allowing such an extreme exoneration for their own negligence or misconduct, beside it was "unfair and unreasonable", would produce the carriers' disinterest in the compliance the most essential duties of the contract. The Supreme Court concluded that the conditions stated in a contract of carriage "ought not to be adverse to the dictates of public policy and morality,"[420] and that such prohibition was based not only in the

[415]19 Wend. 251, 32 Am. Dec. 470.

[416]Sturley, *The History of COGSA…*, 5–6.

[417]4 Sandford, 136.

[418]17 Wall. 357 (1873).

[419]Ibid., 377.

[420]Ibid., 380.

public policy but also in “sound principles of law”.[421] Therefore, the public interest had to be protected and such clauses rejected.[422]

By 1880, when the British carriers increased the use the negligence clauses, most of the State and Federal Courts refused to enforce these “unreasonable” stipulations inserted in the domestic and international bills of lading.[423] When the US courts accepted the application of English law to a case, they were aware of the recognition given by English courts to the negligence clauses. However, they refused to enforce them. Even the other clauses containing special exceptions were construed against the shipowners and in favor of the shipper, and in no case were they accepted to extend such exceptions for the shipowners own negligence.[424] Such rejection became “settled law” and “elementary doctrine”.[425] In the maritime case, *Liverpool and Great Western Steam Company v. Phoenix Insurance Co.* (1889),[426] the U.S. Supreme Court provided again a deeper analysis of the negligence clauses, now in the context of the carriage of goods by sea. The decision quoted extensively and applied the reasoning of *Railroad Co. v. Lockwood.* Such clauses were declared again as “unreasonable, contrary to the public policy and consequently void.”[427] Under this rule, followed by almost all the Federal Courts,[428] the common sea carrier could not contract out his liability for his negligence. Only a few Federal Courts, particularly in New York, had accepted the British practice of excluding

---

[421]Ibid., 368.

[422]Ibid., at 381: “Conceding, therefore, that special contracts, made by common carriers with their customer, limiting their liability, are good and valid so far as they are just and reasonable; to the extent, for example, of excusing them for all losses happening by accident, without any negligence or fraud on their part; when they ask to go still further, and to be excused for negligence –an excuse so repugnant to the law of their foundation and to the public good – they have no longer any plea of justice or reason to support such a stipulation, but the contrary.” *Citing also*: *Express Co. v. Caldwell*, 21 Wall. 264, 268 (1874); *Railroad Co. v. Pratt*, 22 Wall. 123, 134 (1874); *Bank v. Express Co.*, 93 U. S. 174, 183 (1876); *Railway Co. v. Stevens*, 95 U. S. 655 (1877); *Hart v. Pennsylvania Railroad Co.*, 112 U. S. 331, 338, 5 Sup. Ct. Rep. 151 (1884); Phoenix Ins. Co. v. Erie & Western Transp. Co., 117 U. S. 312, 322, 6 Sup. Ct. Rep. 750, 1176, 29 L.Ed. 873 (U.S. Wis. 1886); *Inman v. South Carolina Ry. Co.*, 129 U. S. 128, ante, 249 (1889).

[423]Colinvaux, at 1: “… since 1870’s, the courts resolutely refused to enforce unreasonable conditions in bills of lading.”; Schoenbaum and Yiannopoulos, 365.

[424]In *The Titania* 1883 19 F. 101 (D.C.N.Y.), at 103–104 Judge Brown said: “But although, under the English decisions, it seems to be settled that ship-owners may exempt themselves from damages caused even by their own negligence, provided this intention be unequivocally expressed, yet such causes of special exemption, being inserted for the benefit of the ship-owner, are construed most favorably to the shipper and most strongly against the ship-owner, and will not be held to embrace the latter's own negligence, unless that be specially excepted in connection with the actual cause of the loss.”

[425]Schoenbaum and Yiannopoulos, 365.

[426]129 U.S. 397, 9 S.Ct. 469, 32 L.Ed. 788.

[427]129 U.S. 397, 441–442. Similar statement was made later in: *The Kensington*, 183 U.S. 263, 268 (1902). *See* also Wheeler, 77; Colinvaux, 1.

[428]Wheeler, 82.

total liability in some specific cases.[429] Clauses limiting their liability had to be reasonable, and excluding liability for their own negligence was not considered so.[430] Hence, those clauses had no effect when inserted in contracts governed by the law of the United States, or when the carriage was performed entirely or partially within the United States.[431] It had neither effect in a contract made abroad or through an express clause establishing the application of another law.[432]

2.1.3.3.3 The Enactment of the Harter Act of 1893

(1) **Historical Context**

Despite the non-recognition by most American courts of the clauses imposed by British carriers excluding liability for their own negligence, such a practice continued. By that time, there existed a monopoly of some 20 British liner companies, owning the most suitable ships and moving the majority of the American exportations.[433] Because of the jurisdiction clauses included in their bills of lading, cargo owners of the United Sates were obliged to file their claims in the English courts and under English law, where the negligence clauses were recognized.[434] The situation pushed the powerful cargo interest in the United States to demand statutory regulation to prevent shipowners from such abusive practices.[435] It drove shippers of cargo, underwriters and bankers in the United States to force the settlement of a statutory solution. For this same purpose, the House of Representatives of the United States passed a bill in 1885 where the clauses of exoneration of liability were prohibited.[436] However, this bill never passed the Senate Commerce Committee approval,[437] leaving the problem still pending a solution.

---

[429]Knauth, at 119 cites: "*Rubens vs. Ludgate Hill S.S.Co.* (1892), 1st Dept., 65 Hun 625, 20 N.Y. Supp. 481 at 185–186, affirmed without opinion (1894), 143 N.Y. 629. *See Robertson vs. National S.S.Co.* (1891), 139 N.Y. 416 and *Gleadhill vs. Thompson* (1874), 56 N.Y. 194."

[430]Colinvaux, 2.

[431]Ibid., 2.

[432]Ibid., 2.

[433]Knauth, 119.

[434]D. C. Frederick, 'The Political participation and legal reform in the international maritime rulemaking process: from the HR to the Hamburg rules' (1991) 22, Journal of Maritime Law and Commerce, 81–117, 84, *citing* S. Dor, Bill of Lading Clauses and the Brussels International Convention of 1924 (Hague Rules) 17 (2d ed. 1960); Sturley, *The History of COGSA...*, 10–11.

[435]Knauth, 119.

[436]Sturley, *The History of COGSA...*, 12.

[437]Ibid., 12.

(2) **The Original Project**

As response to the aforementioned problem, the U.S. Congressman Michael R. Harter of Ohio presented a new project in 1892.[438] The first two sections of the original project expressly prohibited clauses of exoneration of liability for loss or damages arising for fault or negligence. For the third section, the original draft showed an absolute obligation to provide a seaworthy ship:

> Section 3. That if any vessel transporting merchandise or property between ports in the United States of America and foreign ports shall, on starting her voyage, be in all respects seaworthy and properly manned, equipped and supplied, neither the vessel, her owner or owners, agent, or master shall become or be held responsible for damage or loss resulting from error of judgment in navigation or in the management of said vessel, if navigated with ordinary skill and care, from the time of her leaving her usual place of loading on her intended voyage until her arrival at the usual place of discharge at her port of delivery.[439]

According to the original proposal, the carrier was exempted of liability for damage or loss occurring from error of judgment or management of the vessel, subject to two conditions: (1) if the vessel was previously in "all respect seaworthy, properly manned, equipped and supplied"; and, (2) "if navigated with ordinary skill and care". This original project was subject to large discussions in the Senate Committee on Commerce, and later presented to the Senate with extensive amendments.[440] The reasons for such amendments are not documented or at least not available because the report of the committee was never printed.[441]

(3) **The Final Version**

The final version, instead of the absolute warranty of providing a seaworthy vessel, reduced the obligation to the practice of due diligence in making the ship seaworthy before the beginning of the voyage. It is similar to the proposed model in the Liverpool Conference Form of 1882. With this formula, if the carrier showed that he practiced due diligence for such a purpose, he was relieved from liability for any loss of or damage to the cargo. The carrier is also relieved of liability for faults or errors in navigation or in the management of the vessels. The obligation to navigate with skill and care was also suppressed and some additional exculpatory causes were included.[442] This version of the Harter Act became Law in the United States, with the presidential signature on the 13th of February of 1893, applying for carriage within national as well as foreign ports.

---

[438]Knauth, 120.

[439]As cited by Knauth at 121.

[440]Ibid., 121.

[441]Ibid., 121.

[442]Harter Act 1893 section §192: "…danger of the sea or other navigable waters, acts of God, or public enemies, or the inherent defect, quality, or vice of the thing carried, or from insufficient package, or seizures under legal process, or for losses resulting from any act or omission of the shipper or owner of the goods, his agent or representative or attempting to save life or property at sea, or from any deviation in rendering such service."

As one of the main objectives of the Act was to regulate the practices developed by British shipowners of excluding contractually all kinds of liability, it introduced for the first time a statutory limitation to the generally accepted principle of freedom of contract. Sections 1 and 2 expressly prohibit any "clause, covenant or agreement" that relieved carriers of liability for their own negligence in the execution of the contract.[443] This was the main purpose for the enactment of this statute, as explained in *The Delaware* (1896),[444] the first case where the Supreme Court of the United States analyzed the Harter Act. Although a collision case, the US Supreme Court paid special attention to the Act and interestingly quoted large paragraphs of the aforementioned petition of the Glasgow Corn Trade Association to the Marquis of Salisbury,[445] to explain the reasons for its enactment.

The effect, however, was not to completely relieve the shipowners from their obligation to provide a seaworthy ship, but to limit it to the practice of due diligence.[446] The common carriers liability was to be judged on negligence. In *The Irrawaddy* (1898),[447] the Supreme Court stated that the main purpose of the Act was to relieve the shipowners from liability for lack of seaworthiness caused by latent defects not discoverable by the utmost diligence.[448]

Other than the mentioned restrictions, the Act did not place major or heavier obligations on the carrier in favor of cargo interest as in the original project. The exception for faults or error in navigations was a special concession to shipowners attending to the novelty of steamships intercontinental navigation, as exposed in the International Law Association meeting in Liverpool of 1882.[449]

It has to be observed that the exclusion of liability, subject to the practice of due diligence, is for "faults or errors in navigation or in the management of the vessels", and not for faults in the fulfillment of other duties. On this matter, the description and separation of duties concerning the cargo and vessel are very important. With regards to the cargo, the carrier is responsible for "proper loading, stowage, custody, care, or proper delivery of any and all lawful merchandise or property

---

[443]Harter Act of 1893 at 46 U.S.C. §§190–191.

[444]161 U.S. 459, 16 S.Ct. 516, 2010 A.M.C. 1803, 40 L.Ed. 771, at 471 or 1813: "It is entirely clear, however, that the whole object of the act is to modify the relations previously existing between the vessel and her cargo. This is apparent not only from the title of the act, but from its general tenor and provisions, which are evidently designed to fix the relations between the cargo and the vessel, and to prohibit contracts restricting the liability of the vessel and owners in certain particulars connected with the construction, repair, and outfit of the vessel, and the care and delivery of the cargo."

[445]*See supra* c) Results of these Conferences.

[446]*See The Carib Prince*, 170 U. S. 655, 42 L. Ed. 1181, 18 Sup. Ct. Rep. 753. (1898).

[447]171 U. S. 192, 193, *sub nom. Flint* v. *Christall*, 43 L. Ed. 132, 18 Sup. Ct. Rep. 833.

[448]At 192–193 the Court said: "Plainly, the main purposes of the act were to relieve the shipowner from liability for latent defects, not discoverable by the utmost care and diligence, and, in event that he has exercised due diligence to make his vessel seaworthy, to exempt him and the ship from responsibility for damage or loss resulting from faults or errors in navigation or in the management of the vessel."

[449]*See supra* note 384.

committed to its or their charge…", and must "carefully handle and stow her cargo and to care for and properly deliver same". Regarding the vessel itself, the carrier must "properly equip, man, provision, and outfit said vessel, and to make said vessel seaworthy and capable of performing her intended voyage…"[450]

With the enactment of this Act, the long-standing and sometimes praised absolute standard of liability for common carrier was finally abolished by statute.[451] The common sea carrier's liability is based on its diligence in the fulfillment of these duties. The Harter Act was judged as "fair and practical in operation".[452] But the absence, or rather, the impossibility of a dogmatic definition for due diligence to be provided by the courts, created uncertainty and frequently shipowners failed in fulfilling this duty, depending on the specific analysis of each case.[453]

(4) **International Impact**

The solution found in the United States and set out in the Harter Act initially had little impact abroad. At the beginning of the 20th century, only three countries followed the model proposed by the act: Australia (1904), New Zealand (1908) and Canada (1910).[454] With England still the major shipping nation, the situation did not change much. The Act faced a problem of international private law: although the negligence clauses were not enforceable in the United States, they were still so in England, were the act had no major validity.[455] Bills of lading governed by English law, containing negligence clauses, were accepted and enforced by the courts, regardless of the "the American principle of public policy against unreasonable conditions."[456] The English courts interpreted "the Harter Act clauses" included in the American bills of lading, with no further relevance over the other clauses on them stated, and continued the full recognition of the unlimited freedom of contract.[457] Even in cases where not negligence clause was stated in the contract, but the Harter Act applied, the English courts continued to recognize warranty of

[450]Harter Act of 1893 at 46 U.S.C. §§190–191.

[451]In the *Martin v. Southwark* 191 U.S. 1, 24 S.Ct. 1, at 2 or 7: "Section 3 must be read with §2 to effectuate the purpose of the act, and shows an intention upon the part of Congress to relax, in certain respects, the harshness of the previous rules of obligation upon shipowners,…".

[452]Knauth, 122.

[453]Ibid., 122.

[454]Ibid., 122.

[455]Colinvaux, 4.

[456]Ibid., 4. *See RE Missouri S.S. Co.* (1889) 42 Ch. D. 321; *Jones v. Oceanic S.N. Co.* [1924] 2 K. B. 730.

[457]Knauth, at 122: "The situation in England, where much of the damage litigation naturally centered because the preponderant marine insurance and ship-owning interest, did not, however, develop in a manner equally satisfactory to the owners and underwriters of cargo."

seaworthiness as absolute.[458] Nevertheless, the Harter Act formula was the model used in further attempts for international unification, which concluded in the adoption of the Hague Rules.

2.1.4 *International Regulations for the Carriage of Goods by Sea*

2.1.4.1 The Hague Rules—1924

2.1.4.1.1 History of Its Adoption

By the end of the 19th century, Great Britain was doubtless a commercial and colonial empire and the dominant power of the sea.[459] In 1897, the British Register listed 2453 sailing vessels of over 100 tons, and 6665 steamships over 100 tons, which, together with other vessels, and vessels owned in British colonies, meant that Britain alone owned just over a half of the world's tonnage, measured in that year at 26.5 million grt.[460] At the beginning of the 20th century shipowners were still politically powerful in the British Empire, but the cargo interest remained more influential in the United States, Canada, Australia and New Zealand, where legislation favoring cargo owners were enacted.[461] Notwithstanding, such legislation did not have widespread international impact as it only regulated the domestic and outbound bills of lading. Importers were still subject to bills of lading issued in countries that continued favoring carriers,[462] particularly in England, where such foreign legislation had no major impact.[463] The cargo interests of the overseas British dominions pressed the Imperial Government for regulation on the subject.[464]

Given the success of the rules on Collision and Salvage adopted in 1910 by the Comité Maritime International, there were also discussions on risk allocation for the

---

[458]In *McFadden v. Blue Star Line* (1905) 1 K.B. 697, Judge Chanell said at 706–707: "Therefore, it seems to me that I must hold that the defect was a substantial one, and that as it existed before and at the time of the loading of the goods it amounted to a breach of the warranty, unless the incorporation of the Harter Act in the bill of lading makes any difference. Then does the incorporation of that Act make any difference? For the purposes of this question I will assume that there was no negligence in the packing of the valve-chest, though I do not decide that point one way or the other. Is the absence of negligence material? In other words, does the incorporation of the Harter Act have the effect of cutting down the absolute warranty of fitness to an undertaking to exercise due diligence to make the ship fit? In my opinion it does not."

[459]Gold, 96.

[460]Ibid., 96. *See* Zimmerman, E. W., *Ocean Shipping*, New York: Prentice-Hall, 1922, 220 ff.

[461]Sturley, *The History of COGSA...*, 18.

[462]Ibid., 18.

[463]Frederick, 85.

[464]Sturley, *The History of COGSA...*, 19.

carriage of goods by sea under bills of lading. These discussions were suspended at the outbreak of the First World War (1914–1918). During the war, commercial progress experienced a decline.[465] The shipping industry was severally affected by the submarine warfare, blockades and nationalizations.[466] The power of the British Empire, though victorious, was diminished, and the British carriers suffered a weakened financial situation that pushed them to make certain concessions.[467] On the other side, the lack of tonnage in ocean shipping pushed the US government to reconsider their position on defending cargo interests.[468] There was a need to make the shipping business more attractive for private investors and increase tonnage.

Still at the beginning of the 1920s, the British shipowners owned the largest merchant fleet of the world and consequently, lead a very influential lobby.[469] They were very powerful in the Parliament[470] and had a strong influence also in the judiciary system.[471] They radically refused to change their 19th century policy of averting Parliamentary interference in the practices they developed under the principle of contractual freedom.[472] As aforementioned, since the 19th century they had interpreted this principle as "freedom for powerful carriers" to impose terms on the commercially disadvantaged shippers.[473] But by this time, it was recognized that such practices were not really concordant to the concept of freedom of contract, as the shipowners dictated unilaterally all the conditions of the bill of lading.[474] Shipowners continued to impose unilaterally these clauses, without any possibility for negotiation by the part of shippers, who were practically obliged to take it or leave it.[475] Shippers complained, arguing over the high insurance premiums they had to pay because of the carrier's immunities, as well as that the inflated premiums for bearing the risk of carrier negligence.[476] However, until the 1920s, neither the

---

[465]Knauth, 124.

[466]J. C. Sweeney, 'The UNCITRAL Draft Convention on Carriage of Good by Sea (Part I)' (1975–1976) 7, Journal of Maritime Law and Commerce, 69–125, 71.

[467]Frederick, 86, 107. This situation was used by the cargo owners of Canada, Australia and New Zealand "to press for equity in commercial maritime transactions."

[468]Basedow, 41.

[469]A. Q. Diamond, 'The Hague-Visby Rules' ([1978]) 2, Lloyd's Maritime and Commercial Law Quarterly, 225–66, 227.

[470]Knauth, 120.

[471]Frederick, 85.

[472]Diamond, 227; Frederick, at 90: Sir Norman Hill who represented the Liverpool Steamshipowners Association and was the leader spokesman during various of the discussions that preceded the HR, expressed "'his personal conviction' that the 'mutual rights [of the parties] to make their bargains without legislative interference' served the best interest of all." *See* Report of the Thirtieth Conference of the International Law Association at xxxix, reprinted in *The Legislative History...*, Vol. 1, 94.

[473]Clarke, 3.

[474]Diamond, at 227: "The only freedom of the shipper was to take the bill of lading or to leave it. And in view of the Conference system even the latter freedom was often illusory".

[475]Ibid., 227.

[476]Clarke, 4. *citing* Markianos, 22–23, Cole, 11, Guyon no. 5.

English Parliament nor the judges attended to those complaints.[477] Another reason that contributed to the debate was the creation of the CIF and CF terms in the contract of sales of goods. Such terms required the endorsee of the bill of lading to absolutely bear the consequences of the bill of lading clauses without having any previous opportunity to negotiate them, because they were simple accepted by, or imposed on the shipper.[478] Over this point, a new lobby was created integrating insurers and bankers to support shippers.[479] This lobby introduced into the debate their concern for keeping and assuring the value of the bills of lading as commercial documents.[480]

In 1921, due in part to the aforementioned events, the Imperial Shipping Committee issued a recommendation for the unification of laws in the British dominions, based primarily on the Canadian Water Carriage of Goods Act of 1910, to be then introduced also in the British legislation.[481] The British government led the unification of laws through the revival of the work of the International Law Association. Shipowners came to support the idea of a uniform international regulation. They preferred to have the same regulation worldwide, rather than facing different national laws in every port of call.[482] But the interest of the United Kingdom government was not limited to a unified regulation for the territories that composed the British Empire. Additionally, they sought such regulation "that British shipowners would not be at a disadvantage as compared with those of the rest of the world."[483] Diamond even affirms that it was the main objective of the British Government to get involved in the discussion for the enactment of an international convention on the subject.[484] In fact, carriers claimed that the enactment of domestic regulation enforcing higher liabilities on them would require an increase of freights, and this would force shippers to turn to foreign carriers to save freight costs.[485] That is why carriers were only willing to accept an international regulation, one which would level the playing field.[486]

---

[477] Knauth, 120; Frederick., 83. *See* Report of the Thirtieth Conference of the International Law Association, 52–53, reprinted in *The Legislative History...*, Vol. 1, 158–159.

[478] Diamond, 227.

[479] Clarke, 4.

[480] Ibid., 4.

[481] Sturley, *The History of COGSA...*, 19.

[482] Ibid., 19.

[483] Colinvaux, 7; Diamond, 227.

[484] Diamond, at 227: "It was this later objective which motivated the attempts made by the British Government both through the International Law Association and at a diplomatic level to secure an international treaty whereby as many contracting States as possible would bind themselves to introduce uniform legislation acceptable to all of them. It was the judgment of many people in the mid-1950s that, in securing the Brussels Convention of 1924, the U.K. Government had managed to combine self-interest with the implementation of the greatest good for the greatest number."

[485] Clarke, 4.

[486] Ibid., 4.

The discussions were reassumed in 1921 by the Maritime Law Committee of the International Law Association, which held a conference in London in May of that year. There, a set of rules was proposed following the same model of the Canadian Water Carriage of Goods Act of 1910 and the Harter Act. In October 1922, the Comité Maritime International, during its conference in London, reviewed the topic adopting some changes and proposed the final set of rules for the adoption at a diplomatic conference.[487] The fifth Diplomatic Conference on Maritime Law took place in Brussels that same year, and the rules were discussed but not formally adopted. In 1923, an additional meeting took place of the subcommittee of Bills of Lading of the Diplomatic Conference and further discussions were held to clarify the text adopted the year before. In all these pre-Brussels conferences it was expressly accepted that the delegates there did not represented countries, but economic interests.[488] These delegates, representing the interests of the key sectors of the maritime shipping industry, and producing rules according to their economic needs, facilitated the decision making process to reach an agreement.[489] The formal adoption of the International Convention for the Unification of Certain Rules of Law Relating to Bills of Lading came finally on the 25th of August, 1924, when the document was opened for its signature,[490] coming into effect the 2nd of June, 1931.[491] In the end, it was a "compromise invoking some new laws" to reach a consensual agreement between the different interests of carriers and shippers of that specific time,[492] a kind of *Lex Mercatoria* at the dawn of the twentieth century.

2.1.4.1.2 Scope of Application

The scope of application of the Rules is set out in Article 10 of the Convention. With a very short and concise statement, the Convention establishes that the rules apply to every bill of lading issued in any of the contracting States. The carriage of goods, according to Article 1, paragraph (e), comprises the period that runs from "the time when the goods are loaded on to the time they are discharged from the ship". It is the period called "tackle to tackle".

[487]Sturley, *The History of COGSA…*, 30.

[488]Frederick, 88.

[489]Ibid., 93. The chairman of the 1921 Hague meeting, Sir Henry Duke said on this regards: "In any Convention in which nations were represented they would vote by nations. We represent interests. In any context or in any discussion in which they were represented they would express their views by interests. But the usage of the Association, as I have learned, and the usage of the Committee is, that the votes are given by individuals.*"* *See* Hague Conference Report at 9, reprinted in 1 Legislative History, at. 115.

[490]Sturley, *The History of COGSA…*, 31–32.

[491]Karan, 27. *See* CMI Yearbook (1992), 40.

[492]Tetley, *Interpretation and Construction…*, 37.

2.1.4.1.3 The Standard of Liability

It is clear that under the previous strict standard, the carriers had an absolute obligation to restore the goods carried. The obligation implied a warranty to provide an absolutely seaworthy ship to assure such a result; but in reality, under that standard, whether the carrier fulfilled this duty or not, was irrelevant. Doubtless, it is a fundamental part of it for the efficacy of the contract, but seaworthiness was obviously not the end in itself of the contract, but just a part of it.[493] The carriers were obliged to restore the goods in every case, or to compensate the cargo owner if they were damaged or lost, whatever the cause of damage or loss was, and unseaworthiness may have been only one cause. A breach of this warranty created liability not for the want of seaworthiness in itself, but only if that was the case of the carrier's failure to restore the goods in proper condition.[494]

But as the carriers began using numerous contractual exceptions of liability, the original and basic contractual undertaking was weakening. Then, the concept of seaworthiness became more relevant. The courts tried to keep the essence of the contract by dividing the absolute obligation to deliver safely into specific and essential duties.[495] If those duties were duly accomplished, the result should be the same; it means the cargo should arrive in proper condition at destination. The courts then defined two main obligations or duties. The first was the duty to provide a seaworthy ship, ensuring thus a minimum level of protection to the cargo from external perils similar to the level provided by the absolute standard.[496] The second duty was to exercise care and skill in the stowage and carriage of the cargo. As pointed before, these duties were not new. They were already established in the historical sources of maritime law to which we referred above. The carrier was not supposed to exclude his liability for the lack of fulfilment of these duties, and for this reason they were categorized as overriding duties.[497] With this distinction of specific and fundamental duties, the courts prompted a redefinition of the basis for liability.[498] This was the formula originally proposed in the "Conference form", discussed in the International Law Association in their meeting of Liverpool of 1882. As aforementioned, it was also later included in the US Harter Act of 1893. The difference was that the first duty, to provide a seaworthy ship, was not kept in its absoluteness, but reduced to a practice of due diligence to make the ship seaworthy.

The redactors of the HR took into account the duties defined previously by the courts, which in turn, came from the older legal bodies of maritime law, but as appeared in the Harter Act. To set a statutory prohibition of the negligence clauses,

---

[493]Clarke, at 124: "The warranty was, in fact, superfluous as Ripert observes."

[494]Ibid., 124.

[495]Ibid., 124. *See Propeller Niagara v. Cordes* (1858) 62 U.S. (21 How) 7, 23.

[496]Ibid., 125. *See Steel v. State Line* (1877) 3 App. Cas. 72, 76.

[497]Ibid., 124.

[498]Ibid., 124.

the HR lawmakers replaced the old absolute standard of liability historically applied to common carriers, for an intermediate formula that fulfilled the interests of the negotiating parties. This intermediate formula was: first, the establishment of a minimum of liability based on the negligence incurred by the carrier in the fulfillment of the two aforementioned basic duties; and, second, releasing the carrier for damages caused by fault or negligence in the navigation and management of the vessel, while still holding those liable for want of care of the goods.[499] The minimum duties that the carrier must perform are set in the Article III as follow:

Article III.

1. The carrier shall be bound before and at the beginning of the voyage to exercise due diligence to –

(a) Make the ship seaworthy.

(b) Properly man, equip and supply the ship.

(c) Make the holds, refrigerating and cool chambers, and all other parts of the ship in which goods are carried, fit and safe for their reception, carriage and preservation.

2. Subject to the provision of Article 4, the carrier shall properly and carefully load, handle, stow, carry, keep, care for, and discharge the goods carried.

…

The provision stated in the first section obliges the carrier to take, adopt and develop preparatory measures to make the ship capable to accomplish successfully the contracted carriage, specifying three main areas of concern:

- The physical condition of the vessel itself;
- The quality, quantity and capability of the crew and the equipment; and,
- The cargo worthiness of the vessel.[500]

Carriers are liable for cargo damages or losses only if they cannot prove that they have exercised due diligence before and at the beginning of the voyage; unless, of course, they warranted a seaworthy vessel.[501] Thus, if the cargo damage or loss occurs because a pre-existing unseaworthy condition in the vessel not discoverable by practicing a reasonable degree of due diligence; or, if the unseaworthiness condition appeared after the beginning of the voyage, the carriers are totally released from liability.[502]

The second provision does not orders expressly to practice due diligence. It states, instead, an obligation of care related to the direct handling of the cargo, from

[499]M. F. Sturley, T. Fujita and G. J. d. van Ziel, *The Rotterdam rules: The UN Convention on Contracts for the International Carriage of Goods Wholly or Partly by Sea* (London: Sweet & Maxwell, 2010), 78.

[500]Wilson, 187.

[501]R. Force, 'A comparison of the Hague, Hague-Visby and the Hamburg Rules: Much Ado About (?)' (1996) 70, Tulane Law Review, 2051–89, 2061.

[502]Ibid., 2061.

the moment of its loading until its discharge. Notwithstanding, the concept of due diligence has been held as equivalent to that of due care and used as synonyms.

A third duty expected from the carrier is the proper and careful navigation and management of the ship during the voyage. But the impossibility in that time of the shipowners to control the decisions and actions of the master, crewmembers and others servants once the ship had left the port, was considered a good argument to exonerate them from liability, when errors of navigation were the cause of damage or loss. An exception of liability in this regard was stated in the Convention.[503] With this exception, 16 other events were also established in a detailed list of defenses to exclude their liability.[504] This long catalog of exceptions of liability listed in Article 4.2 contains many of the exceptions that the British shipowners had commonly inserted in the British liner bill of ladings.[505] They can be grouped in natural incidents, acts of third parties, acts of the cargo owner or inherent defects of the goods and certain reasonable acts of the shipowners or his employees.[506] It has been said that it was the endeavor of the British carriers to keep, as much as possible, the liability system they created through the bills of lading.[507]

In addition, if the carrier was found liable for negligence, they could still limit their liability to a certain minimum amount, of the £100 originally stated, to the current 666.67 units of account per package or unit, or 2 units of account per kilogram, established in the 1979 S.D.R. Protocol.[508] Evident in all these provisions is the influence of lawyers representing the British shipowner's interests, which explains why the HR seems to be more focused on establishing the least amount of liability possible, as a counterpart to the prohibition of the negligence clauses.

2.1.4.1.4 Consequences

(1) **International Standardization**

The Hague Rules were not enacted to be a general and thorough code for the regulation of maritime transportation under bills of ladings.[509] They neither codified the previous or current laws of that time, but they are instead, a mix of civil and common law style statute, called by Tetley as a "codifying statute".[510] The

---

[503]The HR, Article 4, paragraph 2, literal (a): "2. Neither the carrier nor the sip shall be liable for loss or damage arising or resulting from: (a) Act, neglect, or default of the master, mariner, pilot, of the servants of the carrier in the navigation or in the management of the ship."

[504]Sweeney, 72; The HR, Article 4, paragraph 2.

[505]Diamond, 227.

[506]Billah, M. M., *Effects of insurance on maritime liability law: A legal and economic analysis* (Cham: Springer, 2014), 11.

[507]J. O. Honnold, 'Ocean Carriers and Cargo; Clarity and Fairness –Hague or Hamburg?' (1993) 24, Journal of Maritime Law and Commerce, 75–109, 101.

[508]The HR, Article 4, paragraph 5(a).

[509]Tetley, *Interpretation and Construction…*, 37.

[510]Ibid., 37.

Convention mostly attempted to unify certain rules relating to the transport by sea under bills of lading as its title states.[511] In fact, the HR became the first statutory instrument of international standardization of rules to govern the carriage of goods by sea under bills of lading.[512] Several cases reported the intention of standardization of the Hague Rules. One of them was *The Muncaster Castle* (1961) where Viscount Simonds stated that:

> The Hague Rules, as is well known, were the result of the Conference on Maritime Law held at Brussels in 1922 and 1923. Their aim was broadly to standardize within certain limits the rights of every holder of a bill of lading against the shipowner, prescribing an irreducible minimum of the responsibilities and liabilities to be undertaken by the latter.[513]

On a first assessment, it would be said that this purpose was mainly achieved, as these rules have been widely ratified or accessed and applied in the shipping industry to the present. Most of the courts in the world, when dealing with cargo claims, are deciding them according to the rights and duties of the parties established in the HR. This international standardization is without any doubt very positive result. Notwithstanding, the success of this unification has been challenged. Professor Sturley, in a comparative study of the national interpretation of the rules, comes to state that the HR are an "extreme case of international disharmony."[514]

(2) **Compromise for a Fairer Balance**

The second objective of the HR was the statutory prohibition of insertion of negligence clauses. They tried to create a fairer balance between the obligations and risks to be assumed by carriers and cargo owners,[515] by restricting carriers from excluding contractually liability for their own negligence; while, at the same time, replacing statutorily the traditional common law strict liability standard, for a minimum of liabilities.[516] Lord Steyn exposed this purpose in *Effort Shipping Co. Ltd. v. Linden Management, S.A.* (1998):

---

[511] Wilson, 187.

[512] Karan, 27.

[513] [1961] AC 807 at 836, [1961] 1 Lloyd's Rep. 57 at, 67. In the United States *See also The Asturias* 40 F. Supp. 168, 169, 1941 A.M.C. 761, 762 (S.D.N.Y. 1941): "The purpose of the Act is to create international uniformity"; *Senator Linie GmbH v. Sunway Line, Inc.* 291 F.3d 145 at 158, 2002 A.M.C. 1217 at 1232 (2nd Cir. 2002): "One important aspect of the international agreement [The Brussels Convention 1924] and its United States counterpart [COGSA 1936] is the standardization of liability expectations.", as quoted by Tetley, *Interpretation and Construction…*, 57–58.

[514] Honnold, 101; *See* Sturley, "*International Uniform Laws in National Courts*" 27 Va. J. Int'l L. 729, 774–796 (1987).

[515] Frederick, 96.

[516] Diamond, Anthony QC. "*Responsibility for Loss of, or Damage to, Cargo on a Sea Transit: the Hague or Hamburg Convention?*" in P. S. K. Koh, *Carriage of goods by sea* (Singapore: Butterworths, 1986), 110.

> This much we know about the broad objective of the Hague Rules: it was intended to reign in the unbridled freedom of contract of owners to impose term which were 'so unreasonable and unjust in their terms as to exempt from almost every conceivable risk and responsibility'; it aimed to achieve this by a pragmatic compromise between interests of owners and shippers; and the Hague Rules were designed to achieve a part harmonization on the diverse laws of trading nation at least in the areas which the Convention covered.[517]

The abolition of the negligence clauses was reached through a "pragmatic compromise", as described by Lord Steyn, attempting to reach a fairer balance between the parties. American jurisprudence has reported also this intention in *Encyclopedia Britannica v. Hong Kong Producer* (1969):

> The purpose behind Harter, the Hague Rules and COGSA were to achieve a fair balancing of the interest of the carrier, on the one hand, and the shipper, on the other, and also to effectuate a standard and uniform set of provisions for ocean bills of ladings.[518]

On this objective, it must be said that the Convention also succeeded.[519] The rules expressly prohibit clauses exonerating the carriers of liability, or reducing their duties in the performance of the contract. Article 3.8 states the express prohibition of clauses releasing of or diminishing the carrier's liabilities set in the Convention:

> 8. Any clause, covenant, or agreement in a contract of carriage relieving the carrier or the ship from liability for loss or damage to, or in connection with, goods arising from negligence, fault, or failure in the duties and obligation provided in this article or lessening such liability otherwise that as provided in this convention, shall be null and void and of no effect. A benefit of insurance in favor of the carrier or similar clause shall be deemed to be a clause relieving the carrier from liability.

Certainly, the "pragmatic compromise" reflected in the Convention sets a balance,[520] or, better said, a more or less balanced distribution of risks between the shippers and carriers.[521] Such balance, however, may be considered fair, only when it is compared to the former practices based on the unrestricted freedom of contract. Doubtless, a minimum of liability is by far better than no liability at all. But this compromise required shippers to make greater concessions. Under the liability formula stated, if the ship is unseaworthy the carrier will be exonerated simply by proving that he practiced due diligence in making her seaworthy before and at the beginning of the voyage. Such a minimum requirement on the part of the carrier

---

[517] [1998] AC 605 at 621, [1998] 1 Lloyd's Rep. 337, 346, 1998 A.M.C. 1050 at 1065 (HL), as quoted by Tetley, *Interpretation and Construction…*, 62.

[518] 422 F2d. 7, 11, 1969 A.M.C. 1741, 1746(2nd Cir. 1969), [1969] 2 Lloyd's Rep. 536, cert. denied, 397 US 964, 1971 A.M.C. 813 (1970), as quoted by Tetley, *Interpretation and Construction…*, 63. *See also Senator Linie GmbH v. Sunway Line, Inc., see supra* note 513, at 158: "In essence, the purpose of these laws is to allow international maritime actors to operate with greater efficiency under a mantle of fairness."

[519] Diamond, 226, 231–232.

[520] Karan, 27; Tetley, *Interpretation and Construction…*, 62.

[521] Diamond, 226.

means the shipper must bear a higher burden of risk, when his cargo is exposed to danger out of his radius of action, and more in the sphere which only the carrier can properly organize and control.[522] It is clear that shippers will have fewer chances to exercise the control that the carriers have over the vessel's seaworthiness as well as the cargo care during the time it is under the custody of the carrier. This situation is not commonly seen in other types of contracts.

This compromise between carriers and shippers to reach a fairer balance has been suggested by Tetley to be regarded as a principle to orientate the interpretation and construction of the Hague Rules.[523] Paradoxically, in spite of the criticized imbalance disfavoring cargo owners, in the cases referred by the same author, where this "principle of fair balance" has been invoked by some American judges, it has operated against the shippers.[524]

For these and other reasons, the Convention has been criticized as favoring too much the shipowners, who enjoyed strong support from the United Kingdom in the discussions in Brussels.[525] Specifically on the allocation of risks, plus the limitation of liability rules, it is evident that it was tipped in favor of carrier's interests.[526] This criticism of the unfair imbalance was reported in 1971 by Secretariat of the UNCTAD on Bill of ladings.[527] In this report, it was affirmed that due to specific provisions and omissions, the shipowners had *excessive* legal opportunities to elude liability for cargo damages or losses.[528] Others criticized that the low degree of liability assigned to carriers for negligent cargo loss had the added effect of discouraging them to make more investments on safe carriage.[529]

---

[522]Zimmermann, 521.

[523]Tetley, *Interpretation and Construction…*, 62.

[524]Ibid., 62. *quoting Senator Linie GmbH v. Sunway Line, Inc. See supra* note 513. It is about the liability for dangerous goods, and *Caemint Food v. Lloyd's Brasileiro* 647 F. 2d 347, 1981 A.M.C. 1801 (2nd Cir. 1981), on defective packages.

[525]Diamond, at 227: "It was the judgment of many people in the mid-1950s that, in securing the Brussels Convention of 1924, the U.K. Government had managed to combine self-interest with the implementation of the greatest good for the greatest number. The Convention seemed, on the one hand, a diplomatic coup of the first order; on the other hand, the beneficial instrument which had finally brought about the desirable aim of the standardization of bill of lading clauses."

[526]Sweeney, 74. It was even part of the developed countries objections. Such balance in favor of shipowners "must necessarily have affected the cost of insurance, although no compensation is given by way of lower rates for shippers."

[527]UNCTAD, *Bills of Lading: Report by the Secretariat of UNCTAD* (1971).

[528]Ibid., 18, at para. 80: "In part, this arises from the fact that through both specific provisions and omissions, the Hague Rules provide what appears to be an excessive number of opportunities for the shipowner to avoid, legally, liability for loss of cargo and so to reject the claim made by the cargo owner."

[529]Honnold, 107. *See* Kindred, at 613–614; O'Hare, Allocating Risk, 143–144, 152. Hellawel at 363–365; M. F. Sturley, 'Changing Liability Rules and Marine Insurance: Conflicting Empirical Arguments About Hague, Visby, and Hamburg in a Vacuum of Empirical Evidence' (1993) 24, Journal of Maritime Law and Commerce, 119–49.

(3) **Lack of Clarity**

In addition, a more global defect pointed out in the HR is its lack of clarity in many aspects. The reason for this lack of clarity is mainly attributed to a specific fact: to reach an agreement between opposite political and commercial interests, "clarity and consistence of purpose" were, to a considerable extent, sacrificed.[530] Indeed, it is said that the draftsmen of the Convention were more motivated, or perhaps, limited, by the political and commercial interests they represented, that they were more concerned in the economic implications of the rules, than in the language itself.[531] The language of the Convention was conditioned to what the draftsmen were able to concede in order to reach a satisfactory agreement for both parties. Consequently, it seems they preferred wider concepts and imprecise terminology that allow or require further constructions by the courts. The result was an ambiguous Convention that left many aspects undefined and subject to constant confusions.[532] Hence, the rules have been criticized for being poorly drafted[533]; or that they "do not meet the most elementary standards of legal technique, readability and good statutory language."[534]

Although Diamond argues the HR has not raised many problems of interpretation[535]; that is possible to be affirmed only after some decades of vast litigation on the subject and many courts constructions. The reality is that the concept of practicing due diligence can be so vague, wide and imprecise that many controversies on the subject have frequently required litigation. Such ambiguities in the language, plus its incapacity to anticipate the advances made through technology have driven carriers and shippers to long and expensive litigation or arbitration to resolve disputes.[536] These shortcomings and particularly the failure to anticipate economic and technological changes have been pointed out as the most problematic

---

[530]Diamond, 228.

[531]Frederick, 104. Indeed, something that has been noticed by some authors is the great influence of English shipowners in the adoption of this Convention. Diamond reported the concern of the British Government in defending the carriers' interests during the negotiations, *See* supra note 575; Knauth pointed out the influence of the British shipowner in the British Parliament, *See* supra note 561.

[532]Colinvaux, at v, commenting on the incorporation of the HR language into the English COGSA of 1924, he criticizes its ambiguity as follow: "No document gives more scope for ingenuity in its interpretation than a statute which attempts to incorporate into English law the terms of an International Convention. A well drafted enactment like the Sale of Goods Act, 1893, has the effect of crystallizing the law within a few decades; one such as the Carriage of Goods by Sea Act, 1924 puts it into confusion indefinitely. Now, nearly thirty years later, every month sees some new and insoluble problem arising under it."

[533]Honnold, 101. *See* Gröforns, *The Hamburg Rules* 1978 J. Bus. L. 334, 335.

[534]Ibid., 101. Quoting a Norwegian Ministry of Justice. *See* Selvig, *The Hamburg Rules, the Hague Rules and Maritime Insurance Practice*, 12 J. Mar. L. & Com. 299, 302 (1981).

[535]Diamond, 228.

[536]Frederick, 99. *See* Andreani, Revision of the Hague Rules, Activities of UNCTAD and UNCITRAL and Developing Countries, in Studies on the Revision of the Brussels Convention on Bills of lading (F. Berlingeri ed. 1974), 45.

issues that have prompted further attempts for new legislation.[537] Practicing due diligence in making a ship seaworthy is such a complex task, that the issue must be submitted most of the time to judges to determine whether carriers were diligent or not. Therefore, it has required large amounts of litigation to clarify, through court decisions, the many ambiguities contained in this Convention.

These uncertainties generate conflicting results mostly in favor of the carriers, who have taken advantage of them "to delay or defeat claims through tenuous legal technicalities."[538] Due to the jurisdiction clauses that the British carriers continued inserting in their bills of lading, such construction had to be provided by the English Courts, where most of the cargo claims were supposed to be filed.[539] The Convention did not regulate on jurisdiction clauses.

But this is not the only economic consequence of its lack of clarity. In addition, the uncertainties of the Convention relating to some "ill-defined risks" created the need for both, the owner of goods as well as the carriers, to buy insurance against those risks.[540] The legal uncertainty of the rules, as in any other regulation, is recognized as wasteful, responsible for the increase of cost in the delay of settlements and payments of costly legal services, double insurance to cover the same risks and "unnecessary casualty losses when immunity is given to the only party who could guard against those losses."[541]

Despite the criticism to its imbalance, its ambiguity and the obsoleteness of its provisions to regulate the new issues of the shipping industry, the HR continue to be the governing regime for the contract of carriage of goods by sea under bills of lading.

2.1.4.1.5 Amendments

The need for an update of the HR has resulted from the demands of a growing global market and technological advances, both of which have created major changes in the industry. Decades after the adoption of the HR, the interpretations and constructions made in court decisions were not enough to fulfill this need. The results were the adoption of two protocols amending some few and very specific provisions of the original Convention.

(1) **The Visby Amendment—The Brussels Protocol of 1968**

The shortcomings of the original convention became more notorious by the middle of the 20th century. Diamond reports that during the period previous to the adoption of the Visby Protocol, between 1955 and 1963, the HR experienced the greatest

---

[537] Frederick, 105.

[538] Honnold, 101.

[539] Frederick, 85. This author reports that the influence of British Shipowners existed even in the English judiciary.

[540] Frederick, 99.

[541] Honnold, 107.

success.[542] He pointed out that the number of accessions to or ratification of the Convention rose from 17 to a total of 27, and that the number of cases was relatively low.[543] However, in the middle of the 1950s businessmen and lawyers of major maritime member States of the Convention pushed for reforms.[544] This prompted further discussions in the *Comité Maritime International* (CMI) to pursue an update of the rules as a response to many questions arisen due to the vague and incomplete language of the HR. The result of those discussions was the Brussels Protocol of Amendments to the HR, adopted in a Diplomatic Conference on the 23rd of February 1968.[545] One of the main changes introduced was regarding the scope of application of the rules, which was modified by the introduction of a new Article X. Originally, the rules applied exclusively to every bill of lading issued in a contracting party. Now, they also apply in two more situations: (1) when the carriage departs from a port in the contracting State; and, (2) when the contract contained or evidenced in the bill of lading provides that the rules of the Convention or the legislation of a state party applies to the carriage. This new article also clarifies that the application of the rules in the cases listed is regardless of the nationality of the ship, the parties, or any interested person.[546]

One of the controversial proposals presented by the British delegation intended to overrule the construction made in the famous case *The Muncaster Castle*.[547] The court held the carrier liable for the negligence in making the ship seaworthy when this duty was carried out by their subcontractors, or the subcontractor's servants. This decision established the non delegability of the obligation. Such a proposal, however, did not garner international support and was not incorporated into the protocol.[548]

The Visby amendment decreased some carrier's responsibilities[549] and increased others, especially in terms of the limitation of liability. The balance established in the HR was updated here "with a slight list towards shippers".[550] However, the amendments did not affect Article 3, sections 1 and 2 of the original Convention. It did not affect directly nor substantially the original carrier´s duties established in

542Diamond, 232.

543Ibid., at 232: "The greatest test cases on the Rules had mostly been determined and were seen to be very few in number, as compared with the great volumes of cargo claims where the Rules could be satisfactorily applied without the need for litigation."

544Frederick, 94.

545Sweeney, 73. Since 1959 the amendment to the rules were discussed in the Plenary Conference of the CMI in Rijeka. The process took 4 year of discussion until the adoption of the final draft in the CMI conference of 1967 in Stockholm.

546The Visby Protocol Article V(b) and (c).

547[1961] A.C. 807, [1961] 1 Lloyd's Rep. 57; Diamond, 231.

548Ibid., 231.

549It expands the time bar limitation of one year for claims based in deviation, fraud or fundamental breach; the total amount to be recoverable is based in the value of the goods in the port of destination; extension of the carrier´s rights to agent and servants, etc.

550Tetley, *Interpretation and Construction…*, 62.

those provisions. The original Hague Rules with the amendments made by this protocol are observed as a second regime known as the Hague-Visby Rules. Some 28 States, including The United Kingdom and its Overseas Territories, have ratified this amendment.[551] Though intended as an attempt to update the HR, this amendment did not introduce the necessary changes required by the new global economy even at the time of its enactment. As a consequence, it was later required the adoption of the Hamburg Rules.[552]

(2) **The S.D.R. Protocol—The Brussels Protocol of 1979**

By 1979, another amendment was introduced to the Convention. This time, the currency used to calculate the limitation of liability was changed. The former Visby protocol established a limitation of liability to be estimated in French Francs. The new protocol changed this to "units of account", establishing a general limitation of liability to 666.67 units of account per package or unit or 2 units of account per kilogram of gross weight of the goods lost or damaged. Each units of account is calculated using the Special Drawing Right, a value set by the International Monetary Fund and exchangeable with any national currency on which the cargo is valuated. Once again, the carriers obligations regarding the ship and the cargo stated in Article 3 remained intact. The protocol has been ratified or accessed by some 24 States.[553]

2.1.4.2 The Hamburg Rules—1978

2.1.4.2.1 Adoption and Generalities

As a consequence of the fast growth in international trade, the shortcomings of the Hague/Hague-Visby Rules came to be more obvious during the second part of the 20th century. The rules were (are) not suitable for the new realities of a modern carriage of goods. This has prompted the necessity to introduce the amendments that we briefly mentioned above. As those amendments did not go far enough, by 1970, a new movement pursued further regulation on the matter through the discussion of a totally new convention.[554] One of the matters of concern on the HR reported by the Secretariat of the UNCTAD in 1971 was precisely the uncertainty created in the interpretation of the terms "due diligence" and "properly and carefully", among others.[555] In addition to the language problems of the HR, criticism was further raised on the incapacity of those rules to "anticipate economic and

[551]CMI Yearbook 2013, 605–606. Including also the Special Administrative Region of Hong Kong.

[552]Tetley, *Interpretation and Construction…*, 62.

[553]CMI Yearbook 2013, 607.

[554]Diamond, 234.

[555]UNCTAD, *Bills of Lading: Report by the Secretariat…*, 17, para. 73 (d).

technological developments."[556] By the 1960s, the use of containers was increasingly popular; the shipping industry already knew and applied steel ships powered by steam or oil, precise and complete marine charts, radio beams or satellite communication, and navigation aids such as radars or sonar, and weather stations at the carrier's offices.[557] An additional problem involved the conflictive exceptions of the governing regime over nautical fault, established far before the introduction of the aforementioned technological developments.[558] Such technological advances created doubts around the convenience and fairness of keeping this exception. New technological aids allow shipowners to have uninterrupted communication with the vessel while on the sea, and significantly reduce the associated risks of the operation.

Therefore, a process was lead pursuing the enactment of a more complete regulation for carriage of good by sea under bill of lading, and the adjoining liability rules. Some of the participants in the discussions went further to challenge the financial or economic aspect of the HR, demanding the allocation of much higher liabilities on shipowners.[559] A key role for the adoption of a new Convention was played by developing countries, which claimed participation in the creation of an international regime that at that time affected them in a higher extension.[560] The claims of the developing countries made the discussion, led by UNCITRAL, more politically oriented than economic or commercially oriented in contrast to the Hague Convention.[561] Governments, in general, had a much more active role in the drafting of this Convention, although their participation was sometime seen as an attempt to introduce "irrelevant, badly defined and often ill-informed notions of public interest into commercial affairs."[562] Some bodies representing shipowners, shippers and underwriters expected from this Diplomatic Conference to merely put into legal terms their commercial agreements regarding the risks distribution between the parties involved in the maritime adventure.[563] Such an approach put aside that the carriage of goods by sea has an impact that goes beyond the mere interests of the parties involved. Therefore, governments were obliged to intervene also as a response to the frequent ill-conceived notions of freedom of contract implemented in the past.

---

[556]Frederick, 105.

[557]Honnold, 78.

[558]Ibid., 78.

[559]Diamond, 234.

[560]Sweeney, at 73: "Dissatisfaction of the developing world stems essentially from the belief that the operation of traditional maritime law (along with other aspects of international trade law) impairs the balance of payments position of developing states so as to insure continued poverty and perpetual under-development in an industrial age."; Diamond, at 234: "The Hague Rules, it was so said, were a device of the colonialist powers designed to impoverish the developing world." *See* also UNCTAD, *Bills of Lading: Report by the Secretariat...*, 30 para. 172–175.

[561]Frederick, 117.

[562]Diamond, 232.

[563]Ibid., 232.

After almost ten years of discussion carried out by the UNCTAD and UNCITRAL, the United Nations Convention on the Carriage of Goods by Sea was adopted through a diplomatic conference held in Hamburg, Germany on the 30th of March in 1978.[564] It created a mandatory frame of liability for carriers based again on negligence, similar to that of the previous Brussels Convention.[565] Known as the Hamburg Rules, this Convention is a kind of "mini-code" with regulations not only covering the main "problem areas" addressed in the HR, but also a more extensive regulation for other issues surrounding the carriage of goods by sea.[566] The attempt was made in these rules to overcome some of the inconsistencies and ambiguities of the H/H-VR.[567] It came into force on the 1st of November of 1992, following the adherence of 20 States.[568]

2.1.4.2.2 Scope of Application

This Convention, as well as the H-HVR applies exclusively for carriage of goods by sea under bills of lading. The scope of application is provided in Article 2(1). The rules apply to carriages performed: a) from port of loading; or b) to port of discharge; or c) to an optional port of discharge contractually agreed upon where discharge is actually performed; if they are located in one of the Contracting States. It also applies to, d) bill of ladings issued in one of those Sates; or, e) when the parties agree to the application of the rules or the legislation of any State giving effect to them and such an agreement is provided in the bill of lading or any other document evidencing the contract of carriage.

2.1.4.2.3 Period of Responsibility

The period of responsibility for the carrier differs from the H/H-VR. While the former regime covers "from tackle to the tackle"[569]; the Hamburg rules provide a more extensive coverage. In the time when the HR was adopted, it was possible for shippers to deliver the cargo up to the side of the ship and the carrier could immediately upload it.[570] The same procedure happened at the port of discharge. But with the increase of international trade, ports became crowded and ships bigger; the shipper must consequently leave their cargo at the carrier's warehouses until the

---

[564]Honnold, 80.
[565]Koh, 116.
[566]Force, 2055.
[567]Wilson, 217.
[568]Honnold, 80.
[569]The HR Art.1(e) "Carriage of goods covers the period from the time when the goods are loaded on to the time they are discharge from the ship."
[570]Honnold, 81.

vessel is ready for loading; the same happens to consignees who must wait until the cargo is ready to be delivered.[571] That can take many hours or even days, and these periods, as well as the time while in an intermediary port, are excluded from the "tackle to tackle" rule set out in the HR.[572] Because of this new situation, the goal of international unification was lost. Carriers were subject to the national laws of every port for these specific periods, when, typically, most of the losses from weather or theft occur.[573] The situation moved to the adoption of a more extensive rule. The period of responsibility of the carrier "covers the period during which the carrier is in charge of the goods at the port of loading during the carriage and at the port of discharge".[574] This rule makes the carrier responsible for the entire period under which he has the custody of the goods, regardless of whether they are on board the vessel or not. It solves the problem in the determining of when "before or after" is, as set out in the H/H-VR.[575] The provision also mirrors the Harter Act which similarly establishes responsibility from the time the carrier receives the cargo until its delivery.[576] On this aspect, the introduction of a more extensive period of responsibility is doubtless very positive.[577]

2.1.4.2.4 Liability Rule

The liability regime is stated in an affirmative rule of responsibility stated within Article 5.1, as follows:

> The carrier is liable for loss resulting from loss of or damage to the goods, as well as from delay in delivery, if the occurrence which caused the loss, damage, or delay took place while the goods were in his charge as defined in Article 4, unless the carrier proves that he, his servants or agents took all measures that could reasonably be required to avoid the occurrence and its consequences.

The rule is based on the principle of "presumed fault".[578] It does not make any reference to an obligation to provide seaworthiness or any differentiation of the specific duties. They are implied in the rule as the general obligation is not to cause

---

571 Ibid., 82.

572 Ibid., 82.

573 Ibid., 81–82. *See* Kindred, From Hague to Hamburg: International Regulation for the Carriage of Goods by Sea, 7 Dalhousie L.J. 585, 595.

574 The Hamburg Rules Art. 4.1.

575 Wilson, 216.

576 Force, 2059.

577 Honnold, 83.

578 The Annex II of the same Hamburg Rules clearly states that: "It is the common understanding that the liability of the carrier under this Convention is based on the principle of presumed fault or neglect. This means that, as a rule, the burden of proof rests on the carrier but, with respect to certain cases, the provisions of the Convention modify this rule."

loss of or damage to cargo due to negligence.[579] The previous test of practicing due diligence in making the ship seaworthy before and at the beginning of the voyage and properly caring for the cargo is reduced to a uniform test of liability based on fault. This formula simplifies the long-standing problem of identifying the origin or character of the cause of damage or loss to assign liability.

Now, it is irrelevant whether the cause of damage or loss was due to lack of seaworthiness or want of care; both are judged under the same principle, regardless of when negligence occurred.[580] The carrier now has an obligation to exercise due diligence and care during the entire period when he is responsible for the goods.[581] Although it is not specifically expressed, because the carrier, as well as his servants, is under a duty to take all reasonable measures required to prevent the occurrence of damages or losses, this automatically obliges him to make and keep the ship seaworthy during the whole period of the voyage for each cargo concerned.[582] Hence, the carrier's obligation has increased. A new problem will be the interpretation by the courts of the expression "all measures that could reasonably be required to avoid the occurrence and its consequences", included in the last part of the transcribed article.[583] The ambiguousness of the term will certainly warrant more litigation.[584]

Another consequence of this presumed fault rule is the abolition of the list of exceptions contained in the Article 4.2 (d) to (q) of the Hague/Hague-Visby Rules. As none of them involves fault on the part of the carrier, it was not necessary to include them. Carriers, of course, can still allege these causes as such in trials.[585] Particularly clear is that the same principle stated in the exception contained in paragraph 4.2 (q) of the HR, requiring proof of the lack of negligence on the part of the carrier, remains in the Hamburg Rules.[586] However, the US courts ruled that in regards to the immunity set out in this paragraph of the HR, the carrier must prove not only the absence of his negligence, but also the cause of the damage, resulting in the fact that the Hamburg Rules may be more favorable to the carrier on this subject as they must prove only lack of negligence.[587]

One of the major differences introduced by these rules was the upholding of liability for error or negligence in navigation and management of the ship. The long discussion regarding the fairness of such an exception brought finally to its abolition in this Convention. The technological advances applied in navigation today render occurrences of real errors in navigations and management of the ship very

---

579 Force, 2063.

580 Wilson, 217.

581 Force, 2064.

582 Tetley, *Marine Cargo Claims,* Vol. 1, 896.

583 Wilson, 217.

584 Frederick, 113. *See* Hellawell, Allocation Risk Between Cargo Owner and Carrier, 27 Am. J. Comp. L. 363 (1979).

585 Force, 2066.

586 Ibid., 2067.

587 Ibid., 2067.

uncommon.[588] As expected, the carrier's interest complained against it, alleging the possible consequence of freight rates increasing if they were held liable for negligence of their servants.[589] This change, doubtless, facilitates the determination of liability for want of cargo care, which has been frequently hard to determine whether it is a violation of the duty of care or a fault or negligence of master and the crew in managing the vessel. On this point, the system proposed by these rules for one single standard of liability based on the "fault or neglect that cause the loss or damage of the goods", instead of the H/H-VR system based in the interplay of duties and immunities, has made the assignation of liability simpler.[590] The rules however, state some causes of exceptions in case of damage caused by fire;[591] or carriage of live animals and transport on deck, leaving in both cases the opportunity for the parties to contract their own terms.[592]

The order of proof in the Hamburg rules is also simpler that in the H/H-VR. The cargo claimant must submit prima facie evidence that the damage or loss occurred while the goods were under the custody of the carrier, and the latter must then show that the proximate cause of loss or damage falls within the general provision of Article 5(1).[593]

In general, and from a practical point of view, it could be said that the regimes in the H/H-VR and in the Hamburg Rules are practically the same. Beside the differences noted, the rest of the provisions are quite similar in content.[594] Under the Hague system the carrier is obliged to practice due diligence before and at the commencement of the voyage, but if a unseaworthiness condition arises after the voyage begins, the cargo interest may still allege that there was a lack of continuous duty in caring for the cargo. The carrier, in such a case, might be held liable unless he proves that he and his servants were diligent enough to solve the problem in a timely manner.[595] This condition is similar to the Hamburg regime where the carrier will be presumed at fault when the damage or lost occurred during the time of his custody.

2.1.4.2.5 Objections

The main objections for the ratification of these Rules are based mostly on the argument that the increase of liability in favor of the cargo interest may bring consequently higher insurance premiums and freight rates.[596] Nevertheless, Sturley

---

[588]Honnold, 105.

[589]Wilson, 217.

[590]Honnold, 98.

[591]The Hamburg Rules 5.4.

[592]The Hamburg Rules Articles 5.5 and 9.

[593]Tetley, *Marine Cargo Claims,* Vol. 1, 936.

[594]Force, 2085.

[595]Sturley, Fujita and van Ziel, 8.

[596]Force, 2087.

points out that there is not yet any real evidence of such a consequence, and there is no possibility for an accurate assessment without empirical evidence.[597] In its absence, he suggests that such an argument should be abandoned and other approaches should be used to evaluate the carrier's liability.[598]

Other critics have pointed out that the introduction of a new regime will produce additional "litigation, confusion and expense", against the almost "well-established" international regime contained in the HR, especially given the voluminous case law created since its inception.[599] The carrier's interests have objected that the introduction of this new convention will cast aside the "expensive litigation" arisen from the Hague Rules.[600] Honnold contends that perhaps the carriers do not want to abandon the large case law revolving around their immunity for negligence in navigation, management of the ship and fire; that the value of this case law does not lie in its clarity, but in its "ambiguity as a basis for plausible but questionable resistance to cargo claims."[601] Though there are advantages offered by this new regime, since coming into force, they have not reached yet significant application.

2.1.4.3 The Rotterdam Rules—2009

2.1.4.3.1 Adoption and Generalities

After the failure in ratification of the Hamburg Rules, the Comité Maritime International undertook the discussions of a new set of rules in 1998. Later, in 2002, the UNCITRAL Working Group in Transport Law continued the discussions and drafted the final project of a new Convention on the topic.[602] The result was "The United Nations Convention on Contracts for the International Carriage of Good Wholly or Partly by Sea", also known as the Rotterdam Rules, opened for signature on the 23rd of September 2009.[603] At the preset, these rules have been ratified only by Spain and Togo.

One of the main objectives of these rules is to "modernize and harmonize" the rules governing the international carriage of goods by sea, addressing specifically the "technological and commercial development since the adoption of The Hague,

---

[597] Sturley, *Changing Liability Rules…*, 149.

[598] Ibid., 149.

[599] Honnold, 81.

[600] Ibid., 101. *See* a 1998 Bulleting of the Baltic and International Maritime Conference (BIMCO), where in addition it was stated that the Hamburg Rules "must be resisted at every opportunity". Quoted by Waldron in The Hamburg Rules, 1991 J.B.L. 305, 318.

[601] Ibid., 102.

[602] T. Nikaki, 'The Carrier's Duties Under the Rotterdam Rules: Better the Devil You Know?' (2010) 35, Tulane Maritime Law Journal, 1–44, 3.

[603] *United Nations Convention on Contracts for the International Carriage of Goods Wholly or Partly by Sea* (Vienna: United Nations, 2009).

Hague-Visby and Hamburg Rules.[604] This Convention applies almost the same obligations already stated in the HR in relation to the duties of seaworthiness and due care. It repeats most of the same wording of many provisions of the HR, as an effort to maintain the large jurisprudence developed on the H/H-VR,[605] yet introducing some slight changes that make considerable differences with the previous Hague Rules regime.

The Convention introduces in Article 1 some basic definitions that are worthy of examination. First, the "contract of carriage" is defined using the classic description of the activity of a carrier who for the payment of freight, undertakes to carry goods from one place to another. But the definition is not limited to that; it adopts a wider extension to include the land legs of a multimodal carriage, which these rules are also intended to regulate.[606] The rules introduce also the distinction between the mere contractual carrier and the actual carrier, described under the terms of performing party[607] and maritime performing party,[608] which is a person other than the carrier assigned to perform some of the carriers' obligations.[609]

2.1.4.3.2 Period of Responsibility

The period of responsibility also differs from the current H/H-VR. Taking distance of the "tackle to tackle" formula, it is closer to the Hamburg Rules. It establishes in Article 12 that the period of carrier responsibility covers the time from when "the carrier or the performing party receives the goods for carriage and ends when the goods are delivered". The carrier is responsible during the entire period he is in possession of the goods, as the Hamburg Rules state. But it intends to cover also the multimodal transport, in accordance with Article 11, which expressly states the main obligation of the contract, which is to "carry the goods to the place of destination and deliver them to the consignee." The rule was made contemplating the carriage in the door-to-door modality, using the expression "place of

---

[604] Nikaki, 5.

[605] Sturley, Fujita and van Ziel, 82.

[606] Article 1.1.

[607] Article 1.6(a) "'Performing Party' means a person other than the carrier that perform or undertakes to perform any of the carrier's obligation under a contract of carriage with respect to the receipt, loading, handling, stowage, carriage, care, unloading or delivery of the goods, to the extent that such person acts, either directly or indirectly, at the carrier's request or under the carrier's supervision or control."

[608] Article 1.7 "'Maritime performing party' means a performing party to the extent that it perform or undertakes to perform any of the carrier's obligation during the period between the arrival of the goods at the port of loading of a ship and their departure from the port of discharge of a ship. An inland carrier is a maritime performing party only if it performs or undertakes to perform its services exclusively within a port area."

[609] Articles 1.5; 1.6 (a) and 1.7.

destination" instead of "port of destination".[610] Notwithstanding, paragraph 2 of the same article clarifies the exclusion of the carrier's responsibility during the period the goods are under the custody of an authority or third party appointed thereto by law or regulation of the place of reception or delivery. This is the time, for example, when the goods are at the customs clearance or quarantine services, frequently located at the ports facilities to make inspections on inbound/outbound cargo, and where the carrier is obliged to collect or deliver the cargo, in compliance with certain local regulations.

2.1.4.3.3 Liability Rules

The main duties for the carriers are established in articles number 13 and 14. It changes the order of the HR by stating first the specific obligations for protecting the cargo, the duty of care. This foresees that the carrier has responsibility from the reception of the goods, which occurs before to its upload onto the vessel. Second, it states the specific obligations of practicing due diligence to make the ship seaworthy.

(1) **Duty of Care**

The duty of care of the cargo is stated in Article 13 and reads as follows:

Article 13: Specific obligations

1. The carrier shall during the period of its responsibility as defined in Article 12, and subject to Article 26, properly and carefully receive, load, handle, stow, carry, keep, care for, unload and deliver the goods.
2. Notwithstanding paragraph 1 of this article, and without prejudice to the other provision in chapter 4 and to chapter 5 to 7, the carrier and the shipper may agree that the loading, handling, stowing or unloading of the goods is to be performed by the shipper or the consignee. Such an agreement shall be referred to in the contract particular."

The provision makes clear that the obligation is to be exercised during the entire period of the carrier's responsibility. The provision excludes its enforcement when the carrier is not undertaking different legs of transport. The clarification is for cases of multimodal transport, or for the period before they receive the goods or after its delivery.[611] Article 26, thereon referred to, covers the period before loading onto and after discharge from the ship, and the liabilities based in local laws or regulation that covers such periods. The convention also grants to the carriers the possibility to be released from liability for damage or loss during such periods when the shippers partially undertake these functions.[612]

---

[610]Girvin, 445.

[611]Sturley, Fujita and van Ziel, 82.

[612]Article 23 paragraph 2: "Notwithstanding paragraph 1 of this article, and without prejudice to the other provisions in chapter 4 and to chapters 5 to 7, the carrier and the shipper may agree that the loading, handling, stowing or unloading of the goods is to be performed by the shipper, the documentary shipper or the consignee. Such an agreement shall be referred to in the contract particulars."

The rest of the provision states practically the same obligations as the HR with the inclusion of two new points: the carrier must now properly and carefully *receive* and *deliver* the goods. The same responds to the new period of responsibility that covers these two operations. It replaces the term "discharge" for "unload", but this is purely a stylistic change. The latter refers to any means of transport, while the former is more commonly used for maritime transportation.[613] Again it provides that the carrier must perform these actions "properly and carefully". The invariability of these terms responds to the interest in preserving the long-standing jurisprudence related to the H/H-VR where both terms were analyzed.[614]

The second paragraph of the article resolves the problem raised by the FIO's clauses that grant the opportunity to the carrier to contract out the performance of the tasks described in paragraph 1, and the shipper to assume those tasks. It expressly admits the possibility that is recognized in English law but not in other jurisdictions.[615]

(2) **Seaworthiness**

On the obligation to provide a seaworthy ship, the rules state the follow:

Article 14: Specific obligations applicable to the voyage by sea

The carrier is bound before, at the beginning of, and during the voyage by sea to exercise due diligence to:

(a) Make and keep the ship seaworthy;

(b) Properly crew, equip and supply the ship and keep the ship so crewed, equipped and supplied throughout the voyage; and

(c) Make and keep the holds and all other parts of the ship in which the goods are carried, and any containers supplied by the carrier in or upon which the goods are carried, fit and safe for their reception, carriage and preservation."

The Convention repeats the same obligation stated in the HR, but introduces two details that make some significant changes. The obligation of practicing due diligence in making the ship seaworthy is not limited to before and at the beginning of the voyage, it is a continuous one. The carrier is obliged to continue exercising it to avoid any cause of unseaworthiness before and during the voyage. The obligation is not limited *to make*, but now also *to keep* the vessel in proper condition. The provision brought great concern and was objected to under the argument that it would alter the overall risk allocations between the carrier and cargo interest.[616] It places a major burden on the carrier that could be translated into higher freight rates.[617] However, this is simply another argument, and as Sturley argued with

[613]Ibid., 82.

[614]Ibid., 82. He suggests the reading of the 9th Session Report para 117, 119.

[615]Girvin, 445–446.

[616]Ibid., 446.

[617]Ibid., 446.

respect to the Hamburg Rules, there is not empirical evidence to justify such a hypothesis.[618]

The obligation of due diligence remains the same as in the HR. Carriers must do what is reasonably possible under the circumstances.[619] The exercise of this obligation depends on where the vessel is and when the unseaworthiness condition is discovered. If it is discovered while the vessel is at port, it may more easily facilitate the carrier in its repair, whereas the same problem will present much more difficulties while at sea.[620] It is obvious that for a carrier it would be more difficult to carry out reparations on the ship during the voyage in high seas as opposed to while in port or at a shipyard. It is not expected that the carrier performs the same activities or actions that would be possible on shore. The carrier therefore, must keep the vessel specially equipped to avoid any possible unseaworthiness condition during the voyage. On this point, technology can and will play a very important role.

An uncertainty in the convention that may bring difficulties regards the extension of the obligation of due diligence in making and keeping the vessel seaworthy, throughout the voyage at sea. When is the voyage at sea over? Is there liability for damages due to lack of seaworthiness during the unloading process?[621] These are new questions to be clarified by the Courts. Another innovation of these rules is the inclusion of provisions on containers. The continuity of the obligation also includes the obligation to make and keep fit and safe every place where the goods are carried, including the holds, but also the containers supplied by the carrier.[622]

The Rotterdam Rules are another effort to update the H/H-VR regime and to create uniform rules for some other areas of the modern carriage of good by sea. Unfortunately, at the moment of this writing, they do not enjoy much acceptance for ratification.

2.2 Conclusion

A reappraisal of historical maritime law shows that since ancient times, and for almost 20 centuries, the standard of liability for the sea carrier has been strict. If not absolute in all systems at all times, at least it was far stricter than the current standard set by the Hague Rules. Since Roman times, public policy appears to have

---

[618] *See supra* note 597.

[619] Sturley, Fujita and van Ziel, 85.

[620] Ibid. "However, at sea, whether the due diligence obligation requires the carrier to instruct the ship to effect repairs or not would depend on several factors, including the extent of the leakage, the possibility of causing damage to the cargo, the delay that may be caused, the weather condition and the availability of skills within the crew to effect such repairs."

[621] Y. Baatz, *The Rotterdam Rules: A practical annotation* (London: Informa Maritime Law, 2009) Michael Tsimplis "Obligation of the Carrier", 39.

[622] Sturley, Fujita and van Ziel, 84.

been the main reason supporting this heavy standard. It is known that Roman legislators grounded their decision on their philosophical thinking while pondering the principles of equity, fairness and justice. With the expansion of the Roman Empire, this strict standard spread throughout Europe and lasted for several centuries.

Despite the implications of such a demanding standard, carriers were able to develop their industry. That happened during a time when maritime transportation was still rudimentary; when carriers were exposed to many risks; when aids to navigation were scarce[623]; and without the benefit of the technological advances that we have now. In some cases, they carried out their business without limitation of liability or marine insurance. Ancient and medieval shipowners were surely under extreme pressure to use all resources available to prevent and avoid risks, and to make the voyage as safe as possible. The flourishing of maritime commerce reported by historians in cities of the Mediterranean Sea and North Sea during the ancient times, the Middle Ages and the modern era, was due, among other factors, to the availability and improvement of maritime transport during those times. Such availability suggests that the business of carriage of goods by sea was still attractive and profitable in spite of the burden of the strict standard.

The possibility to exclude liability for passenger's goods as stated under Roman law could also have played a role in defining the carrier's liability during that and subsequent times. However, any practice of this kind was not documented. Besides the provision on the Digest for the possibility to reduce or exclude liability, in the subsequent bodies of maritime law studied, whether statutes or codification of the law merchant, there is no exact references to such a possibility. This lack of evidence favors the theory that the strict liability was continually enforced. It was not until the 19th century that the issue became evident and subject to judicial attention in England. Previous to that century, the English courts' decisions show the prevalence of the absolute standard. English judges of the 17th and 18th century, adopted same reasoning of the Roman jurists to assign strict liability to common carriers, as a measure of public policy to protect a commercial activity that was essential for the development of the English society.

The evolution of liability to a negligence base standard was the consequence of the converging circumstances surrounding the exploding growth of international trade at the end of the 19th century. The predominant position gained by the British shipowners during this commercial boom, based on the necessity of merchants for carriage, allowed them extreme advantages, where they engaged in strongly criticized abusive practices. In addition to these economic and commercial facts, the new philosophical ideas of liberalism, in vogue at that time, did not exactly favor the values and public policy observed by the Romans, nor those held by Lord Hold and Lord Mansfield. On the contrary, this new philosophy claimed greater individual freedom in business, which became the foundational basis to allow freedom of contract in greater extension. This scenario prompted the introduction of clauses

[623]Mangone, 12.

excluding liability even for their own negligence. The delay in any statutory regulation is attributed, among other reasons, and according to the cited authors, to the shipowners' influences in the British parliament. It took almost half a century to adopt an international convention to create and unify international rules governing the contract of carriage of good by sea under bill of lading, not to mention specifically preventing the imposition of "negligence clauses".

It is understandable that the solution to such a long dispute was also created by the specific needs of the time. The Brussels Convention was obviously not the result of deep legal reasoning based in the leading values and sound principles of law, but a pragmatic solution for a commercial problem and serving the temporal commercial interests of the parties involved. Instead of the distinguished jurists of ancient Rome who shaped the strict standard, or the English judges of the 18th century who applied it, the Hague Rules were created mostly by businessmen. Doubtless, law has to respond and give solutions to business needs, as it is a fundamental part of society. Therefore, one must be careful when judging the lawmakers of the Convention for not doing more. It is obvious that the negotiations between the two main parties of the shipping industry were not, and never have been, an easy task. It is also obvious that many of the current issues resulting from technological change, that today demand statutory attention, could not have been foreseen in that time. It is perhaps more appropriate to raise concern for what we are failing to do today. Once those commercial needs are overcome, as well as many obstacles of that time through the application of technology, why to keep in force a legal instrument that does not reflect the main purposes of law and does not fit the current reality? There is no convincing justification. Of course, the solution is not to go back to the strict liability standard. It would be certainly unfair to hold carriers liable for something that is completely out of their control to prevent. In addition, the situations or risks that motivated to hold carriers as insurers are not a major threat at the present. However, the current reality of the carriage of goods by sea requires something more than the minimum duties established in the Convention.

Because of the insufficiency of the current governing regime and considering the relevance of this contract for the world's economy, the international community, through the UNCITRAL, proposed two new sets of regulations contained in the Hamburg Rules and the Rotterdam Rules. They come as a response to the new necessities of contemporary navigation and trade, especially the impact of technological development. This response to the existence of new technological advances has been precisely an increase in carrier's liabilities. The changes proposed are an attempt to introduce a new and fairer balance of risks between the parties; knowing that these risks can be assumed by the carriers thanks to the new technologies that enable better, safer, and more efficient navigation. The new regimes clearly demand a higher degree of diligence, which is possible through the access to new technological advances that facilitate more control over the maritime adventure, to prevent and avoid risks, and improve their performance and efficiency. Hence, taking into account the special circumstances that drove the adoption of the HR, the way in which it was redacted, and considering that many of the difficulties and hardships faced by shipowners when this Convention was discussed have been long ago overcome, the logic indicates that the "fair balance of risks" must be updated to make it actually fair. The results, however, has been unsuccessful as these Conventions have not, as of yet, achieved extensive approval and the international carriage of goods by sea continues to be governed by the almost century-old and outdated Hague Rules.

To reach a fairer balance, it requires the allocation of more responsibilities on the carriers, and this is where application of new technologies may play a key role. Thanks to the modern technologies, carriers are capable of having more control of the maritime adventure. In the 21st Century the idea of a "risky adventure", does not have the same meaning as it once did in the 19th century.[624] Technology has improved in such a way that many risks at sea simply do not exist anymore. Therefore, the review of historic maritime law suggests that if carriers since ancient times were able to develop their business under such a heavy standard, then, contemporary carriers should be able to assume the additional duties and responsibilities set in the Hamburg and Rotterdam rules. They have more technical resources to foresee, prevent and avoid risks, and given that their liability is still limited to negligence in the compliance of some minimum duties.

But history seems to repeat itself. Ninety years after the adoption of the Hague Rules, the adoption of new conventions show the awareness of the international community of the need for new regulations. Because of the expansion of the global market, countries' economies are strongly dependent on the exchange of products in foreign markets. Their success requires an efficient and safe maritime transportation, to prevent cargo loss, damage and delay as much as possible. At the same time, there is a need to reduce international litigation resulting from cargo claims. Trial between carriers and shippers in foreign courts has become increasingly complicate, lengthy and especially expensive. Hence, we may say it is time for a new evolution in the sea carrier's liability standard. Working with the current normative, the enforcement of new technologies as part of the due diligence duty, may help to reach higher efficiency. In this regard, today's lawmakers must assure a better equilibrium, such as lawmakers during ancient times seem to have achieved in a more efficient way. Perhaps the more expedient approach is not through legislative actions, but through the judiciary. Keeping the same standard based on the practice of due diligence, the courts could and should require the application of certain new and specific technologies as part of this duty. To promote efficiency and achieve the reduction of cargo losses and damages, certain technologies should be enforced. Which technologies are those? This is a question to be answered in the next parts.

---

[624] J.M. Alcantara, *"The Grand Liability Unified Theory"*, Scritti in Onore di Francesco Berlingieri, Il Diritto Marittimo, 2 Vols. (Genova, 2010) Vol. 1, 30.

Chapter 3
The Obligation of Practicing Due Diligence in the Carriage of Goods by Sea

This part deals primarily with the obligations that, according to the Hague/Hague-Visby Rules, every carrier subject to them must practice. It is firstly presented the introduction or adoption of the rules in American, English and German legislations. Then, it develops the construction that has been given to the fundamental concepts of due diligence, seaworthiness and cargo care. The explanation of the same is made in light of the American and English jurisprudence. There is no doubt that these two nations were leaders in maritime transport for most of the last century, so the decisions made by their courts, analising such duties, have come to complement the gaps and uncertainties of the Hague/Hague-Visby Rules. As explained, the rules impose to practice "due diligence". What should be understood by such an obligations? There have been some doubts if said duty should be understood as to perform due diligence or reasonable diligence; or, if there is a difference between diligence and care. In this regard, some contributions are presented to elucidate this apparent conflict. Regarding the concept of seawothiness, this has been widely discussed and fragmented into various aspects, considering the different elements of the ship and the maritime adventure as well. This deals with both legal and technical aspects, paying special attention to the implications that technology has had on them. Finally, passing from the ship to the cargo, it addresses the duty to protect the cargo that under the same regulations every carrier must meet.

3.1 Statutory Frame of the Carriage of Goods by Sea Under Bill of Lading

3.1.1 The Incorporation of the Hague Rules into the National Laws

The Hague Rules are the present international regime governing most part of the carriage of good by sea under bills of lading. Though the criticism against them, they were adopted by the main players of the international trade. In its original form

V.H. Chacón, *The Due Diligence in Maritime Transportation in the Technological Era*, Springer Series on Naval Architecture, Marine Engineering, Shipbuilding and Shipping 5, DOI 10.1007/978-3-319-66002-8_3

or later in its version of the Hague-Visby Rules, they constitute the legal framework within which the carriage of goods by sea is today performed. Currently, over 70 sovereign States have ratified or accessed the Convention.[1] As explained, the HR were conceived to achieve the unification of certain rules relating to the carriage of goods by sea, but not to be an extensive code. They provide a very basic and general mandatory framework, with the purpose of leaving space for further negotiation between the two parties.[2] The Convention does not provide any further indication, nor definitions or clarifications, on the duties set in Article 3. Much less, it could not foresee the impact that new technologies have had in the shipping industry, and especially in the fulfillment of these duties.

The study of the history of the Hague Rules presented in the previous part, showed the predominant role played by England and the United States in the discussions, redaction and adoption of the Convention. Their outstanding activity in international trade since the 19th century and their long maritime tradition drove their corresponding courts to develop large jurisprudence on the subject. The jurisprudence previous to the Hague Rules, particularly those contained in American decisions deciding cases under the Harter Act, helped to construct the meaning and extension of the duties of practicing due diligence in making ships seaworthy and the care of the cargo.

With almost the same basic wording, the Convention was incorporated into the national legislations of the United Kingdom, United States and also in Germany. Though many countries have adopted the Hague Rules or its amended form known as the Hague-Visby Rules,[3] it must be recognized that the first two countries have developed perhaps the largest body of jurisprudence regarding the duties set in Article 3, Sects. 1 and 2 of the rules. Apart from their large maritime activity, this is due to the fact that many carriers continued to include jurisdiction clauses in the bills of lading establishing the courts of these two countries as the exclusive forum to decide cargo claims arising under this regime. Though the Hague rules case-law has also suffered from ambiguities that, it is said, have operated against cargo claims, it has a special value when dealing with the carriers' duties.[4] It was through court decisions from England and the United States, that judges further defined the meaning, extensions, intentions, and proper application of the Convention. The analysis and conclusions in this jurisprudence came to complement and fulfill many of the shortcomings of the Hague/Hague-Visby Rules. Hence, it is possible to say that, in a certain way, the maritime law of carriage of good by sea under bills of lading has been modeled by these jurisdictions. Those constructions have been later followed by others courts in different countries during the following decades. The reasoning expressed in those decisions tried then to adapt the rules to the new realities of the maritime commerce. Nowadays, it can be affirmed that the Hague Rules must necessarily be read in conjunction with the large case-law arisen from it since its adoption over 90 years ago.

---

[1]CMI Yearbook 2013, 599–601.

[2]Wilson, 186–187.

[3]For a larger comparative analysis of some more national laws *See* Ziegler, 41–60.

[4]Honnold, 102.

Otherwise, the observation of the text alone, from different legal systems and national laws, may result in different conjectures, which may or may not suit the main purposes of the Convention. It is in that jurisprudence where we can find the extension of the due diligence obligation, the concept of seaworthiness, the construction of the duty of care of the cargo and the impact of new technologies in these duties.

3.1.1.1 English Law

3.1.1.1.1 The Carriage of Goods by Sea Act of 1924

Prior to the ratification of the HR on the 2nd of June 1930,[5] The United Kingdom passed an Act in 1924, using practically the same wording of the Convention. The section 3 of the English Act holds the HR as implied terms of any contract governed by the Act.[6] This incorporation of the vague wording of the Brussels Convention into the Act was a subject of criticism, for example, by Colinvaux, who pointed out that the act "puts it into confusion indefinitely".[7] In addition, such incorporation carried the risk for English judges to make a constrained interpretation of its provisions, by applying their own precedents to a regulation that pretended to be of international application. They could have been tempted to apply the same principles and conclusions resulting from the long and extensive maritime jurisprudential tradition of England, which may have hampered the intentions of international unification of the regulation. To prevent that, in 1932, Lord Macmillan, in the case *Stag Line LTD v. Foscolo, Mango and Co. Ltd.*,[8] argued that the Carriage of Goods by Sea Act of 1924, needed a construction based not merely on domestic precedents, but rather on broader principles of general application, as to reach the main purpose of uniformity intended in the HR. Lord Macmillan expressly said that:

> It is important to remember that the Act of 1924 was the outcome of an International Conference, and that the rules in the Schedule have an international currency. As these rules must come under the consideration of foreign courts it is desirable in the interest of uniformity that their interpretation should not be rigidly controlled by domestic precedents of antecedent date, but rather that the language of the rules should be construed on broad principles of general acceptation.[9]

This criterion was followed in later leading cases,[10] and it is perhaps one of the reasons that the construction of the HR by the English Courts has been used as a direct reference for global application. Indeed, the rationale set out by Lord

---

[5]CMI Yearbook 2013, 601.

[6]Hughes, 205.

[7]Colinvaux, v. *See supra* note. 532.

[8][1932] AC 328.

[9]Ibid., 350.

[10]This opinion was also cited in *Maxine Footwear Co. Ltd. v. Can. Government Merchant Marine.* [1959] AC 589 (PC); *The Muncaster Castle* [1961] AC 807, 874; *The Rafaela S*, [2003] 2 Lloyd's Rep. 113 (CA).

Macmillan in the aforementioned case has been adopted by several courts in many jurisdictions, including American courts.[11] This Act was in force for over 50 years, until the adoption of a new Act regulating the subject in 1977.

3.1.1.1.2 The Carriage of Goods by Sea Act of 1971

The English COGSA of 1971 keeps the same principle of the previous Act of 1924 in enforcing legislative control over the mutual rights and responsibilities of the parties of the bill of lading.[12] It introduced in a Schedule the provisions of the HR as amended by the Visby Protocol of 1968; though the Protocol was not ratified until October of 1976.[13] In 1977, the United Kingdom denounced the Brussels Convention of 1924, and the Act of 1971, as well as the Visby amendments, came into force on 23rd of June of 1977.[14] The Act provides that the Hague-Visby Rules have the "force of law" in all the circumstances set in the Article X of the Convention.[15] Therefore, they apply as if they were part or directly enacted by statute in the United Kingdom.[16] Other judges for the interpretation of this Act followed the same criterion for the construction of the rules given by Lord McMillan in the aforementioned case of 1932. In *The Hollandia* (1983), Lord Diplock, quoting Lord Macmillan's analysis, pointed the necessity of a purposive rather than a literalistic construction as a way to prevent the evasiveness of the rules in the use of "colourable devices" that are not expressly stated in the rules but also not prohibited.[17] The 1971 Act applies to every shipment from the United Kingdom and any contracting state of the HR, and enforced by the English Courts in case of controversy as the applicable law with the exclusion of any other law.[18]Although there is not an expressed provision regarding the inland carriage, this Act applies to every shipment departing from ports within the United Kingdom,[19] regardless of

---

[11]Tetley, *Interpretation and Construction…*, 55–56, *citing Francosteel Corp. v. Nederlandsch Amerikaansche Stoomvaart-Maatschppij* 249 Cal. App.2d 880, 889, 1897 A.M.C. 2440, 2448 (Cal. App. 1967), cert. denied, 389 US 931 (1967); *Sunkist Growers, Inc. v. Adelaide Shipping Lines, Ltd.* 603 F.2d 1327, 1338, 1979 A.M.C. 2787, 2802-03 (9th Cir. 1979), cert denied, 444 US 1012, 1980 A.M.C. 2102 (1980).

[12]Scrutton and Eder, 401.

[13]W. Tetley, B. G. McDonough and E. B. Nixon, *Marine cargo claims*, 3rd edn. (Montreal, QC, Canada: International Shipping Publications, BLAIS, 1988), 1098.

[14]Ibid., 1098.

[15]Section 1(2).

[16]As stated by Lord Diplock in *The Hollandia* [1983] 1 AC 565 (HL).

[17][1983] 1 A.C. 565, as quoted by Hughes, 205.

[18]Tetley, McDonough and Nixon, at 205: "It has been held that a choice of forum clause indicating a jurisdiction which would apply a lower package/unit limitation (that of the HR) is invalid under Art. III(8) which, like the rest of the Rules, has the force of law: The Hollandia, [1983] 1 A.C. 565. It is not clear whether a choice of law clause affected by this reasoning would be entirely invalid, or merely invalid as regards matters regulated by the Rules."

[19]Section 1(3).

whether or not the carriage is between ports of two different states, and applies also to coastal carriage where there is an issuance of a bill of lading or similar document of title.[20] In the absence of a negotiable document similar to a bill of lading, the Act still applies, if the parties expressly agree thereon.[21] In addition, it applies to shipments where no bill of lading has been issued, but where there is an intention that they be covered by such a document, and the regulations expected to be contained therein will apply anyway.[22] A particularity of this Act not found in similar statutes, is the provision contained in Section 3 that expressly states the abolition of any implied or absolute undertaking of seaworthiness when the rules apply.[23] However, what is understood as seaworthiness remains the same throughout.[24]

3.1.1.2 American Law

3.1.1.2.1 The Carriage of Goods by Sea Act of 1936

The American Carriage of Goods by Sea Act was the result of complaints in the interpretation and application of the Harter Act. By 1933, still under the Harter Act, the Supreme Court of the United States established in *The Isis*,[25] that the shipowner had to prove that he practiced due diligence in making the ship seaworthy in "all aspects" as a pre-condition to access to the exoneration of liability for errors in navigation or management of the ship, regardless of any causal connection between the unseaworthiness and the cargo loss or damage. If the vessel was unseaworthy, the carrier could not be released for errors in navigation. This heavy requirement against shipowners pushed the enactment of the new legislation.

As well as the United Kingdom, the United States enacted its own legislation on the matter, prior to the ratification of the HR. The American Carriage of Goods by Sea Act (COGSA), mostly preserved the concepts of the Harter Act, but introduced the requirement of a causal link between unseaworthiness and the cargo damage when imposing liability on the carriers.[26] It follows the similar wording of the HR

---

[20]Tetley, McDonough and Nixon, 205; Section 1(4).

[21]Section 1(6).

[22]*Pyrene Co. Ltd. v. Scindia Navigation Co. Ltd.* [1954] 2 QB 402.

[23]Girvin, at 421: "There shall not be implied in any contract for the carriage of foods by sea to which the rules apply by virtue of this Act any absolute undertaking by the carrier of the goods to provide a seaworthy ship".

[24]Hughes, 43.

[25]*May v. Hamburg-Amerikanische Aktiengesellschaft*, 290 U.S. 333, 54 S.Ct. 162, 7 L.Ed. 533 (1933).

[26]*Matter of Ta Chi Navigation (Panama) Corp., S.A.* 513 F. Supp. 148, 157, 1981 A.M.C. 2350 (E.D.La.).

and is included in the Title 46 of the US Code, Sects. 30701 *et seq.*[27] This was enacted in 1936, whereas the HR were ratified later on June 27th, 1937,[28] following 12 years of controversies in the Congress over its enactment.[29] Such ratification, however, was subject to the "understanding" that COGSA must prevail to the extent that its text differs from that of the HR.[30] In contrast to the United Kingdom, the United Stated has not ratified the Visby amendment of 1968.

Regarding the scope of application, COGSA did not overrule the Harter Act. It was supplanted greatly but not completely abolished.[31] COGSA applies only to foreign commerce.[32] The Harter act, rather, covers domestic and foreign voyages under a bill of lading.[33] It covers domestic waterborne carriage within the United States, unless the bill of lading expressly incorporates the COGSA into the contract. The interplay of these two Acts is made more evident when defining the period of responsibility of the carrier. Under COGSA, the carrier has an obligation of care of the goods from the time of loading till the time of discharge. It is the so-called "tackle to tackle" period. Under the Harter Act, the carrier is responsible for goods "committed to its or their charge", which means, during the entire period the goods are under his custody.[34] It does not contain any restriction or period of time for its application. Hence, the Harter Act applies in all the areas where COGSA does not. For example, it applies during the time previous to the loading and after the discharge from the vessel.[35] The Harter Act expresses duties as to loading, stowage, custody, care and delivery. The carrier must practice his obligation of care until the cargo in effectively delivered to its consignee.[36] In this aspect, the Harter Act is then applicable for the bill of lading in international trade in the areas or periods that COGSA does not cover. However, this situation is very rare as most of the bills of lading provide that the latter applies throughout all the time the goods are in

---

[27]In *Herd & Co. Inc. V. Krawill Machinery Corp.*, 359 US 297, 301, 1959 A.M.C. 879, 882–83 The US Supreme Court acknowledged that "The legislative history of the Act shows that it was lifted almost bodily from the Hague Rules of 1921, as amended by the Brussels Convention of 1924, 51 Statt. 233. The effect of those Rules was to establish uniform ocean bills of lading to govern the right and liabilities of carriers and shippers *iter se* in international trade."

[28]Tetley, McDonough and Nixon, 1101.

[29]Sturley, C*hanging Liability Rules…*, 120.

[30]Ibid., 120.

[31]Gilmore and Black, JR., 144.

[32]Ibid., 147.

[33]Ibid., 147.

[34]Mangone, 85.

[35]Gilmore and Black, JR., 147.

[36]Ibid. "Cogsa expressly saves Harter from repeal in so far as it applies to the periods prior to loading and after discharge from the ship. Thus, in the absence of stipulation, Harter still applies: (1) To all "coastwide" trade –that is, to bills of lading covering shipments by water from one port of the United States to another. (2) To the period, even in foreign trade, during which the carrier has custody of the goods, before they are loaded on the ship and after they are unloaded from the ship."

custody of the carrier.[37] The Harter Act also covers the carriage of live animals and cargo on deck, which are expressly excluded under COGSA.[38] COGSA allows the freedom of contract in the carriage of goods by sea, but only with respect to the increase of liabilities for the carrier, not to reduce his duties or to introduce further exonerations of liability.[39] This prohibition clearly responds to the historical reasons for the Act and the interest to prevent the disparity of bargaining power between the parties.[40]

The carrier's duties under the HR and the US COGSA are the same.[41] The main provision on this regards, are found in the Section 3(1) and 3(2) of COGSA, stating the affirmative grounds that enable the cargo interest to recover their cargo losses or damages.[42] The obligation of seaworthiness established in the Harter Act is the same under the COGSA. Again, it is not an absolute, but limited to a duty to exercise due diligence in making the ship seaworthy.[43] The only difference is that under the Harter Act, as stated in *The Isis*, the lack of seaworthiness was "a condition of exception", while in COGSA, as Section 4(1) states, there must be a causal connection between the lack of seaworthiness and the damage.[44] Other than that, there are no major differences between the Harter Act and COGSA. Therefore, older cases decided under the Harter Act are frequently cited in the constructions under the COGSA.[45]

Along with the aforementioned regulations, there are two more Acts that have relevance in the carriage of goods. The Federal Bills of Lading Act of 1916, known as the Pomerene Act,[46] and the Carmack Amendment.[47] The first regulates the bill of lading in foreign commerce by land and sea transportation. It differentiates between a "straight" bill of lading addressed to a specific named consignee, and an "order" bill of lading, which does not indicate the name of a specific consignee, but leaves it open, intended for negotiation with potential buyers of the goods.[48] The latter relates to the carriage of goods inland, establishing strict liability for terrestrial carriage, but subject also to the exceptions commonly known, and the possibility to limit its liability through notice.[49]

---

[37]Tetley, McDonough and Nixon, 1101.

[38]Section 1(c).

[39]Gilmore and Black, JR., 145.

[40]Ibid., 147.

[41]Mangone, 84.

[42]Gilmore and Black, JR., 149.

[43]Mangone, 87.

[44]Schoenbaum, 891–892.

[45]Gilmore and Black, JR., 148–149. The concept of seaworthiness, for example, has not a different meaning in the two acts.

[46]49 U.S.C. § 8010.

[47]49 U.S.C. § 11706.

[48]Mangone, 79.

[49]Tetley, *Marine Cargo Claims*, Vol. 2, 2601.

3.1.1.3 German Law

The German maritime law on carriage of goods by sea is contained in the Code of Commerce or *Handelsgesetzbuch* (HGB), book V, Section 476 and subsequent sections. Originally, this contract was also subject to total contractual freedom, until the normative rules stated in the HR became part of the national legislation in August 10th, 1937.[50] The Hague Rules were ratified by the *Deutsche Reich* in 1939, but the regulation of the HGB has held priority over the Convention.[51] The Visby rules have not been ratified, but its provisions were incorporated into German national law in 1986, by the *Zweite Seerechtsänderungsgesetz*.[52] The legal regime for carriage of goods by sea is therefore the Hague-Visby Rules embodied in the internal law. This incorporation, however, did not follow the same legislative structure or writing order of the HR, but its regulations are included in different sections of the HGB. In 2013, a statute was enacted introducing substantial changes to book V of the HGB, but it still preserves the same concepts of the Hague-Visby Rules.

The scope of application follows the wider rule stated in Article 10 of the HVR. It applies to every shipment under a bill of lading issued in Germany or in one of the contracting states of the HVR, or to shipments where a bill of lading is not issued, but where the carriage is from or to Germany or one of those contracting states. It also applies when agreed upon by the parties.[53] Regarding the States member of the original HR Convention, it still respects the Article 10 of the same, and applies that regime in its original form to bills of lading issued in those countries.[54]

Under German law, given its characteristic level of specialization, the carrier liability has been divided into four areas established and developed in different sections of the code of commerce. These are: (1) Liability for See- and cargo worthiness; (2) liability for faulty care of the cargo; (3) Liability for delay; and, (4) Liability for incorrect issuance of the bill of lading.[55] We will refer to the first two that are the object of this study.

The obligation to practice due diligence in making a ship seaworthy, prior to and at the beginning of a voyage has, under German law, a character of public order.[56]

---

[50] R. Herber, *Seehandelsrecht: Systematische Darstellung* (Berlin, New York: de Gruyter, 1999), 30; Ziegler, 52.

[51] Ziegler, 53.

[52] Ibid., 54. This Act was enacted on the 25th of July 1986 amending the previous regulation.

[53] Tetley, *Marine Cargo Claims*, Vol. 2, 2464; Ziegler, 53.

[54] Ziegler, 54.

[55] H. J. Abraham, *Das Seerecht: Ein Grundriss mit Hinweisen auf d. Sonderrechte anderer Verkehrsmittel, vornehml. d. Binnenschiffahrts- u. Luftrecht*, 4th edn. (Berlin, New York: de Gruyter, 1974), 179.

[56] Tetley, *Marine Cargo Claims*, Vol. 2, 876. *See* Bundesgerishtshof, February 28, 1983, [1984] ETL 342, at 349: "Nun liegt hier allerdings ein Fall anfänglicher Ladungsuntüchtigkeit eines Schiffes vor. Von der Haftung hierfür können sich aber der Verfrachter, der Reeder oder der Kapitän durch Allgemeine Geschäftsbedingungen, zu denen auch die vorformulierten Bedingungen einer Charterpartie zu rechnen sind (§ 1Abs. 1 AGBG) nicht wirksam freizeichnen".

The German law demands from carriers identical duties in practicing due diligence in making the ship seaworthy and in taking care of the cargo. The main provision is in the Section 485 of the HGB, which reads as follow:

Section 485

Seaworthiness and Cargoworthiness

The carrier must ensure that the ship is in seaworthy condition and properly furnished, equipped, manned, and sufficiently supplied (seaworthiness); the carrier must also ensure that the holds, including the refrigerating and cooling chambers, are fit for the reception, carriage, and preservation of the goods (cargoworthiness).[57]

The provision correlates the Article 3(1) of the HR. The differentiation between sea and cargo worthiness, as pointed out by Rabe, has no real significance.[58] It is merely a useful instrument for the delimitation, without exception, of the obligation set out in this article, in relation to the obligation established in the former Article 606, now Article 498 as modified by the amendments of 2013, that refers specifically to the cargo care obligation.[59] Under the former Article 606, the carrier was only liable for damage or loss of the goods. That rule was not exclusive to carriage under bills of lading, but applies for every type of contract of carriage.[60] As mentioned, after the amendment of 2013, said rule is contained in paragraph two of Article 498, which reads as follow:

Section 498

Grounds for Liability

(1) The carrier shall be liable for any damage resulting from the loss of or physical damage to the goods occurring between the time the goods are taken over and their delivery.

[57]Commercial Code in the revised version published in the Bundesgesetzblatt (BGBl., Federal Law Gazette), Part III, Section 4100-1, Book 1, as amended by Article 3 of the Act of 22 December 2015 (Federal Law Gazette Part I p. 2567), and Book 5, as amended by Article 1 of the Act of 20 April 2013 (BGBl. I p. 831). Translation from German provided by the Bundesministerium der Justiz und für Verbraucherschutz, available at http://www.gesetze-im-internet.de/englisch_hgb/englisch_hgb.html#p0018. Last visited May 30th, 2017.

[58]D. Rabe, *Seehandelsrecht: Fünftes Buch des Handelsgesetzbuches mit Nebenvorschriften und internationalen Übereinkommen*, 4th edn. (München: Beck, 2000), 405.

[59]Ibid. 405. The former Article 606 related to damage directly to the cargo not necessarily caused by unseaworthiness: "606. Der Verfrachter ist verpflichtet, beim Einladen, Stauen, Befördern, Behandeln und Ausladen der Güter mit der Sorgfalt eines ordentliches Verfrachters zu verfahren. Er haftet für den Schaden, der durch Verlust oder Beschädigung der Güter in der Zeit von der Annahme bis zur Ablieferung entsteht, es sei denn, dass der Verlust oder die Beschädigung auf Umständen beruht, die durch die Sorgfalt eines ordentlichen Verfrachter nicht abgewendet werden konnten".

[60]Ibid. 246, It is comparable to the GENCON 2 charter party model.

(2) The carrier shall be released from liability pursuant to paragraph (1) insofar as the loss of or physical damage to the goods was due to circumstances which could not have been avoided by a prudent carrier exercising due care. If the goods were carried by a ship that was not in seaworthy or cargoworthy condition, and if the facts of the case indicate a likelihood that the goods were lost or physically damaged due to the ship's lack of seaworthiness or cargoworthiness, then the carrier shall be released from liability pursuant to paragraph (1) only if the carrier can prove that the lack of seaworthiness or cargoworthiness could not have been discovered prior to commencement of the journey by a prudent carrier exercising due care.

(3) If the damaged party contributed to the occurrence of the damage, due to its fault or neglect, then the obligation to pay compensation and the amount of the compensation payable shall depend on the circumstances, and specifically on the extent to which the damages were caused primarily by one or the other party.[61]

The *Handelsgesetzbuch* used to contain other provisions relating to "*Seetüchtigkeit*" (seaworthiness). The first of them was established in Article 513 that defined the obligations of the master in preparing the ship for the voyage.[62] The seaworthiness obligation of the carriers was practically the same as the obligation assigned to the captain.[63] But the latter related mostly to the state or capacity of the ship's structure, the perils of the sea to be encountered and uncommon known exceptional peril.[64] The obligation of seaworthiness imposed on the captain referred to the availability of equipment, the accomplishment of the documentation, certificates, etc., required according to the international conventions establishing basic standards for ship construction and equipment. However, this provision was excluded in the amendments introduced to the code in 2013, and does not exist anymore.

[61]Commercial Code in the revised version published in the Bundesgesetzblatt (BGBl., Federal Law Gazette), Part III, Section 4100-1, Book 1, as amended by Article 3 of the Act of 22 December 2015 (Federal Law Gazette Part I p. 2567), and Book 5, as amended by Article 1 of the Act of 20 April 2013 (BGBl. I p. 831). Translation from German provided by the Bundesministerium der Justiz und für Verbraucherschutz, available at http://www.gesetze-im-internet.de/englisch_hgb/englisch_hgb.html#p0018. Last visited May 30th, 2017.

[62]The former Article 513 of the HGB stated that the master had to ensure before the voyage that the ship is seaworthy, properly furnished, equiped, crewed and provisioned, and that all the documents related to the ship, the crew and the cargo are on board: "Der Kapitän hat vor dem Antritte der Reise dafür zu sorgen, dass das Schiff in seetüchtigem Stande, gehörig eingerichtet und ausgerüstet, gehörig bemmant und verproviantiert ist und dass die zum Ausweise für Schiff, Besatzung und Ladung erforderlichen Papiere an Bord sind"; H. J. Abraham, *Das Seerecht in der Bundesrepublik Deutschland: Kommentar und Materialsammlung*, Großkommentare der Praxis, 4th edn. (Berlin [u.a.]: de Gruyter, 1978), 491.

[63]Rabe, 404.

[64]Ibid., 246.

3.2 The Carriers Duties Under the Hague/Hague-Visby Rules

3.2.1 To Practice Due Diligence

3.2.1.1 Definition of Due Diligence

Under the international legal framework for carriage of goods by sea based on the HR, the definition of due diligence is of paramount importance. We saw already that the origin of the concept is found in Roman law, which has been followed not only by the civil law systems, but in common law as well. Diligence is the standard to measure negligence. Its non-observance carries with it the punishment for negligence.[65] As mentioned, one of the shortcomings of the Convention is the lack of at least a general definition for due diligence or a general parameter of what must be understood as due diligence in the context of maritime transportation.[66] Given the complexity of the issue, providing a definition in the Convention was doubtless no simple task. Indeed, Clarke affirms that a precise definition "is discouraged by the variable level of the obligation."[67] The Convention left the due diligence duty to be interpreted according to its general conception of that time. To find out the intention of the HR redactors in introducing this term in the Convention as the measure in assigning liability, we must first determine what their conception for "due diligence" was. The concept was not new. It was already largely studied and applied in the Anglo-American case law. It must be remembered also that the duty was first proposed during the "Conference Form" discussed in the conference of Liverpool of 1882. Then it was used in the Harter Act of 1993 and subsequently adopted by other countries following the same model. The definition provided by the 1910 edition of the Black's Law Dictionary is perhaps a reference of what the then legal community understood as due diligence. Black describes due diligence in the following words:

> A measure of prudence, activity, or assiduity, as is properly to be expected from, and ordinarily exercised by, a reasonable and prudent man under the particular circumstances; not measured by any absolute standard, but depending on the relative facts of the special case.[68]

---

[65]H. C. M. Black, *A Law Dictionary*, 2nd. (St. Paul, Minn.: West Publishing Co., 1910), 405. *See Heaven v. Pender*, (1883)11 QBD 506; *Smith v. Clarke Hardware Co.*, 100 Ga. 163, 28 S. E. 73, 39 L. R. A. 607; *Cleveland, C., C. & ST. L. RY. CO.*, v. *Railroad Co.*, v. *Ballentine*, 84 Fed. 935, 28 C.C.A. 572 (C.A. 7 1898).

[66]Koh, 110.

[67]Clarke, 204, *citing* Mazeaud, Rev. Trimestrielle de Driot Civil 1936. I no. 48.

[68]Black, 369. This definition has been used in some more recent American cases such as *C. Itoh & Co. (America) Inc*.v. *M/V Hans Leonhardt*, 719 F. Supp. 479, 504, 1990 A.M.C. 733, 743 (E.D. La. 1989).

Keeping in mind that the Canadian Water Carriage of Goods Act of 1910 was also taken as a reference for the redaction of the HR,[69] the definition provided by The Canadian Supreme Court in 1918 in the case *Grain Growers Export Co.* v. *Canada Steamship Lines Ltd.* would illustrate what was expected under this concept specifically for the carriage of goods by sea:

> [Due diligence is] not merely a praiseworthy or sincere, though unsuccessful, effort, but such an intelligent and efficient attempt as shall make it so [i.e. seaworthy], as far as diligence can secure it.[70]

The task of providing a definition for due diligence under the HR context has been carried out by the courts, which provide partial descriptions of the same; but still fell short of a dogmatic or detailed and specific definition for due diligence in this context.[71] English judges have not provided a clear definition of due diligence, due to a "traditional reluctance" in providing "unnecessary" definitions in court decisions.[72] They instead equated the concept of due diligence to the duty of reasonable care,[73] a concept that was far more familiar in common law. American courts recently provided an even more succinct definition: "due diligence consists of whatever a reasonable competent vessel owner would do under the circumstances."[74] A more recent edition of the same Black's dictionary also reduces the description of "diligence" to: "a continual effort to accomplish something; care, caution, the attention and care required from a person in a given situation",[75] and "due diligence" as "the diligence reasonably expected, from and ordinarily exercised by a person who seeks to satisfy a legal requirement or to discharge an obligation."[76]

After long years of maritime litigation, the abundant case law has given shape to a description of some general characteristic of this obligation and what is expected from the carriers in performing this duty. In general, due diligence since the time of the redaction of the HR, has been conceived as a positive duty that requires from the carrier an efficient and effective activity in order to reach a proper seaworthiness condition of the vessel. It demands actions, not just lack of omissions, within the reasonableness and limitations that the circumstances present. In simple words, being diligent means that the carrier has to do everything he can reasonable do to assure the proper seaworthy condition of the vessel, and the successful carriage and delivery of the goods.

---

[69]Imperial Shipping Committee issued in 1921 suggested the unification of laws in the British dominions to be based on this Act. *See* supra a) History of its Adoption; note 481.

[70]43 O.L.R. 330, 344–345 (Ont. S.C. App. Div.), upheld (1919) 59 S.C.R. 643 (Supr. C. of Can.).

[71]Knauth, 122.

[72]Clarke, 205.

[73]Ibid., 205; Wilson, 188. *See The Amstelslot* [1963] 2 Lloyd's Rep 223, 235.

[74]*See* M/V *Tuxpan Lim. Procs.*, 765 F. Supp. 1150, 1179, (S.D.N.Y. 1991). Also refers to *General Foods Corp. v. The Troubador*, 98 F. Supp. 207, 210 (S.D.N.Y.1951).

[75]Garner and Black, 523.

[76]Ibid., 523.

3.2.1.2 Characteristics of the Obligation

As mentioned, the doubts resulting from the ambiguities and shortcomings of the Convention came to be fulfilled by the Courts. Not exactly with diaphanous clarity, but it is what we have. As the definition of due diligence was left to the general conception of the term, the 90 years of Hague case-law has come to clarify some general characteristics of the obligation.

3.2.1.2.1 Due or Reasonable

Because the definition for practicing "due diligence" stated in the HR is based on the previously mentioned sources that were originally written in English language, some discussion arose from its translation into French. The official version of the HR was redacted originally in French. The term was translated from English into French as "*diligence raisonable*". Its translation into English was not literally "*reasonable diligence*", but it fell back to the exact terminology used in the Harter Act. The variation in the translation from French into English from "reasonable" to "due", might have presented initially some difficulties in its construction. "*Due*" was defined at that time as "just, proper, regular, lawful, [and] sufficient".[77] This description suggests a stricter accomplishment of the necessary, extensive or sufficient measures to reach a similar degree of seaworthiness as the previous absolute undertaking demanded. The obligation could then be seen as absolute.[78] Given a hypothetical incongruence of the translated terms, some French commentators have objected to the use of the term "*reasonable*" precisely for its vagueness, and suggested instead the phrase "*soins appropiées*", which is the literal translation of "due" into French.[79] An early American case decided under the Harter Act, *The Irrawaddy* (1898), argued that the purpose of the Act was to relieve the shipowner from liability for latent defects, not discoverable through the "utmost care and diligence".[80] "Utmost" suggests that the original intention of the Harter Act was the practice of a very thorough or exhaustive work of diligence.[81]

---

[77] Black, 400.

[78] Tetley, *Marine Cargo Claims*, Vol. 1, 370.

[79] Clarke, 205. He refers to Aubrun and Audouin. He points also that however Aubrum remarked that it is not a meticulous and extraordinary diligence. This opinion was also followed by Marais and Rouen.

[80] 171 U. S. 192, 193, *sub nom. Flint* v. *Christall*, 43 L.Ed. 132, 18 Sup. Ct. Rep. 833. (1898).

[81] Indeed in *The Friesland* 104 F. 99 (D.C.N.Y. 1900) two cases were cited where the level of inspection required as part of the due diligence obligation seemed higher, at 100: "In the case of The Edwin I. Morrison, 153 U.S. 199, 215, 14 Sup.Ct. 823, 38 L.Ed. 688, it was held to be the duty of the owner to make such necessary and proper inspection from time to time as might give assurance of the seaworthiness of the vessel; and in The Phoenicia (D.C.) 90 Fed. 116, affirmed in 40 C.C.A. 221, 99 Fed. 1005, the absence of a thorough and careful inspection, it was held, made the ship chargeable with the loss."

However, this initial and apparently more demanding character of the duty was diminished later. It is perhaps one of the aspects where the imprecision of the Convention has worked in favor of carriers. Most of the courts adopted a construction closer to the meaning provided by the original French version. "Reasonable" is defined as "ordinary" or "usual",[82] and requires the attention to specific features of the issue subject to diligence. This construction would be more appropriate for the assessment of the obligation. The broad scope of different characteristics and elements involved in maritime transportation make it obviously difficult to set concrete general standards that can apply to every vessel; and the latter construction offers more flexibility that fits more adequately the implications of making a ship seaworthy. Therefore, the courts, even before the Convention, already understood this and most of the decisions on cargo claims in the jurisdictions under study followed the "reasonability" of the obligation in order to measure it. The English Court of Appeal has provided one of the most recent expositions explaining the reasonableness of this obligation:

> [W]ether the carrier, its servants, agents and independent contractors have exercised all reasonable skill and care to ensure that the vessel was seaworthy at the commencement of its voyage, namely, reasonably fit to encounter the ordinary incidents of the voyage.[83]

Further, in *The Eurasian Dream* (2002),[84] it was reaffirmed that the exercising of due diligence is equivalent to the "exercise of reasonable care and skill."

American Courts on their part have followed the same construction of the English Courts as clearly seen in cases such as *The Troubador* (1951),[85] and the M/V *Tuxpan Lim. Procs.* (1991).[86] In *The President of India v. West Coast Steamship Co., (The Portland Trader)* (1963),[87] still evaluating the obligation under the absolute standard, the District Court established that:

> Although the duty to furnish a seaworthy ship is absolute and is a specie of liability without fault, limited neither by concepts of negligence nor by those which might be contractual in nature… the obligation does not require the owner to furnish a ship or gear beyond that which is reasonable fit for the use intended. In other words, the standard is not an accident free ship, nor an obligation to provide a ship or gear which might withstand all conceivable hazards. In the last analysis, the obligation, although absolute, means nothing more or less than the duty to furnish a ship and equipment reasonably suitable for the intended use or service.[88]

---

[82]Black, 994.

[83]Tetley, *Marine Cargo Claims*, Vol. 1, 876. *The Kapitan Sakharov* [2000] 2 Lloyd's Rep. 255.

[84][2002] 1 Lloyd's Rep. 719.

[85]98 F. Supp. 207, 210 (S.D.N.Y.1951).

[86]765 F. Supp. 1150, 1179, (S.D. N.Y. 1991).

[87]213 F. Supp. 352, 1963 A.M.C. 649 (D.C.Or. 1962).

[88]Ibid., 356, for this statement the court also cited the following cases: *Mitchell v. Trawler Racer, Inc.*, 362 U.S. 539, 80 S.Ct. 926, 4 L.Ed.2d 941 (1960); *Boudoin v. Lykes Bros. S.S. Co., Inc.*, 348 U.S. 336, 75 S.Ct. 382, 99 L.Ed. 354 (1955); *Michalic v. Cleveland Tankers, Inc.*, 364 U.S. 325, 81 S.Ct. 6, 5 L.Ed.2d 20 (1960).

One of the reasons for this "amelioration" of the duty is due to the sometimes unavoidable existence of latent defects not discoverable by due diligence. A too strict construction of the duty would hold carriers liable for even undiscovered latent defects, coming directly into conflict with exoneration of liability as stated in Article 4 Section 2(p) of the HR. According to English courts to achieve this exoneration, the defect must be so, that there are no possibilities for them to have been discovered.[89] The plaintiff can overcome this defense by proving the existence of a test that could reveal the latent defect.[90] Here, the courts must determine if such a test demanded by plaintiffs can be reasonably expected on the part of the carriers in practicing due diligence. This issue was subject to analysis by an English court in The *Hellenic Dolphin* (1978),[91] which presented an explanation for this construction. The cargo interest alleged that the vessel required a detailed examination by the superintending engineer at the turn-around port of a round trip voyage. According to the court, it seemed too pretentious to require the carrier to perform measures that, although they may have helped to discover problems on the ship, were neither commonly nor reasonably practiced. The court rejected the argument because they considered it a demand for perfection, which goes beyond the owner's due diligence obligation.[92]

Therefore, recent definitions of the concepts of due, common and reasonable diligence in the general context have made them synonyms.[93] With this in mind, any abstract and theoretical distinction between due and reasonable has surrendered to a unification of the criteria for an even more flexible standard. The challenge here is that the conception of "reasonable under the circumstances" makes the obligation more relative and more difficult to assess when trying to measure it. It leaves little room for the configuration of at least some minimum objective parameters in measuring the carrier's obligation. To the question what is reasonable? The answer depends on the knowledge, evaluation, and critical thinking of every Judge in each case.

3.2.1.2.2 Diligence or Care

The terms of *due, proper, ordinary, adequate and reasonable care* have been defined as all equivalent to "a test of liability for negligence, the degree of care that a prudent and competent person engaged in the same line of business or endeavor would exercise under similar circumstances."[94] In the same regards, Jowwit's

---

[89]Clarke, 210, *quoting Dimitrios N. Rallias* (1922) 13 Ll.L.R. 363, at 366: "Even assuming that (the surveyors) were not negligent, if this *could* have been discovered by ordinary care then the defendant does not come within the protection of the bill of lading".

[90]Ibid., 210.

[91][1978] 2 Lloyd's Rep. 336.

[92]Tetley, *Marine Cargo Claims*, Vol. 1, 929.

[93]Garner and Black, 523, 574, 1380.

[94]Ibid., 240.

Dictionary of English law describes *diligentia* as a synonym for care.[95] Scrutton states also that "the standard imposed by the obligation to exercise due diligence appears to be equivalent to that of the common law duty of care."[96] Hence, English courts have interpreted the concept as equivalent to the common law duty of reasonable care.[97]

Such equivalence of these concepts was evident in *The Muncaster Castle* (1960), where Lord Willmer said that "an obligation to exercise due diligence is to my mind indistinguishable from an obligation to exercise due care—A concept not unfamiliar in English law."[98] However, this construction has not been accepted without objections. It was expressly denied by Lord Radcliffe under the argument that "the measure of the carrier's duty is somewhere between that of reasonable care and that of a duty to see that care is taken."[99] Notwithstanding, later cases continued to maintain this terminological equivalence.[100] A few years later, Lord Devlin reaffirmed it in the *Union of India v. N.V. Reederij Amsterdam* (1963):[101] "no lack of care" is equivalent to no lack of due diligence.[102] More Recently, *The Eurasian Dream* (2002) also confirmed such a construction.[103] Therefore, any distinction between diligence and care, has not had a meaningful effect in the interpretation of the duty of practicing due diligence. Diligence and care in the context of the maritime transportation are treated as synonyms.

3.2.1.2.3 The Degree of Diligence

The degrees of diligence extracted from the Roman texts have been observed in the Anglo-American modern legal theory, though there are some disputes as to whether or not the doctrine of the three degrees of negligence in English law is, in fact, derived from Roman law.[104] Although not expressly stated as three degrees of negligence, there are indeed three degrees of care. These being both correlative terms, means de facto, there are three degrees of negligence.[105] Notwithstanding,

---

95 W. A. Jowitt and J. Burke, *Jowitt's dictionary of English law*, 2nd edn. (London: Sweet & Maxwell, 1977), 614.

96 Scrutton and Eder, 424.

97 Girvin, 423.

98 [1960] I QB 536, 581.

99 Clarke, 203–204, *citing The Muncaster Castle* [1961] A.C. 807, 862; Jolowics 1960 Camb. L. J. 17, 1961 Camb. L.J. 165. Riska pp. 94, 102. *Cf.* Villareal 2 J. Mar. L. & Com. 763.

100 Ibid., 204. *See Leesh River Tea Co. v. B.I.S.N.* [1967] 2 QB 250, 277.

101 [1963] 2 Lloyd's Rep. 223.

102 Ibid., 235.

103 *Supra note* 84; Girvin, 423.

104 Note, 'Three Degrees of Negligence', at 655: "Prior to the publication of the essay of Sir William Jones it has no existence in the English Law, except as directly and wholly traceable to the Roman law."

105 Ibid., 668.

there are more references to two degrees of diligence. The first is the *exacta diligentia* or the utmost diligence, also known as *bonus pater familias,* such as is expected to be observed by a good and thoughtful man. The second one is the *quantum in suis rebus adhiebere solitus est*, or the diligence which a man usually employs in his own affairs.[106] The latter degree of diligence is the one required to sea carriers, not only with the introduction of the HR, but also prior to it.[107] The standard required by the law is that of what an ordinary careful owner would demand for his own ship.[108] Hence, the obligation of due diligence is expected to be exercise by a "*prudent shipowner*",[109] or a "*competent vessel owner*".[110] This "*prudent man*" or a prudent vessel's owner must make a genuine, competent and reasonable effort in making the ship seaworthy.[111]

Now, it is also criticized that the standard expected by the judges does not exactly correspond to the reality. The judiciary expects that the ordinary careful owner is "a paragon of watchfulness, attentiveness and responsiveness, who keeps abreast of all technical development, tolerates little that is less than perfect, takes only the best advice, plans ahead with meticulous care, works in harmony with classification societies, employs skilled and competent inspectors to supervise everything done on his behalf."[112] From this point of view, the standard demanded of shipowners or carriers, in general, is indeed very high, but it is not

---

[106]Jowwit, Earl and Walsh, Clifford. "Jowwitt's Dictionary of English Law". 2nd. Edition, Sueer & Maxwell Limited, 1977, 614, 669.

[107]*See McFadden v. Blue Star Line* (1905) 1 K.B. 697, at 706 Judge Channell quoted a passage of Carver on Carriage by Sea (s. 18) stating: "...the vessel 'must have that degree of fitness which an ordinary careful and prudent owner would require his vessel to have at the commencement of her voyage having regard to all the probable circumstances of it.'".

[108]Editorial, 'Seaworthiness—the ilussion of the Hague compromise' (2006) 12, The Journal of International Maritime Law, 87–8, 87. *See* FC *Bradley & Sons v. Federal Steam Navigation Co.* [1926] 24 Ll.L.Rep. 446, at 454: 'The ship must have that degree of fitness which an ordinary careful owner would require his vessel to have at the commencement of her voyage having regard to all the probable circumstances of it. Would a prudent owner have required that it (sc the defect) be made good before sending his ship to sea, had he known of it?".

[109]*See MDC Ltd. v. NV Zeevaart Maatschapij Beursstraat* (1962) [1962] 1 Lloyd's Rep. 180, at 186: "...and the most common test is: Would a prudent shipowner, if he had known of the defect, have sent the ship to sea in that condition?".

[110]*See* M/V *Tuxpan Lim. Procs.* 765 F. Supp. 1150, 1179, (S.D.N.Y. 1991). *See* also *General Foods Corp. v. The Troubador*, 98 F. Supp. 207, 210 (S.D.N.Y. 1951).

[111]Tetley, *Marine Cargo Claims*, Vol. 1, 876. Tetley cites the Canadian case *Grain Growers Export Co. v. Canada Steamship Lines Ltd.* (1918) 43 O.L.R. 330 at 344–345 (Ont. S.C. App. Div.), upheld (1919) 59 S.C.R. 643 (Supr. C. of Can.); Longley, 62, *citing The Southwark* 191 U.S. 1, 15, 24 S.Ct. 1, 48 L.Ed. 65 (1903); *The Edwin I. Morrison*, 153 U.S. 199, 215, 14 S. Ct. 823, 38 L.Ed. 688 (1894); *The Esso Providence*, 112 F. Supp. 631, 1953 A.M.C. 1317 (S.D.N. Y.).

[112]Editorial, 87.

the degree applied in the commercial reality.[113] Given this expectation, the courts rarely ignore defects found in vessels and liability for them is consequently assigned.[114]

3.2.2 *The Ship's Seaworthiness*

3.2.2.1 Definition of Seaworthiness

3.2.2.1.1 Conceptualization for the Maritime Insurance Contract

There are two conceptions for seaworthiness, depending on the type of contract. In both the marine insurance conception and in the carriage of goods conception, seaworthiness is conceived as "a comprehensive and relative term."[115] The main difference is that in the case of maritime insurance industry, seaworthiness is the reasonable fitness of the ship in all aspects to encounter the ordinary perils of the sea during the adventure insured,[116] whereas in the carriage of goods by sea, the concept includes this same requirement, but goes further to include the capabilities to transport safely the contracted cargo. Another major difference is found in the consequences of lack of seaworthiness. In the case of maritime insurance, the seaworthiness is still a warranty, an implied condition for every voyage policy. If the vessel is found unseaworthy, unless exceptions are stated, the insurance policy is held as void.[117] In the carriage of goods by sea, we explained already that the obligation is relative, and reduced to the practice of due diligence.

3.2.2.1.2 Conceptualization for the Carriage of Goods by Sea

In addition to providing no definition for due diligence, the HR provides none for seaworthiness. However, this concept had been underlined, used and applied since long before the HR, mostly in the common law. One of the oldest American cases dealing with the concept of seaworthiness is the *Dupond v. Vance* (1856).[118] There, the Supreme Court of the United States defined seaworthiness as follows:

---

[113]Ibid., 87.

[114]Ibid., 87.

[115]Gilmore and Black, JR., 152.

[116]UK Marine Insurance Act 1906, s. 39 (4).

[117]Ibid., 154.

[118]19 How. (60 U.S.) 162, 167, 15. L.Ed. 584 (1856).

> To constitute seaworthiness of the hull of a vessel in respect to cargo, the hull must be so tight, staunch, and strong, as to be competent to resist all ordinary action of the sea, and to prosecute and complete the voyage without damage to the cargo under deck.[119]

Under English law, one of the oldest definitions of seaworthiness was configured in *Kopitoff v. Wilson* (1876).[120] There the Queen Bench Division said that the vessel must be "…fit to meet and undergo the perils of the sea and other incidental risks to which of necessity she must be exposed in the course of the voyage".[121]

Both definitions having been construed in the middle of the 19th century, seem to limit the concept to the condition of the vessel's hull and in regards to the perils of the sea. Once the Harter Act came into force, the Supreme Court of the United States included in its concept the other necessary components of the ship for the proper and safe navigation and the protection of the cargo. Some cases described it in an even simpler way. Seaworthiness was "the fitness of the vessel to carry the undertaken cargo."[122] In *The Southwark* (1903),[123] Justice Day introduced a more elaborate description, partially quoting the definition proposed by the Bouvier′s Law Dictionary:

> Bouvier's Law Dictionary defines 'seaworthiness' to be: 'In maritime law, the sufficiency of the vessel in materials, construction, equipment, officers, men and outfit for the trade or services in which it is employed.' …In the case of The Sylvia, 171 U.S. 462, Mr. Justice Gray said: ′The test of seaworthiness" is whether the vessel is reasonably fit to carry the cargo which she has undertaken to transport.' This is the commonly accepted definition of seaworthiness. As seaworthiness depends not only upon the vessel being staunch and fit to meet the perils of the sea, but upon its character in reference to the particular cargo to be transported, it follows that a vessel must be able to transport the cargo which it is held out as fit to carry or it is not seaworthy in that respect.[124]

Bouvier introduced other aspects of the vessels such as equipment, crew, etc., as part of the seaworthy condition that are obviously fundamental for the success of the maritime adventure and the care of the cargo. His definition emphasized the relative nature of the seaworthy condition in relation to the specific cargo to be carried. This same meaning, with the description of its relation to the cargo to be

---

[119] Ibid., 167.

[120] (1876) 1 QBD 377; Wilson, 9.

[121] Ibid., 380.

[122] *See The Sylvia*, 171 U. S. 462, 43 L. Ed. 241, 19 Sup. Ct. Rep. 7. (1898).

[123] 191 U.S. 1, 24 S.Ct. 1, 48 L.Ed. 65 (1903).

[124] Ibid., 8–9, *See* Bouvier Law Dictionary, 2nd Edition, Vol. 2, 506. Also in H. R. Baer, *Admiralty Law of the Supreme Court*, 3rd edn. (Charlottesville, Va: Michie Co., 1979), 500.

carried, continues to be the standard in cases decided under the Hague Rules.[125] From these definitions and the jurisprudential development, some legal and technical characteristics of the obligation have been underlined.

3.2.2.2 Legal Aspects

3.2.2.2.1 An Overriding Obligation

We mentioned previously that the courts since before the adoption of the HR distinguished this duty as an overriding one. The HR continued the same construction. It has special effect in dealing with concurrent causes of damage that involves the exculpatory cause of errors in navigations.[126] This prioritization is also particularly evident when the damage is caused by both lack of seaworthiness and one of the exculpatory causes listed in Article 4(2). If there is concurrence of both causes, the carrier is liable for the entire loss, as having not fulfilled the obligation of practicing due diligence in making the ship seaworthy and prevents any possibility to escape liability.[127] Lord Somervell expressly made this point in the already cited case *Maxine Footwear Co. Ltd. V. Can. Government Merchant Marine* (1959).[128] In the decision the Court said that:

> Article III, rule 1, is an overriding obligation. If it is not fulfilled and the non-fulfillment causes the damage the immunities of Articles IV cannot be relied on. This is the natural construction apart from the opening words of Article III, rule 2. The fact that that rule is made subject to the provisions of Article IV and rule 1 is not so conditioned makes the point clear beyond argument.[129]

---

[125] *See Empresa Cubana Importadora de Alimentos Alimport v. Iasmos Shipping Co. S.A.* (*The Good Friend*). [1984] 2 Lloyd's Rep 586, at 592 defines de concept of seaworthiness as: "used in the ordinary meaning, and not in any extended or unnatural meaning. It means that the vessel – with her master and crew—is herself fit to encounter the perils of the voyage and also that she is fit to carry the cargo safely on the voyage". Also in: *Great China Metal Industries Co. Ltd. V. Malasyan International Shipping Corp. Berhard* (*The Bunga Seroja*) [1999] 1 Lloyd's Rep 512 (HCA), at 86; *The Gang Cheng* (1998) 6 MLJ 488; *Actis Steamship Co. Ltd. v. The Sanko Steamship Co. Ltd.*, (*The Aquacharm*) [1982] 1WLR 119 (CA); Tetley, *Marine Cargo Claims*, Vol. 1, at 877: "The state of the vessel in such a condition, with such equipment, and manned by such a master and crew, that normally the cargo will be loaded, carried, cared for and discharge properly and safely on the contemplated voyage."

[126] Tetley, *Marine Cargo Claims*, Vol. 1, 903. *See The Temple Bar* 45 F. Supp. 608 (D. Md. 1942), at 617, 1942 A.M.C. 1125, at 1139: "…if the facts in any case disclose unseaworthiness resulting from the vessel owner's failure to exercise due diligence to make the vessel seaworthy, which concur with negligent navigation in causing the loss, the owner will be liable. That is to say, unseaworthiness cannot be transformed into bad seamanship for the purpose of avoiding responsibility for loss of vessel or cargo."

[127] Ibid., 903. *See The Eurasian Dream* [2002] 1 Lloyd's Rep. 719, 737.

[128] [1959] A.C. 589, [1959] 2 Lloyd's Rep. 105 (P.C.).

[129] Ibid., 602–03 or 113.

The same criterion has been maintained in subsequent cases such as in the *Mediterranean Freight Services Ltd. v BP Oil International Ltd (The Fiona)* (1993).[130] Here, Judge Diamond, Q.C. stated that the exercise of this overriding obligation is a pre-condition for access to the defenses set in Article 4(6).[131] Similar construction was given by the high Court of Australia in the case *Great China Metal Industries Co. Ltd. v Malaysian International Shipping Corp. Bhd (The Bunga Seroja)*(1999).[132] In both cases, the decisions were grounded citing the leading opinion of Lord Somervell in *The Maxine Footwear Co.* The reason for such a construction seems to be grounded in the fact that the Article 3(1), is not subject to Article 4, as Article 3(2) is.[133]

3.2.2.2.2 Not Delegable

With the complexity of large maritime operations, shipowners frequently rely on third parties to inspect, repair and prepare the vessel for the voyage. However, due diligence to make the ship seaworthy is an "inescapable personal obligation".[134] English and American courts have established that this obligation is extended to all shipowner's officers, subordinate employees, or employees of independent contractors.[135] Delegating the obligation to well reputed agents does not exonerate carriers from liability if want of due diligence is found. This principle was long ago established in the American case *The Colima* (1897),[136] decided under the Harter Act. The carrier is held responsible for the work performed by his servants, agent, independent contractor, or any other person assigned with a necessary task on the ship.[137] A few years later this principle was reaffirmed in *International Navigation*

---

[130][1993] 1 Lloyd's Rep. 257, affirmed by the Court of Appeal (Civil Division) at [1994] 2 Lloyd's Rep. 506.

[131]Tetley, *Marine Cargo Claims*, Vol. 1, 882, *quoting The Fiona* at 286: "The exceptions in art. IV, r. 6 are clearly in my judgment subject to the performance by the carrier of his *overriding obligation* set out in art. III, r. 1. So also in my judgment is the right to an indemnity conferred by the first paragraph of the rule" (Emphasis by Tetley).

[132][1999] 1 Lloyd's Rep. 512.

[133]Girvin, 429.

[134]Scrutton and Eder, 423.

[135]Baer, 503.

[136]82 F. 665 (S.D.N.Y. 1897), at 678: "This section [Section 3 of the Harter Act] has been in several cases adjudged to require due diligence, not merely in the personal acts of the owner, but also on the part of the agents he may employ, or to whom he may have committed the work of fitting the vessel for sea. The act requires in other words, due diligence in the work itself."

[137]Also in *Federazione Italiana Dei Consorzi Agrari v. Mandask Compania De Vapores, S.A.*, 284 F. Supp. 356, *aff'd in part, rev'd in part*, 388 F.2d 434 (2nd Cir.), *cert. denied*, 393 U.S. 828, 89 S. Ct. 92, 21 L.Ed.2d 99 (1968); *In re Marine Sulphur Transport Corp.*, 312 F. Supp. 1081, 1105 (S. D.N.Y.1970), *aff'd in part, rev'd in part*, 460 F.2d 89 (2nd Cir.), *cert. denied*, 409 U.S. 982, 93 S. Ct. 318, 34 L.Ed.2d 246 (1972); *Complaint of Tecomar S.A.* 765 F. Supp. 1150, 1179 (S.D.N.Y. 1991).

Company v. Farr & Bailey Manufacturing Company (1901).[138] This construction has been maintained in American law up to the present.[139] Accordingly, it holds the carrier liable even when the shipper chooses the marine surveyor by agreement with the carrier.[140]

In line with the reasoning of *The Colima* case, and citing it, the House of Lords in the famous case *Riverstone Meat Co. Pty. v. Lancashire Shipping Co. (The Muncaster Castle)* (1961)[141] clearly stated that regardless the quality and prestige of the shipyard and the classification society hired to repair and inspect the vessel, respectively; if due diligence was not exercised, the shipowner is liable.[142] The shipowner must respond for the want of due diligence in the work of repair, whomsoever has done it.[143] Lord Keith grounded the decision on the following analysis:

> It would, however, be a most sweeping change if [the Hague Rules] had the result providing carriers with a simple escape from their new obligation to exercise due diligence to make ship seaworthy… But in dubio the courts should, in a charge of the suggested dimensions, lean to the other way. The language of the Hague Rules does not, I think, lead to the result contended for the respondents. The carrier will have some relief which, weighed in the scales, is not inconsiderable when contrasted with his previous common law position.[144]

This case became the leading decision on this point. This issue was the object of large consternation on the part shipowners, who proposed later possible amendments to the HR through the Brussels Protocol of 1968 (Visby Amendment), all of which were finally rejected.[145] *The Amstelslot* (1963)[146] complemented the rule exposed in *The Muncaster Castle*. There, the Lloyd's Register of Shipping

---

[138]181 U.S. 218, 45 L.Ed. 830, 21 S. Ct. 591. (1901), at 225: "We do not think that a shipowner exercises due diligence within the meaning of the act by merely furnishing proper structure and equipment, for the diligence required is diligence to make the ship in *all respects* seaworthy, and that, in our judgment, means due diligence on the part of all the owner's servants in the use of the equipment before the commencement of the voyage and until it is actually commenced."

[139]Schoenbaum, 893. *See Cerro Sales, Corp. v. Atlantic Marine Enterprises Inc.*, 403 F. Supp. 562, 1976 A.M.C. 375 (S.D.N.Y. 1975).

[140]Ibid., 893, note 15. *See The Point Brava*, 1 F. Supp. 1150, 1991 A.M.C. 2432 (S.D.N.Y.).

[141][1961] A.C. 807, [1961] 1 Lloyd's Rep. 57. Baer, at 506: "[this case] was said to be the first decision in which the Hague Rules were discussed by an English court since their adoption in the Australian Carriage of Goods by Sea Act."

[142]Longley, at 65: "As the result of the negligence of one of the shipyard employees, when the vessel was returned to her owner she had a defect which, on the subsequent voyage, caused damage to cargo."

[143][1961] AC 807, 844.

[144]Ibid., 872.

[145]Girvin, at 424: "…it was one of the issues for revision considered by the CMI at its Conference at Rijeka in 1959 and at its Stockholm Conference in 1953…. However, at the later Diplomatic Conference at Brussels in 1967 and 1968, the amendment was rejected…" *See supra* (1). The Visby Amendment –The Brussels Protocol of 1968.

[146]*See Union of India v. N. V Reedrij Amserdam* [1963] 2 Lloyd's Rep. 223 (H.L.).

performed an inspection of the vessel without finding that the gear had a crack. To be released from liability, Lord Reid explained, the carrier must prove not only that he and his servants had exercised due diligence, but that Lloyds Register had also done so.[147] The House of Lords found that Lloyd's took reasonable care in doing the inspection and consequently, the obligation was accomplished.[148] In later cases, it was added that the owners or managers were not allowed, in the practice of due diligence, to "wash their hands...of all questions... or to leave everything to the unassisted discretion of the masters."[149] It is not enough to be diligent in selecting the inspector, or sub-contractor or any third party hired to practice due diligence in their behalf. They must, in fact, be diligent. A consequence of this rule is that the surveyor's certificate is not a valid proof in and of itself that due diligence was applied. In such cases, the best proof is the testimony of the surveyor who personally made the inspection and prepared and signed the certificates.[150] What specifically the inspector or surveyor has done is what can be assessed as diligent or not.

3.2.2.2.3 Liability Depends on Direct Control of the Vessel

In the case The *Waterville Victory* (1951),[151] an American court ruled that when a defect resulting from original construction rends the vessel later unseaworthy and previous tests were done where the defect was not discovered, the carrier is not liable. This ruling was similarly made by the English courts in the already commented *The Muncaster Castle* (1961),[152] where Lord Keith also said:

> He [the carrier] will be protected against latent defects, in the strict sense, in work done on his ship, that is to say, defects not due to any negligent workmanship of repairers or others employed by the repairers, and, as I see it, against defects making for unseaworthiness in the ship, however caused, before it became his ship, if these could not be discovered by him, or competent experts employed by him, by the exercise of due diligence.[153]

It happens when the ship, being new, comes with a defect of construction or it is chartered or purchased from another party with defects that are not discoverable by exercising due diligence at the time of the takeover.[154] Liability for the carrier arises

---

[147]Wilson, 190.

[148]Ibid., 190.

[149]*See The England* [1973] 1 Lloyd's Rep. 373 at 383; approved by Lord Brandon in *The Marion* [1984] 1 AC 563, 577.

[150]Tetley, *Marine Cargo Claims*, Vol. 1, 928. *See The Niagara*. 1944 A.M.C. 1307, 1312 (B.C.S. C.).

[151]1951 A.M.C. 320 (S.D.N.Y.), *aff'd* per curiam, *Continental Ins. Co. v. United States* 195 F.2d 527, 1952 A.M.C. 669 (2nd Cir.).

[152][1961] A.C. 807, [1961] 1 Lloyd's Rep. 57.

[153]Ibid., 872.

[154]Wilson, 189.

only when the ship is under his control.[155] If the seaworthiness defect occurs during the time previous to the delivery of the ship, before the carrier has control over the ship, and it was not discoverable by practicing a customary inspection once received, the carrier will be released of liability. However, the carrier cannot rely on protection of the certificate of a Lloyd's surveyor or any classification society, if the defect could have been discovered through a reasonable inspection of the vessel at the time that he took the vessel over.[156] This was confirmed in *Parson Corp. v. CV Scheepvaatonderneing Happy Ranger* (*The Happy Ranger*) (2006).[157] The shipowner was held liable for not practicing due diligence on a vessel that in her maiden voyage had a defect in one of her crane's rams horn hooks which broke in the loading operation causing damage to a process vessel.[158] The court agreed that the ship had not come 'within the shipowner's orbit' until her delivery and the carrier cannot be liable for any negligence prior to delivery.[159] Nevertheless, such a defect could have been found by practicing an adequate testing of the cranes and hooks after taking delivery, which the defendant did not diligently do.[160]

3.2.2.3 Technical Aspects

The courts have determined some general characteristics of the seaworthiness condition. It has an objective and a relative aspect, requires foreseeability, it evolves with the pass of time and it has been fragmented in the different important components of the ship.

3.2.2.3.1 Objective Aspect

Within the concept of seaworthiness, it is possible to distinguish two aspects: the objective and relative seaworthiness.[161] The first one refers to the objective conditions of every ship to navigate properly and safely, and its capability to confront the navigational risks and perils of the sea.[162] This objective standard has been set by the statutory regulations in order to assure a minimum of ship's safety. These regulations have come into force mostly in recent decades through international conventions enacted by the International Maritime Organization (IMO), such as the International Convention for the Safety of Life at Sea Convention (SOLAS) of

---

155 Ibid. 189.

156 Ibid., 189–190; Girvin, 425.

157 [2006] EWHC 122 (Comm); [2006] 1 Lloyd's Rep. 649.

158 Girvin, 425.

159 Ibid., 425.

160 Ibid., 425.

161 Sánchez Calero, Vol. 170, 323–324.

162 Ibid., 323–324.

1974; the International Management Code for the Safe Operation of Ships and for Pollution Prevention (ISM Code) of 1993; or the International Convention on Standards of Training, Certification and Watchkeeping for Seafarers (STCW) of 1978. The carrier must observe and apply the technical standards set out in these international conventions, the regulations set out by the coast guard, and the general rules of classification societies relating to vessel's construction, equipment and manning.[163] If the vessel does not fulfill all of these basic requirements, the vessel is to be held unseaworthy.[164] On this aspect, shipowners have the help of flag state or port state authorities who, by statute, must perform regular inspections on the vessels, to assure its navigability and safety.[165] The absence of the mandatory certificates may arise in liability, as it indicates that the regulations were not adhered to and that is evidence of a lack of due diligence.[166] However, the mere certificates of these authorities or those issued by classification societies do not alone establish the practice of due diligence in meeting these standards.[167] These regulations are of course not enough to assure that a specific vessel is suitable for the carriage of a specific cargo. They set very basic standards to grant safe navigation, and mostly do not address standards for cargo protection. This brings us to the second aspect of seaworthiness, or the relative aspect relating to the specific conditions of the voyage and the goods to be carried.[168]

3.2.2.3.2 Relative Aspect

From the time the obligation of providing a seaworthy vessel was absolute, the courts in the common law assessed the seaworthiness condition attending to the special circumstances of the contract and the specific conditions of the maritime adventure. Frequently cited as one of the leading cases in English law on the matter,

---

[163]*See The Black Gull*, 250 F.2d 777, 1958 A.M.C. 277 (2nd Cir.)*; second appeal*, 269 F.2d 68, 1960 A.M.C. 163 (2nd Cir.); Tetley, *Marine Cargo Claims*, Vol. 1, 899, *see The Kapitan Sakharov* [2000] 2 Lloyd's Rep. 255, 266 (C.A.).

[164]Longley, 43.

[165]Sánchez Calero, Vol. 170, 324.

[166]Ibid., 324.

[167]Ibid., 324; Longley, 43 *citing The Folmina*, 212 U.S. 354, 29 S. Ct. 363, 53 L.Ed. 546 (1909); *The Pinellas*, 45 F.2d 174, 1930 A.M.C. 1875 (4th Cir.); *The Gonzenheim*, 36 F 2d. 869, 1930 A. M.C. 122 (5th Cir.); *The Negus*, 298 Fed. 749, 1923 A.M.C. 625 (S.D.N.Y.); *The Archer*, 29 F2d. 134, 1928 A.M.C. 1615 (E.D.N.Y.); *The Georgian*, 4 F. Supp. 718, 1933 A.M.C. 1540 (S.D. Fla.), *aff'd*, 76 F.2d 550 (5th Cir. 1935). Also in the *Complaint of Tecomar S.A.*, 765 F. Supp. 1179 (S. D.N.Y. 1991); *Navegacion Castro Riva, S.A. of Panama v. The Nordholm*, 178 F. Supp. 736, 741 (E.D.La.1959) ("classification societies often continue vessels in class long after their highest and best use would be as scrap"), *aff'd*, 287 F.2d 398 (5th Cir.1961); *Bank Line v. Porter*, 25 F.2d 843, 845 (4th Cir.), *cert. denied*, 278 U.S. 623, 49 S.Ct. 25, 73 L.Ed. 544 (1928), ("certificate of surveyors disregarded where vessel found unseaworthy in fact").

[168]Sánchez Calero, Vol. 170, 324.

Lord Blackburn stated in the *Steel v. State Line Steamship Co.* (1877),[169] that the question if a ship is reasonably fit to carry her cargo must be "determined upon the whole circumstances and the whole evidence".[170] The specific characteristics of the cargo, the time or season when the voyage is taking place, the condition of the ports, the area of the globe,[171] among others, are key elements to be considered when assessing the vessel's seaworthiness. Based on the English COGSA of 1924, the court clearly stated in *FC Bradley & Sons v. Federal Steam Navigation Co.* (1927):

> In the law of carriage by sea neither seaworthiness nor due diligence is absolute. Both are relative, among other things, to the state of the knowledge and the standard prevailing at the material time.[172]

Accordingly, in *MDC Ltd. v. NV Zeevaart Maatschapij Beursstraat* (1962), it was further argued:

> The test in a case of this kind, of course is not absolute: you do not test it by absolute perfection or by absolute guarantee of successful carriage. It has to be looked at realistically, and the most common test is: Would a prudent shipowner, if he had known of the defect, have sent the ship to sea in that condition?[173]

For their part, the American courts have determined that due diligence involves the care which the situation demands.[174] *The Sylvia* (1898)[175] echoed the same reasoning previously made by the English courts. The test of seaworthiness is whether the vessel is reasonably fit to carry the cargo which she has undertaken to transport.[176] By the time of this decision, the Harter Act was already enacted, and one of the problems it presented was whether the cause of the damage was due to unseaworthiness or to the negligence of the crew. This situation obliged the courts to base their reasoning not on a general standard or set of rules, but in the specific conditions of every case. For example, water entering through port holes in a ship might render different results in each case. In *The Sylvia*, the cause of cargo damage was seawater seeping into the hatches. The shipowner was relieved of liability because the entrance of water was caused by careless management of the vessel and not due to lack of seaworthiness. A few years later, the Supreme Court dealt again with the same cause of damage in *International Navigation Company v. Farr &*

[169] L.R. 3 App. Cas. 72.

[170] J. D. Lucas, *Cases and materials on admiralty*, University casebook series, 2nd edn. (Mineola N.Y: Foundation Press, 1978), 652.

[171] Clarke, 125.

[172] (1927) 27 Ll.L.Rep. 395, 396.

[173] [1962] 1 Lloyd's Rep. 180, 186.

[174] *See The Waterville Victory*, 1951 A.M.C. 320 (S.D.N.Y.), *aff'd* per curiam, *Continental Ins. Co. v. United* States 195 F.2d. 527, 1952 A.M.C. 669 (2nd Cir.); *See* also *The Pocone*, 159 F.2d 661, 665, 1947 A.M.C. 306 (2nd Cir.); *Mamiye v. Barber*, 360 F.2d 774 (2nd Cir. 1966).

[175] 171 U.S. 462, 19 S.Ct. 7, 43 L.Ed. 241 (1898).

[176] Lucas, 651.

Bailey Manufacturing Company (1901).[177] This time, the vessel deemed unseaworthy before breaking ground for not properly closing the glass and iron covers of the portholes.

The countless number of circumstances surrounding every case and the particular conditions of the marine adventure makes clear that this obligation cannot be based in an absolute, unique or general standard.[178] It is a relative duty and its assessment depends on the elements of the specific carriage contracted, taking into account, among other aspects: the specialties and nature of the voyage undertaken, the cargo and its stowage,[179] the type of vessel,[180] the conditions of the ports involved,[181] area in the world, and the season of the year. A vessel might be seaworthy for navigation in summer, but not during winter.[182] Similarly, a vessel may be seaworthy for a specific cargo, but not for another.[183]

With the exceptions of the minimum standards provided by international conventions, this relative aspect makes it more difficult to establish general parameters or standards regarding the protection of cargo which may be exactly followed in order to make the ship seaworthy. Rather than an abstract and immutable legal concept, this duty is of variable character and intensity, depending on the circumstances of each case and each time.[184]

---

[177] 181 U.S. 218, 45 L.Ed.830, 21 S.Ct.591. (1901); Lucas, at 652: "The Court went on to reject the argument that despite the fact of unseaworthiness the condition arose from an error in the management within the Harter Act exemption, and similarly to reject the argument that the owner is responsible for the provision of a proper structure and equipment, and for the diligence of his "shore" personnel, but is not responsible for the negligence of his "sea" personnel through the negligence result in unseaworthiness before breaking ground."

[178] *See Philippine Sugar C. Agency v. Kokusai Kisen*, Etc. (*The Naples Maru*) 106 F.2d 32, 1939 A.M.C. 1087 (2nd Cir.), at 34–35: "The standard of seaworthiness, like so many other legal standards, must always be uncertain, for the law cannot fix in advance those precautions in hull and gear which will be necessary to meet the manifold dangers of the sea."

[179] Longley, 43.

[180] *See Complaint of Tecomar S.A.*, 765 F. Supp. 1178 (S.D.N.Y. 1991); *Tug Ocean Prince, Inc. v. United States*, 584 F.2d 1151, 1155 (2nd Cir.1978), *cert. denied*, 440 U.S. 959, 99 S. Ct. 1499, 59 L.Ed. 2d 772 (1979).

[181] *See Tidmarsh v. Washington Fire & Marine Ins. Co.*, 4 Mason 439, 441; 23 F. Cas. 1197, 1198 (C.C. Mass.).

[182] Baer, 501.

[183] Ibid., 501; *See The Sagamore* (1924), 300 Fed. 701, 1924 A.M.C. 961 (2nd Cir.), at 704: "The vessel may be perfectly seaworthy for cargo-carrying purposes around the harbour, and not seaworthy for oceanic carriage; and she may be seaworthy for the carriage of a load lumber, and not be seaworthy for a load of steel rails. The law implies a warranty by owner … of seaworthiness for the purposes for which the 'Sagamore' was chartered."

[184] Tetley, *Marine Cargo Claims*, Vol. 1, 899. *See Mobil Shipping Transp. Co. v. Wonsild Liquid Carriers*, 190 F. 3d 64, 69, 1999 A.M.C. 2719, 2710 (2nd Cir. 1999).

Notwithstanding, it is important to remark, as pointed out by Clarke, that once the vessel seaworthiness is ascertained, the obligation becomes absolute with respect to that specific cargo and voyage.[185] The obligation assumed by the carrier is not to carry any cargo to any destination; but rather, he assumes the obligation to carry a particular cargo to a particular destination by a particular route.[186] This means that when it is determined what a vessel needs for the proper performance of a specific carriage, the carrier must comply with those relative standards, which then creates an absolute obligation regarding that particular contract.[187]

3.2.2.3.3 Evolution

In close connection with the relative aspect of the obligation, seaworthiness has an evolving character, requiring the assessment of this obligation to be done within a specific timeframe. Since the 19th century, judges have kept to almost the same definition for seaworthiness. It was noted that the vessel´s sufficiency could not be based on static positive legal rules, because the means and modes of navigation and cargo carriage are always subject to changes and improvements towards making them more efficient and safer.[188] Back in 1827, in the case of *Tidmarsh v. Washington Fire & Marine Ins. Co.*,[189] Story C. J. argued that:

> Much argument has been employed at the bar upon the question of the nature and extent of seaworthiness. It has been properly remarked, that the standard of seaworthiness has been gradually raised within the last thirty years, from a more perfect knowledge of ship-building, a more enlarged experience of maritime risks, and an increased skill in navigation. In many ports, sails and other equipment would now be deemed essential, which at an earlier period were not customary on the same voyages.[190]

This same reasoning was echoed in *The Titania* (1883)[191] and later, under the Harter Act, in *The Southwark* (1903).[192] In the latter, Justice Day also remarked specifically this evolving character.[193] He referred to "improvements in navigation require new implements or new forms of old ones" that, initially, were not necessary or regarded as basics for the vessel's seaworthiness, but through the passage

---

[185]Clarke, at 125: "Secondly, the warranty is said to be absolute: once the meaning of seaworthiness is ascertained, (relative, as above), in any particular case, the obligation of the carrier to ensure that his ship meets that relative standards is absolute." *See* Carver No, 107; Scrutton, 80 ff.

[186]Ibid., 126.

[187]Ibid., 126.

[188]J. Bouvier, *Bouvier's Law Dictionary and Concise Encyclopedia*, 3 rev. (being the eigth Edition) by Francis Rawle, 2 vols. (Kansas City, Mo.: Vernon Law Book Company, 1914), II, 3028.

[189]4 Mason 439, 23 F. Cas. 1197 (C.C. of Mass.).

[190]Ibid., 441 or 1198.

[191]19 F. 101 (D.C.N.Y 1883).

[192]191 U.S. 1, 24 S.Ct. 1, 48 L.Ed. 65 (1903).

[193]Ibid., 3 or 8. *See supra* note 124.

of time, have become part of the seaworthiness condition.[194] Baer, commenting on the quotation made by justice Day on *The Southwark*, points out an elemental conclusion: "vessels must be kept up to date", "what would qualify as a seaworthy vessel in 1900 may well not do so in 1970,"[195] and even less today. Under that syllogism, one can immediately understand, for example, why the use of radars or wireless communications has become part of the seaworthiness condition of a vessel.[196]

Observing this evolving character together the relativeness and reasonableness of the obligation, an American District Court in the case *Electro-Tec Corp.* v. *S/S Dart Atlantica* (1984) stated another basic conclusion in this regards: "what was unreasonable yesterday may be reasonable today."[197] Tetley points out that this statement is an explanation for why courts, especially American courts, have been willing to analyze the impacts of new technologies in the carriage of goods by sea.[198] Indeed, more American courts have analyzed the implications of new technologies in the seaworthiness condition of the vessel than their English peers.

Within the English law, the case law regarding this evolutionary character is not as clear. English authors on the topic express that the temporal character of seaworthiness demands attention to "the state of knowledge and the scientific progress at the time of the contract."[199] This was recently acknowledged in *The Eurasian Dream* (2002): "Seaworthiness must be judged by the standards and practices of the industry at the relevant time, at least so long as those standards and practices are reasonable".[200] This begs the question: what is reasonable today? This question will be discussed in more detail in the third part.

3.2.2.3.4 Foreseeability

The shipowner or carrier when preparing the vessel for voyage must consider the normal perils and vicissitudes he may encounter during the sea adventure.[201] Foreseeable conditions include, for example, the weather during the different seasons of the year. The prudent shipowner must take the proper measures and prepare his vessel for both fare conditions in summer, but also for such conditions that may

---

[194]Bouvier, II, 3028, *See The Northern Belle*, 9 Wall. (U.S.) 526, 19 L.Ed. 746; *Palmer v. Ins. Co.*, 116 N.Y. 599, 23 N.E. 5.

[195]Baer, 500.

[196]Ibid, 500.

[197]598 F. Supp. 929, 934, 1985 A.M.C. 1606, 1612 (D.Md. 1984).

[198]Tetley, *Interpretation and Construction...*, 64.

[199]Clarke, 125, *citing* Carver, Carriage by Sea (12th ed.), London 1971, No. 109.

[200][2002] 1 Lloyd's Rep. 719, at para. 126.

[201]Longley, 44, *citing The Ciprya*, 137 F.2d 326, 1946 A.M.C 947 (2nd Cir); *The Gonzenheim*, 36 F.2d 869, 1930 A.M.C. 122 (5th Cir); *Barge B.A.* 1401. (*Hampton Roads Carriers, Inc. v. Allied Chemical Corp.*), 329 F2d. 387, 1964 A.M.C. 2458 (4th. Cir. 1964); *The Roseina*, 1937 A.M.C. 359 (S.D.N.Y.); *The Anastasia*, 11 F. Supp. 314, 1935 A.M.C. 579 (E.D.Va.).

arise from winter hardships.[202] Such foresight is especially needed when dealing with dangerous cargo.[203]

Storms are a typical case of peril of the sea. Courts have denied exoneration of liability when storms are foreseeable. Determining if a storm constitutes a peril of the sea, depends on whether or not, the storm was expectable.[204] When expectable, heavy weather conditions are not considered peril of the sea.[205] So, if the direction and force of the winds show the possibility of a storm, the carrier must take measures to assure the avoidance of such phenomenon or have his vessel well enough equipped to overcome the consequences of the ill weather. Here is another point where technology plays a paramount role. The new devices and systems for weather forecasting should be available and used by carriers before the commencement of the voyage to foresee the weather conditions that may be encountered during the voyage.

3.2.2.3.5 Fragmentation

The shipowner's obligations in making the ship seaworthy, has been initially measured along two main areas. The first is the overall condition of the vessel itself, as well as its equipment and crew. The second is its cargo worthiness, or its capacity to receive and carry safely the contracted cargo.[206] The case law has analyzed the different components required by a vessel for the successful performance of the maritime adventure and the compliance of the contract. This separate analysis has created a kind of fragmentation of the concept of seaworthiness in order to enable a better study of the different causes that might produce damage to or loss of the cargo. Such fragmentation was markedly clear in the Anglo-American maritime law, which divided the concept in proper manning, equipping, supplying and the fitness in receiving and caring for the cargo.[207] This distinction is in response to its express mention in the Hague Rules. The international character of maritime transport demanded the express mention of these aspects to explain the

---

[202]Ibid., at 48: "In *The Pennsylvania*, 1956 A.M.C. 1810 (D. Ore), it was held that a welded vessel was unseaworthy for winter voyage across the Pacific on which she was bound to be subjected to low temperatures which would accentuate the danger of fracture of the steel plates of her hull which were notch sensitive at low temperatures."

[203]Ibid., 66.

[204]*See Complaint of Tecomar S.A.*, 765 F. Supp. 1175 (S.D.N.Y. 1991); *J. Gerber & Co. v. SS Sabine Howaldt*, 437 F.2d 580, 596 (2nd Cir. 1971); *New Rotterdam Ins. Co. v. The S. S. Loppersum*, 215 F. Supp. 563, 567 (S.D.N.Y.1963).

[205]*Complaint of TecomarS.A., see supra* note 204.

[206]Wilson, 12.

[207]Gilmore and Black, JR., 151. *See The Southwark*, 191 U.S. 1, 24 S. Ct. 1 (1903); *The Benjamin Noble*, 244 F. 95, 97 (6th Cir. 1917) affirmed *sub nom. Capitol Transp. Co. v. Cambria Steel Co.* 249 U.S. 334, 39 S.Ct. 292 (1919); *R. T. Jones Lumber Co. v. Roen S.S. Co.*, 270 F 2d. 456, 1960 A.M.C. 46 (2nd Cir. 1959).

comprehensiveness of the duty, especially for countries that might have a different concept of seaworthiness.[208] The main aspects included in the vessel's seaworthiness are explained as follows.

(1) **Vessel Worthiness**

This refers to the condition of the vessel itself, the structural condition of the hull and the state and functionality of the machinery. The vessel must be sufficiently tight, staunch and strong to resist the ordinary perils of the sea.[209] Under common law, this was the classic definition of seaworthiness.[210] Some concrete examples within this aspect include the following: the engines have to be in good conditions,[211] the plates must be tight and capable of resisting the weather,[212] the shell plating must be intact without fractures,[213] the steering gear must be in optimal condition,[214] and deck port holes must be sufficiently fastened.[215] In addition, the following must be avoided or repaired as part of the due diligence obligation: cracks in bullhead plating,[216] corrosion of the steelwork,[217] and leaking rivets.[218] The lack of proper equipment for the ordinary process of loading and discharging the cargo will also rend the vessel unseaworthy.[219]

The vessel′s design and structural composition make up most of the objective character of seaworthiness. Nowadays, there are standards established by international conventions and classification societies. The SOLAS Convention provides also detailed specifications of some vessels features to be mandatorily accomplished. They create, at the same time, assurance for the vessel's stability, functions and proper sailing. These regulations are meant to ensure a safer journey, and thus the safety of the goods carried. Vessel-worthiness is often determined by the vessel

---

[208]Ibid., 151.

[209]Bouvier, II, 3027. *Sanford & Brooks Co. v. Columbia Dredging Co.*, 177 Fed. 878, 101 C.C.A. 92 (4th Cir. 1910).

[210]*The Muncaster Castle* [1961] 1 Lloyd's Rep. 57; *The Amstelslot* [1963] 2 Lloyd's Rep 223; *The Hellenic Dolphin* [1978] 2 Lloyd's Rep. 336.; *The Friso* [1980] 1 Lloyd′s Rep. 469; *The Theodegmon* [1990] 1 Lloyd's Rep. 52.

[211]*See Hong Kong Fir Shipping Co., v. Kawasaki Kisen Kaisha* [1962] 2 QB 26; *The Amstelslot* [1963] 2 Lloyd's Rep. 223.

[212]Longley, 49.

[213]Girvin, 386, *citing The Toledo* [1995] 1 Lloyd's Rep. 40.

[214]Ibid. *See The Turret Crown*, 297 Fed. 766, 1924 A.M.C. 253 (2nd Cir.); *The Seeger*, 104 F.2d 167, 1939 A.M.C. 792 (2nd Cir.); *The Ionian Pion*eer, 236 F.2d 78, 1956 A.M.C. 175 (5th Cir.); *The Naiwa*, 3 F.2d 385, 1925 A.M.C., 85 (4th Cir.); *Erie & St. Lawrence Corpotaration v. Barnes Ames Co.*, 52 F.2d 217 (W.D.N.Y. 1931).

[215]Ibid., 386. *See Steel v. State Steamship Co.* (1877) 3 App. Cas. 72.

[216]*See Federazione Italiana v. Mandask Compania* 388 F.2d 434. 1968 A.M.C. 315 (2nd Cir. 1962).

[217]*The Torenia* [1983] 2 Lloyd's Rep. 210, 230 and 234.

[218]Ibid., 386. *See Charles Brown & Co. v. Nitrate Producer Steamship Co.* (1937) 58 LI.L.R. 188.

[219]Ibid., 386. *See Hang Feng Shipping & Trading Co. Ltd. v. Mullion & Co. Ltd.*, [1966] 1 Lloyd's Rep. 511, 524.

construction. As mentioned, if there are latent structural defects originated in its construction and not discoverable through due diligence once the shipowner takes control of the ship; this relieves the carrier's liability for unseaworthiness.[220]

(2) **Human Worthiness**

aa) Numbers, Qualifications and Capabilities of the Mariners

The human element on the ship is a fundamental part of the vessels seaworthiness condition. The quantity, qualifications and behavior of the people managing the ships have been subject of many court decisions when assessing this obligation. It has been clearly stated and continually repeated that the master, mates and the crew must be sufficient in number and competent to attend the vessel's operation.[221] The matter is of such importance that the international conventions governing the labor at sea and the safety of life at sea, have established a minimum number and capacities of the crew members on board.[222] The violation of statutory regulations regarding the number and the capacity of the crewmembers rends the vessel unseaworthy.[223] Thus, the carrier is held liable for damages caused by an incompetent engineer or other officers.[224] Therefore, due diligence must be exercised in the selection of the personal; this means the master and the crew.[225] On this subject, The *Eurasian Dream* (2002)[226] provided two guidelines the carrier should take into account when exercising due diligence in the selection of personnel:

(1) The appointment of a generally competent master/crew (e.g. by inspecting the seaman's documents, interviews and inquiries from previous employers to ensure that the person is reasonable fit for the position he is hired)[227]; and,
(2) The specific competence of the master in relation to the vessel and the voyage in question.[228]

---

[220]Tetley, *Marine Cargo Claims*, Vol. 1, 880.

[221]*See Orient Mid-East Lines v. A Shipment of Rice*, 496 F.2d 1032 (5th Cir. 1974); *Matter of Ta Chi Navigation (Panama) Corp., S.A.* 513 F. Supp. 148, 1981 A.M.C. 2350. (E.D.La.).

[222]SOLAS, STCW, Maritime Labor Convention of 2006.

[223]Longley, 54, *citing The E. Madison Hall*, 140 F.2d 589, 1944 A.M.C. 202 (4th Cir.); *The Nancy Moran*, 80 F. Supp. 623, 1948 A.M.C. 1609 (S.D.N.Y.); *The Claribel*, 1964 A.M.C. 957 (E.D. La.), 341 F.2d 956, *aff'd* 1965 A.M.C. 535 (5th Cir.); but *See The Spartan*, 47 F.2d 189, 1931 A.M.C. 1 (2nd Cir.).

[224]Wilson, 11. *See The Makedonia* [1962] 1 Lloyd's Rep 316; *Hong Kong Fir Shipping Co., v. Kawasaki Kisen Kaisha*, [1962] 2 QB 26; The Farrandoc [1967] 2 Lloyd's Rep. 276; *Heinrich C. Horn v. Cia de Navegacion* Fruco [1969] A.M.C. 1495.

[225]Schoenbaum and Yiannopoulos, 388.

[226]*See Papera Traders Co. Limited & Others v Hyundai Merchant Marine Co. Limited, The Keihin Co. Limited*, [2002] 1 Lloyd's Rep. 719.

[227]Also in *The Makedonia*, [1962] Lloyd's Rep. 316, 337.

[228]Tetley, *Marine Cargo Claims*, Vol. 1, 892. *See Robin Hood Flour Mills Ltd. v. NM Paterson & Sons Ltd. (The Farrandoc)*, [1967] 2 Lloyd's Rep. 276, 282.

Merely relying on the certificates of competence or licenses a mariner holds is not enough to fulfill this obligation.[229] The crewmember must be properly interviewed. Courts have paid special attention to the length and nature of the interview, as evidenced in *The Makedonia* (1962),[230] where an interview of 5 min with a chief engineer was considered too short to make a diligent assessment of his character.[231] The shipowner must inquire into his experience,[232] examine his seaman's book[233]; and discuss his present psychological and health condition.[234] The language of the crew, specifically their lack of proficiency in English has also been considered cause of unseaworthiness.[235] These requirements demonstrate that this obligation, at least in this aspect, is very strict. However, the master and crew must not be "perfect".[236] The temporal absence of crewmembers is neither a cause of lack of seaworthiness.[237] In general, the incompetence of the captain and crew might have several origins or reasons. Tetley lists some of them:

- A disabling want of skill;
- A disabling want of knowledge;
- An inherent lack of ability;
- A lack of adequate training or instruction;
- A lack of knowledge about a particular vessel and/or its system;
- A disinclination to perform the work properly; or,
- A physical or mental disability or incapacity (drunkenness or illness).[238]

The disabling want of knowledge is considered the same as the disabling want of skills, according to an English court in *Standard Oil Co. of New York v. The Clan Line Steamers Ltd.* (1924).[239] Both situations make the master unfit to command the vessel.[240] If certain parts of the ship require a special attention or operation, the shipowner must hire more qualified and skilled seafarers who have the necessary

---

[229]Ibid., 892.

[230][1962] 1 Lloyd's Rep. 316.

[231]Clarke, 216.

[232]Longley, 55, *citing The Pinellas*, 45 F.2d 174, 1930 A.M.C. 1875 (4th Cir.). *See* also *The Hong Kong Fir*, [1962] 2 QB 26.

[233]Clarke, 216, *citing The Makedonia*, *supra* note 230, at 336.

[234]Longley, 55, *citing The Mountoswald*, 15 Lloyd's List L. Rep. 144 (C.A.1923), about a master well experienced and with years of license was frequently intoxicated, situation widely known and because of that, the vessel was held unseaworthy.

[235]*See The Condor*, *The Nordpol*, 84 F.2d 3, 1936 A.M.C. 1010 (2nd Cir.); *in re Pacific Mail S. S. Co.* 130 Fed. 76 (9th Cir. 1904).

[236]Girvin, 387. *citing State Trading Corp. of India v. Doyle Carrier Inc.* (*The Jute Express*) [1991] 2 Lloyd's Rep. 55; *Rio Tinto Co. Ltd. v. The Seed Shipping Co. Ltd.* (1926) 24 Ll. L.R. 316.

[237]Bouvier, II, 3028. *See Draper v. Commercial Ins. Co.*, 4 Duer (N.Y.) 234; *Holland v. Seven Hundred & Twenty-Five Tons of Coal*, 36 Fed. 784 (D.C.E.D.Wis).

[238]Tetley, *Marine Cargo Claims*, Vol. 1, 891.

[239]*(The Clan Gordon)* [1924] AC 100, 120–21.

[240]*The Makedonia*,*see supra* note 230.

skills needed for that specific part or equipment of the vessel. For example, in the case of the *Hong Kong Fir Shipping Co., v. Kawasaki Kisen Kaisha* (1962),[241] the ship's engines were very old and the owners were informed of that condition. Such condition made necessary to engage an engine-room staff "of exceptional ability, experience and dependability."[242] The specific knowledge of the vessels characteristics, her systems and equipment require obviously, that the master and crew must have the skills to operate them. Any lack of knowledge, which hinders them in successfully carrying out their duties on board rends the ship unseaworthy.[243] This point is more relevant when dealing with new features or uncommon design, characteristics or special needs of a vessel.[244] The shipowner will be liable if he does not make his crew aware of such special features on his vessel.[245] Hence, shipowners must make the master and the crew aware of the new technological devices or systems used on the vessel and then confirm that they are capable enough to manage them.[246]

bb) Differentiating Between Want of Human Worthiness and Nautical Fault or Errors in Navigation

All these aspects of incompetence and inefficiency of the master and crew are matters of facts. The incompetence of the crew has to be determined by the occurrence of an incident, but not always necessarily a series of incidents. On this point, *The Makedonia* (1962)[247] argues also that the occurrence of one or more mistakes does not immediately constitute incompetence.[248] Incompetence must be then differentiated from mere mistakes or negligence.[249] Notwithstanding, the occurrence of a mistake or negligence on the part of the master or the crew, does not immediately release the shipowner from liability.[250] Under American law, errors in navigation might be so extreme and the faults so gross that they then raise the

---

241[1961] 1 Lloyd's Rep. 159.

242Ibid., 168. As quoted by Clarke, 216, *citing also The Farrandoc* [1967] 2 Lloyd's Rep. 276.

243*Standard Oil Co.v. The Clan Line (The Clan Gordon), see supra note* 239; *The Farrandoc, see supra* note 242; *The Star Sea*, [1997] 1 Lloyd's Rep. 360, 374, (operation of the CO^2 fire-fighting system).

244*The Elkton* 49 F.2d 700, 1931 A.M.C. 1040 (2nd Cir.).

245*See Standard Oil Co. v. The Clan Line (The Clan Gordon), see supra* note 239.

246In the *The Farrandoc*, [1967] 2 Lloyd's Rep. 276, at 282, the Canadian Exchequer Court held on this aspect that: "Even after making such inquiries he would, in my opinion, inquire how far the man's experience fitted him for the service in the particular ship and take steps to see that the man was adequately instructed with respect to any features of the particular ship with which it was necessary for him to be familiar to properly discharge the duties of his position and to avoid damage to the ship and ger cargo."

247*See supra* note 230.

248*See The Star Sea*, [1997] 1 Lloyd's Rep. 360, 374.

249Tetley, *Marine Cargo Claims*, Vol. 1, 891.

250*See Matter of Ta Chi Navigation (Panama) Corp., S.A.*, 513 F. Supp. 148, 158, 1981 A.M.C. 2350 (E.D.La.).

presumption of incapacity or incompetence of the master or crew.[251] A mistake or error caused by incompetence of the master or a crew member, in the absence of further proof that the shipowner made the proper inquiries when hiring them, will result in the shipowner's liability in not practicing due diligence in manning the ship.[252] A different situation is the disinclination to perform the job properly, when the master or crewmember, having the skill and knowledge, simply has a disable lack of will and does not perform the job required of him.[253]

(3) **Equipment Worthiness**

Proper equipment allowing the ship to navigate safely is doubtless a very important part of its seaworthiness.[254] This is perhaps the area where technology has a major impact, since new devices are discovered and implemented every year. The case-law has made clear that the vessel must be supplied with a "compass, sextants and sounding apparatus",[255] "charts, light books, pilot books, list of radio beacons and notices to mariners",[256] cranes on board to perform loading and unloading of cargo,[257] among others. Practicing due diligence in this regards means to have all these equipment with all its components and in proper working condition.

The compass is one of the oldest and clearest examples of the equipment that a vessel must have on board and working properly. A defective compass renders the vessel unseaworthy. It was held since the end of the 19th century in the case *Richelieu & O. Nav. CO. v. Boston Marine Ins. Co.* (1890).[258] This was an action upon an insurance policy for the stranding of the steamer *The Spartan*, a Canadian vessel navigating in Canadian waters. The stranding, although covered by the policy, was excluded if the same was caused by want of ordinary care and skill in navigation and of seaworthiness. The master, the mates, and two wheelsmen stated in the protest, that the cause of the accident was the foggy weather and a defective

[251]*See The Cygnet*, 126 F. 742 (1st Cir. 1903); *In McGill v. Michigan Steamship Co.*, 144 F. 788 (9th Cir. 1906); *Matter of Ta Chi Navigation (Panama) Corp., S.A.* 513 F. Supp. 148, 159. 1981 A.M.C. 2350 (E.D.La.).

[252]*Matter of Ta Chi Navigation (Panama)Corp., S.A.*, *see supra* note 251.

[253]*The Makedonia,see supra* note 230 at 335.

[254]Longley, 50.

[255]Ibid., 50. *See The Claribel*, 1964 A.M.C. 967 (E.D. La.), 341 F.2d 956, 1965 A.M.C. 535 (5th Cir.). Also in *U.S. v. Ultramar Shipping Co., Inc.*, 685 F. Supp. 887, 1988 A.M.C. 527 (S.D.N.Y., 1987), quoting at 899 2A Benedict on Admiralty § 67.

[256]Ibid., 50. *See The W.W. Bruce*, 94 F.2d 834, 1938 A.M.C. 232 (2d Cir.); *The Maria*, 91 F.2d 819, 1937 A.M.C. 934 (4th Cir.); *The Maria*, 15 F. Supp. 745, 1936 A.M.C. 1314 (S.D.N.Y.); *Trinidad Shipping Co. v. Frame Alston Co.* 88 Fed. 528 (S.D.N.Y.1898); The Glenville, 1962 A. M.C. 2311 (S.D. Tex.); but *See The Silverway*, 15 F.2d 648, 1926 A.M.C. 1645 (5th Cir.). Also in *U.S. v. Ultramar Shipping Co., Inc. see supra* note 255, *quoting* at 899 2A Benedict on Admiralty § 67.

[257]Girvin, 387, *citing Parsons Corp. v. CV Scheepvaatonderneming (The Happy Ranger)* [2006] EWHC 122; [2006] 1 Lloyd's Rep. 649.

[258]136 U.S. 408, 10 S.Ct. 934, 34 L.Ed. 398.

compass. These two conditions were considered by the Supreme Court of the United States the predominant and proximate cause of the accident, concluding that a defective compass rendered the vessel unseaworthy.[259] The shipowner had the obligation to provide a good and reliable compass kept in proper condition for safe navigation. If there were defects in the compass, the court said,-known or unknown-that caused the stranding, then the vessel was not seaworthy. English law holds the same consideration regarding defective compass.[260]

Defective lights in a control panel of a seawater cooling system[261] or a defective fathometer may rend the ship unseaworthy as well. In *The Chickasaw*- (1966),[262] the master knew of the defect before the voyage and he was the only person who could have ordered its reparation. The cause of the damage also deprived the carrier to limit his liability.[263] However, in another case, the lack of a fathometer did not render the vessel unseaworthy as other instruments were on hand and made possible the fixing of the position of a vessel involved in lightering.[264]

Measuring equipment of humidity and temperature in the holds as well as fans or mechanical ventilation are also required. The *Associated Metal & Minerals Corp. v. M/V Olympic Mentor* (1997)[265] stated that the carrier was liable for the lack of such equipment on the vessel, when a cargo of steel coils was damaged by sweat. The vessel was held unseaworthy for this type of cargo.[266]

Adequate navigation charts are another piece required on board the vessel. They do not necessarily have to be the latest updated chart, provided, that they have the necessary information.[267] A similar criterion applies for lists of radio beacons. In *The American Smelting v. Irish Spruce* (1977),[268] the ship was held seaworthy though it did not have a new list of radio beacon on board. The list onboard dated from 1969, although there was a newer version from 1971, but with the same

[259]Ibid., at 425 or 938: "If the compass on the new iron vessel was not sufficiently protected to traverse correctly, the vessel was as little sea worthy as if she had no compass, and this should have been carefully ascertained before she started on her voyage. If there was no fault in the compass, then it is very evident that the officer, who is 30 or 40 miles wrong in his calculation, and driving through a thick fog with a full head of steam, and first discovers his true position by running on an island, a cape, or a continent, has neither the skill nor the prudence to be intrusted with such a command; and for want of such an officer the vessel is not seaworthy."

[260]*See Paterson Steamship Ltd. v. Robin Hood Mill*s (1937) 58 Ll.L.R. 33.

[261]*See Louis Dreyfus Corp. v. 27,946 Long Tons of Corn*, 830 F.2d 1321, 1324, 1326. (5th Cir. 1987).

[262]265 F. Supp. 595, 1966 A.M.C. 2219 (S.D. Cal. 1966), upheld in 414 F.2d 724, 1969 A.M.C. 1682, [1970] 1 Lloyd's Rep. 437 (9th Cir. 1969).

[263]Tetley, *Marine Cargo Claims*, Vol. 1, 913.

[264]*See U.S. v.Ultramar Shipping, see supra* note 255, at 899–900 or 545.

[265]1997 A.M.C. 1140, 1155 (S.D.N.Y. 1955).

[266]Tetley, *Marine Cargo Claims*, Vol. 1, 914.

[267]*See Temple Bar* 45 F. Supp. 608, 815, affirmed 137 F. 2d 293 (4th Cir. 1943); The *Heddernheim* 39 F. Supp. 558, 1941 A.M.C. 730 (S.D.N.Y. 1941); *The Aakre* 122 F.2d. 469, 141 A.M.C. 1263 (2nd Cir. 1941).

[268]548 F.2d. 56, 1977 A.M.C. 780 (2nd Cir. 1977).

information. The cause of the incident was that the older list was not consulted by the master, nor by an officer. It was deemed as an error of navigation and not an error for lack of equipment.[269]

As mentioned before, it is in this area of seaworthiness where technology today has a major impact. The shipping industry now has a variety of devices and systems that enable the performance of major efficiency and safety during the maritime adventure. These applications are frequently not regulated statutorily and mostly left to the market, until they reach a certain point of general use where they become mandatory. One example of this is the use of lorans and radars. In 1962, in the *President of India v. West Coast S.S. Co. (S.S. Portland Trader)*,[270] the Judge argued that, at that time, there was no worldwide or American practice using radar and loran. Under this argument, a vessel without such equipment was not held unseaworthy. However, once these devices became a common use in navigation, judges remained in their reluctance to recognize its absence on board as a cause for unseaworthiness.[271] They were only recognized as so, when on board, they were defective.[272] Following these cases, Tetley concludes that modern equipment is not per se a requirement for seaworthiness, but when installed, they must be properly installed and maintained.[273] With the pass of time, some equipment has become mandatory by international conventions. This is exactly the case with regards to magnetic compasses, radars, among others, which were made mandatory in the SOLAS convention for some vessels constructed after the 1st of July 2002.[274]

(4) **Cargo Worthiness**

Because cargo is the main reason of the shipping industry, every part of the ship must be fitted to protect it from the normal and predictable risks. There are two aspects to consider: the condition of the vessel itself for the cargo and the stowage of the cargo on her.

[269]Tetley, *Marine Cargo Claims*, Vol. 1, 912.

[270]213 F. Supp. 352, 356–357, 1963 A.M.C. 649 (D.C.Or. 1962).

[271]*See Irish Spruce (Irish Shipping Ltd. Lim. Procs.)* 1975 A.M.C. 2259, 2568 (S.D.N.Y. 1975), reversed in appeal on other grounds, 548 F.2d. 56, 1977 A.M.C. 780 (2nd Cir. 1977).

[272]*See Texaco North Dakota, Lim. Procs.* 570 F. Supp. 1272, 1289, 1985 A.M.C. 1650, 1674 (E. D.La. 1983).

[273]Tetley, *Marine Cargo Claims*, Vol. 1, 915.

[274]*SOLAS: Consolidated text of the International Convention for the Safety of Life at Sea, 1974, and its Protocol of 1988: Articles, annexes, and certificates*, 5th edn. (London: International Maritime Organization, 2009) Chapter V, Regulation 19.

aa) The Condition of the Vessel Itself for the Cargo

This first aspect contemplates issues such as any prior fumigation and cleaning of the hatches and other places where cargo will be stowed until its upload[275]; the existence on the ship of proper pumps to drain surplus water from the cargo[276]; and, proper refrigeration systems,[277] all of which are measures to make the ship cargo worthy. Perishable cargo requires refrigeration. One of the oldest cases on this issue in the American jurisprudence was also *The Southwark* (1903).[278] The case related to a cargo of meat that required refrigeration. The vessel was held unseaworthy for not providing a working cooling system capable of transporting such a perishable cargo and for not stowing it properly to allow the air to circulate through the cargo. The case dealt with the use of a refrigerating apparatus for a cooling chamber, a technology relatively new at that time. The arguments there established a framework for the analysis on the use of new technologies on ships that will be discussed at a later point in this work. The English Courts have argued this aspect in cases such as *The Good Friend* (1984),[279] where Lord Staughton held that:

> The obligation to make a ship seaworthy itself includes an obligation to see that the ship is fit for cargo service. Where the particular service is specified in the contract, it is an obligation to see that the ship is fit to carry the specified cargo on the specified voyage.[280]

bb) The Stowage of the Cargo

In the second aspect, the cargo worthiness is not limited to the ability to carry the specific cargo, but also covers to its proper stowage on the vessel.[281] An overload of the vessel that disables her to safely sail may be held as lack of seaworthiness.[282] In this case, the observation of the Plimsoll marks becomes fundamental. Loading over the respective load line makes the vessel unseaworthy.[283] The draft or the

---

[275]Wilson, 12. *See Tattersall v. National Steamship Co.* (1984) 12 QBD 297; *The Tres Flores* [1973] 2 Lloyd´s Rep. 247.

[276]Ibid., 12; *see Stanton v. Richardson* (1874) 9 CP 390.

[277]Ibid., 12; *see Cargo per Maori King v. Huges* [1895] 2 QB 550.

[278]191 U.S. 1, 24 S. Ct. 1, 48 L.Ed. 65 (1903).

[279]*Empresa Cubana Importadora de Alimentos Alimport v. Iasmos Shipping Co. S.A.* [1984] 2 Lloyd's Rep. 586, 592.

[280]Similar criterion was held in *The Aquacharm* [1982] 1 Lloyd's Rep. 7, 9 (C.A.); *The Gudermes* [1991] 1 Lloyd's Rep. 456, 472; *The Subro Valour* [1995] 1 Lloyd's Rep. 509, 516.

[281]Bouvier, II, 3027. *Harloff v. Barber & Co.* 150 Fed. 185; *Kish v. Taylor* [1912] AC 604; *The Friso* [1980] 1 Lloyd´s Rep. 469.

[282]Ibid., 3027.

[283]Longley, 54, *citing The Sagamore*, 300 Fed. 701, 1924 A.M.C. 961 (2d Cir.); *The Indien*, 5 F. Supp. 349, 1933 A.M.C. 1342 (S.D. Cal.), 71 F.2d 752, 1934 A.M.C. 1050 (9th Cir.); *Wellesly v. Hooper*, 185 Fed. 733 (9th Cir. 1911); *The Vestris*, 60 F.2d 273, 1932 A.M.C. 863 (S.D.N.Y.); *Petition of T.J. Howard*, 53 F. Supp. 556, 1943 A.M.C. 1263 (E.D.N.Y.), 155 F.2d 780 (2nd Cir. 1946); *The Sea Rambler*, 83 Lloyd's List L. Rep. 84 (K.B. Div. 1949).

stowage plan must be proper in order to preserve the stability of the vessel, assuring proper trim without substantial list.[284] A bad stowage may be the cause of instability of the ship or damage to the cargo wrongly stowed or to other cargo.[285] If cargo is stowed improperly, and damages arise from the vessel's instability, the carrier is held liable for not practicing due diligence, if such an instability results from improper stowage done in the same port or in a previous port. When the instability arises from cargo wrongly stowed subsequently and in a different port, then the fault is considered an error in the nautical management of the ship and the carrier will be exculpated from liability.[286] In addition, incorrect stowage may cause damage to the cargo or other cargo adjacent to it that is wrongly stowed, which will also hold the carrier liable.[287] Notwithstanding, it must be pointed out that such damage could be considered under the obligation of cargo care set out in Article 3 (2).[288] In any case, to hold a carrier liable, the lack of cargo worthiness must be connected directly to the cargo damage. There must be a defect of the vessel that threatens the cargo.[289]

(5) Container Worthiness

Since the first use of containers in 1956 by a North American company, containers have become the predominant method of liner shipping operations.[290] They were developed by the shipping industry to improve efficiency, allowing for faster and safer cargo movement throughout the oceans.[291] The use of containers brought up concerns about their relation to the vessel and carrying out due diligence in making

---

[284] Ibid. *See The Horaisan Maur*, 73 F.2d 526, 1925 A.M.C. 96 (2nd Cir.).

[285] Tetley, *Marine Cargo Claims*, Vol. 1, 918. *See The Friso*, [1980] 1 Lloyd's Rep. 469, 476.

[286] Ibid., 918.

[287] Ibid. *See Nuzzo* v. *Rederi A/S Wallenco* 304 F.2d 506, 508, 1962 A.M.C. 1871, 1874 (2nd Cir. 1962); *Blommer Chocolate Co.* v. *Nosira Sharon Ltd.* 776 F. Supp. 760, 776, 1994 A.M.C. 1807 (S.D.N.Y. 1991), aff'd 963 F.2d 1522 (2nd Cir. 1992); *Sogem-Afrimet, Inc.* v. *M/V Ikan Selayang*, 951 F. Supp. 429, 442, 1998 A.M.C. 1366, 1384 (S.D.N.Y. 1996).

[288] Ibid., 919. Tetley mentions here that the obligation in the Article 3(2) of the HR is absolute, but in his chapter 26 properly dealing with this provision, says at page 1321 that this obligation "although a stringent one, is not absolute."

[289] Ibid., 899. *See The Apostolis*, [1997] 2 Lloyd's Rep. 241, 257 (C.A.). *See* also *R.T. Jones Lumber Co. v. Roen Steamship Co.* 270 F.2d 456, 458, 1960 A.M.C. 46, 49 (2nd Cir. 1959); *Mobil Shipping and Transp. Co.* v. *Wonsild Liquid Carriers Ltd.* 190 F.3d 64, 69, 1999 A.M.C. 2710 (2nd Cir. 1999).

[290] M. D. Booker, *Containers: Conditions, law and practice of carriage and use* (London: Derek Beattie, 1987) Vol. 1, at xxvii: "Sealand were the pioneers in 1956 and by 1958 were operating six converted ships between New York and Puerto Rico. By April 1966 the first North Atlantic services had started between Boston and Rotterdam and the first shore-based cranes for transhipping containers had been installed."; Talley, W. K., 'Ocean Container Shipping: Impacts of a Technological Improvement' (2000) 34, Journal of Economic Issues, 933–48, 933.

[291] *See Leather Best, Inc. v. S.S. Mormaclynx*, 313 F. Supp., 1373, 1970 A.M.C. 1310 (E.D.N.Y. 1970), at 1376: "Containers are provided primarily for the convenience of the carrier, since they cut down handling time and can save as much as 90% of the time required for unloading and reloading a vessel."

a vessel 'cargo worthy'. Much of the discussion has been about whether the containers are part of the ship or not. Depending on the answer, it will be possible to determine if the obligation stated in the Article 3(1) of the HR, is extensive to them. Containers were obviously not considered in the Hague Rules, as they had not been invented yet. First, to determine which treatment must be given to containers, it is necessary to define what a container is. It has been defined as a "very efficient handling, loading, stowing and unloading device"[292]; a "removable compartment of the ship" where goods are stowed by the shippers and unstowed by the consignee[293]; or, a "permanent reusable article of transport equipment".[294] The latter includes the definitions given by the International Convention on Safe Containers (CSC) of 1972.[295]

Under American law, most of the cases dealing with the concept of containers related to the limitation of liability. The question has been whether the container, not its content, might be considered a package or not.[296] One of the first American cases on this topic was the *Leather Best, Inc. v. S.S. Mormaclynx* (1971).[297] There, the Federal Appellate Court stated that containers were "functionally part of the ship." Being part of the ship, when they are provided by the carrier, it was concluded that the carrier must exercise diligence in making the container cargo worthy.[298] The same conception was adhered to later in other cases.[299] One of them

---

[292]S. Simon, 'Latest Development in the Law of Shipping Containers' (1973) 4, J. Mar. L. & Com., 441–54, 442, 447.

[293]S. Simon, 'The Law of Shipping Containers: Part I: A Look at the Factual and Legislative Background, and a Critique of a New Judicial Formula.' (1974) 5, J. Mar. L. & Com., 507–32, 512.

[294]Ibid., 513.

[295]Article II(1). " 'Container' means an article of transport equipment:

(a) of permanent character and accordingly strong enough to be suitable for repeated use;
(b) specially designed to facilitate the transport of goods by one or more modes of transport, without intermediate reloading;
(c) designed to be secure and/or readily handled, having corner fittings for these purposes;
(d) Of size such that the area enclosed by the four outer bottom corners is either:
(i) at least 14 sq.m. (150sq.ft.) or
(ii) at least 7 sq.m. (75 sq.ft.) if it is fitted with top corner fitting;

the term 'container' includes neither vehicles nor packaging, however, containers when carried on chassis are included."

[296]*See Standard Electrica, S.A. v. Hamburg Sudamerikanische Dampfschifffahrts-Gessellschaft*, 373 F.2d 943, 1967 A.M.C. 881; *Encyclopaedia Britannica Inc. v. The Hong Kong Producer*, 1969 A.M.C. 1741.

[297]*Leather's Best, Inc. v. S. S. Mormaclynx*, 451 F. 2d 800, 815, 1971 A.M.C. 2383, 15 Fed. R. Serv. 2d 651 (C.A.2, 1971).

[298]Tetley, *Marine Cargo Claims*, Vol. 1, 921.

[299]*See Mitsui & Co., Ltd. v. American Export Lines, Inc.* 636 F.2d 807, 816, 1981 A.M.C. 331 (C. A.N.Y. 1981); *Binladen BSB Landscaping v. M.V. Nedlloyd Rotterdam*, 1985 A.M.C. 2113, 2120, 759 F. 2d 1006. 1012 (2nd Cir.).

was *Houlden & Co.* v. *S.S. Red Jacket* (1977).[300] An eight year-old container, used in some 20 or 30 voyages, sustained major structural damage that rendered it unseaworthy. The carrier did not practice due diligence as he permitted the container to be loaded on board.[301] The US District Court of the Southern District of New York held that: "The standard of reasonable fitness applies to all of the ship's equipment, including containers supplied to shippers for the purpose of 'house to house' shipments."[302] This statement seems to limit this consideration exclusively to containers provided for a "*house to house*" shipment. Notwithstanding, differing from the previous cases, the container was not deemed as part of the ship, but as equipment of the same. Held as equipment, the carrier must then provide a container which is fit and suitable for its purpose in every voyage and in whatever the modality of the shipment. Later, in *Eastman Kodak Co.* v. *S.S. Sealand Voyager* (1991),[303] the container was held again as "a functional part of the ship", and the carrier was also liable for not practicing due diligence in making it seaworthy or cargo worthy prior to the commencement of the voyage.

In other contexts, the US Courts have also conceptualized the containers as parts of the ships. For example, in contracts of container leasing;[304] or in a labor claim, where the Supreme Court of the United States stated that "the container is a modern substitute for the hold of the vessel."[305] Hence, defective reefer container,[306] improper stowed, lashed or dangerously stacked, compromising the cargo and the ship stability, renders the ship unseaworthy and makes the carrier liable for want of due diligence.[307]

Similar analysis is not clearly found in the English jurisprudence.[308] Regarding maritime liens, the House of Lords rejected to enforce a maritime lien over a vessel, holding that there was no possibility to ascertain the exact vessel for which the

[300]1977 A.M.C. 1382, 1401–1402 (S.D.N.Y. 1977), aff'd without opinion, 582 F.2d 1271 (2nd Cir. 1978).

[301]Ibid., 1398, 1401.

[302]Ibid., 1402. The carrier was held liable despite an improper packing of the cargo that caused damage to the container's skin. Also the damage occurred during a storm in January in the North Pacific which was not considered peril of the sea because those are usual weather conditions.

[303]1991 A.M.C. 2356, 2361 (S.D.N.J.1991), aff'd 958 F.2d 362, 1992 A.M.C. 1520 and 1816 (3rd Cir. 1992).

[304]P.-J. Bordahandy, 'Containers: a Conundrum or a concept?' (2005) 11, The Journal of International Maritime Law, 342–71, 354. *See CTI Container Leasing Corporation v. Oceanic Operations Corp.* 652 F.2d 377, 1982 A.M.C. 2541 (2nd Cir. 1982).; *Flexivan Leasing, Inc. v. M/V C.C. San Francisco* 1986 A.M.C. 333.

[305]*See Northeast Marine Terminal Co., Inc. v. Caputo*, 432 U.S. 249, 270, 97 S. Ct. 2348, 2360, 53 L.Ed.2d 320 (1977).

[306]*Binladen BSB Landscaping v. M.V. Nedlloyd Rotterdam* 593 F. Supp. 546 (D.C.N.Y. 1984).

[307]Tetley, *Marine Cargo Claims*, Vol. 1, 922.

[308]N. J. Margetson, 'Liability of the carrier under the Hague (Visby) Rules for cargo damage caused by unseaworthiness of its containers' (2008) 14, The Journal of International Maritime Law, 153–61, 157.

containers were provided.[309] Thus, it can be deduced that in the English Law, at least regarding maritime liens, containers are not considered "part of the ship".[310] Notwithstanding, English law leaves the liability for defective containers to the contractual stipulation between the parties, allowing the exclusion of liability for damages caused by faulty containers.[311] Courts in other countries, such as in the Netherlands, have also held that containers should be cargo worthy and the obligation in the HR is extended to them.[312] Australia seems to follow same approach although with less clarity.[313]

The opinion that holds containers as part of the vessel is challenged by the argument contending that they are most of the time out of the vessel and out of the scope of action or control of the carriers.[314] Certainly, the shipowner may not have the same control over a container as he has over the vessel itself. But neither does the shipper have enough control over it as to require him to bear the risk resulting in the use of containers. The argument fails further when the transport is multimodal, now offered more and more often by the same sea carriers in the "house to house" modality. The application of the HR to a later invention such as the container can only be achieved through an analogy using the regular means of transport or stowage commonly used at that time. The carrier using a break-bulk or a reefer ship must practice due diligence in making the holds and cooling chamber fit and safe for reception and carriage of goods as the Article 3(1)(c) of the HR states. Now, in a containership, the container serves practically the same functions of the "holds" or "cooling chambers" of the ship. The main difference is that these "holds" are movable and transferable to different means of transport. In this sense, a containership comes to be a ship whose holds are divisible or individualized into as many containers as it can carry. Furthermore, the design fits perfectly the structure of containerships which are deckless vessels especially built for that purpose.[315] If not deemed as a "hold" of the ship, it could be held as "equipment" according to Article

[309]Bordahandy, 356. *See The River Rima* [1987] 3 All ER 1 [1987] 2 Lloyd's Rep 106: [1988] 1WLR 758 [1988] 2 Lloyd's Rep 193, HL.

[310]Ibid., 356.

[311]Margetson, 157.

[312]Ibid., at 153. He refers to the decision of the Supreme Court of Netherlands in the *NDS Provider* of the 1 of February of 2008, nr C06/082 HR, published in RvdW 2008, 177.

[313]Bordahandy, at 357: "Thus the NSW[New South Wales] Court of Appeals in the *TNT Express* appears to have been inconsistent in both conceptualizing a container as 'part of the ship', thus implicitly extending to containers the carrier's personal duty of seaworthiness and also requesting that shippers proceed to a reasonable inspections (for seaworthiness) of the container before loading."

[314]Margetson, 161.

[315]*See Du Pont de Nemours Inter, S.A. v. S.S. Mormacvega* 493 F.2d 97, 102 1974 A.M.C. 67, (C. A.N.Y. 1974); *Konica Business Machines, Inc. v. Vessel "Sea-Land Consumer"* 153 F3d. 1076, 1078 (C.A. 9 (Cal.) 1998), 1998 A.M.C. 2705, 98 Cal. Daily Op. Serv. 6955, 98 Daily Journal D. A.R. 9595, citing the judgment of the first instance district court at 1996 WL 468770; *Insurance Co. of North America v. Blue Star (North America), Ltd.* Not reported in F. Supp., 1997 WL 345235 (S.D.N.Y. 1997), 1997 A.M.C. 2434.

3(1)(b) or as "other parts of the ship in which goods are carried", under Article 3(1)(c). In both cases, the carrier must exercise due diligence in making them "equipment-worthy" or cargo worthy for the voyage. This is obviously an analogy made as a necessity to determine the carrier's obligation under the HR with regards to containers. Otherwise, the matter of containers would be without regulation, or subject to the national regulation of every state, which goes against one of the main purposes of the HR to create uniformity of the regulations. However, the HR also attempted to create a fairer distribution of risks between carriers and cargo interest. It would be against this objective to totally release the carrier of liability for a device developed and offered by them, with which they have much more control than shippers. In *Mitsui & Co., Ltd. v. American Export Lines, Inc.* (1981),[316] the United States Court of Appeals acknowledged this reality.[317]

In resume, when carriers provide the container, they must practice due diligence on it.[318] The carrier will be exculpated however, when the container is provided and loaded by the shipper or loaded with dangerous goods without notification of its condition and with no reasonable possibility to detect any harmful content that may be held in the container.[319] In the end, responsibility for a defective container rests on the party who provided the container.

(6) **Supply Worthiness**

The carrier must provide enough bunkers for the whole voyage or at least have arranged for further supplies at the ports to be called in on the route.[320] Bunkers,[321] medicines, and food for the crew are additional components in determining the ship's seaworthiness.[322] Lack of sufficient bunker causing loss of power with any consequent fluctuation in the temperature needed for the cargo, (perishable cargo for example) renders the ship unseaworthy.[323]

---

[316]636 F.2d 807, 816, 1981 A.M.C. 331 (C.A.N.Y. 1981).

[317]Ibid., at 816: "… containers are typically supplied by the carrier, must be returned to the carrier by the consignee, and are used and reused hundreds of times. Many ships, including the S.S. Red Jacket, are so constructed that shipments must be made in containers. The shipper normally pays for the weight of the pallet but not for that of the container." *See also supra note* 291 and accompanying text.

[318]Tetley, *Marine Cargo Claims*, Vol. 1, 921.

[319]Ibid., 922.

[320]Ibid., 896.

[321]*See McIver v. Tate Steamer* [1903] 1 K.B. 362; *Northumbrian Shipping Co. v. Trimm* [1939] AC 397; *Fiumana Societa di Navigazione v. Bunge & Co. Ltd.* [1930] 2 K.B. 47.

[322]Bouvier, II, 3028.

[323]Girvin, 387, *citing Rey Banano del Pacifico CA v. Transportes Navieros Ecuatorianos* (*The Isla Fernandina*) [2000] 2 Lloyd's Rep. 15.

(7) **Document Worthiness**

The carrier must comply with all national and international regulations and have on board the necessary certificates from the port health authorities.[324] It is necessary to have on board the compulsory documentations such as the certificates required by the ISM and ISPS Codes.[325]

3.2.2.4 When Must Due Diligence Be Practiced?

3.2.2.4.1 Before and at the Beginning of the Voyage

Another inexact issue in the Rules is the period of time when the obligation of due diligence must be exercised. Article 3 Section 1 states "before and at the beginning of the voyage." The expression offered some difficulties in determining exactly when this period begins and ends. It has been described as the period from the moment of loading until the ship puts to sea.[326] This description was stated in English Law in *Maxine Footwear Co. Ltd. v. Can. Government Merchant Marine* (1959).[327] Lord Somervell delimited this period as follows:

> […] 'before and at the beginning of the voyage' means the period from at least the beginning of the loading until the vessel starts on her voyage. The word 'before' cannot in their opinion be read as meaning 'at the commencement of the loading'.
>
> On that view the obligation to exercise due diligence to make the ship seaworthy continued over the whole of the period from the beginning of loading until the ship sank. There was a failure to exercise due diligence during that period. As a result the ship became unseaworthy and this unseaworthiness caused the damage and loss of the appellant's goods.[328]

Logically, explains Lord Somervell, it would be nonsense if the duty were to be practiced just before the loading, to then appear later again just before the beginning of the voyage.[329] Any unseaworthiness condition arising in the vessel after she sails cannot be claimed as a want of due diligence, unless the cause of the damage was discoverable under a diligent inspection in the mentioned period.

The moment at which the loading of the goods onto the ship starts may not present any problem in precision. But determining the beginning of the voyage might be more difficult. Schoenbaum says that the duty ends when the vessel "breaks ground."[330] Tetley explains that that moment when the ship starts her

[324] *See Cheikh Boutros Selim El-Khoury v. Ceylon Shipping Lines Ltd.* (*The Madeleine*), [1967] 2 Lloyd's Rep. 224, 241.

[325] *See The Eurasian Dream* supra note 84.

[326] Clarke, 127.

[327] [1959] AC 589 (PC).

[328] Ibid., 602.

[329] Ibid., 604.

[330] Schoenbaum, 894.

voyage might be a little bit more problematic. To determine the exact moment when the voyage begins the following elements must be observed: the weight of the ship's anchor or slips,[331] battening down of the hatches, visitors sent ashore, orders from the bridge so that the ship actually moves under its own power, by tugs, or both.[332] In addition, if the beginning of the voyage is when the vessel is put to sea, then the period comprised in a river stage between the port of loading and the sea would be included within the period of the obligation.[333] Taking as a reference the exception contained in the Article 4(2)(a), on damages caused by the negligence in navigation and management of the ship, which is based in the incapacity of the carrier to practice his supervision and control on the master and crew, Clarke suggests that this should be the criterion to determine when the voyage starts.[334]

In some cases, some preparation for the journey may be done when the vessel is already underway, and when it is properly planned and carried out after leaving the port, the vessel will not be considered unseaworthy.[335] The carrier at this point, must assure, through the practice of due diligence, that the vessel is in sufficient condition to navigate and accomplish the contractual obligation. This corresponds to the common law principle establishing that "a thing supplied should be reasonable fit for the purpose for which it is intended," which then demands the maximum duty of the supplier.[336]

Although the argument made by the courts in the *Maxine Footwear Co.*, was limited to a specific period of time, it is obvious that to prepare a vessel sufficiently in all the aspects described above, requires much effort well in advance to the beginning of the loading period. The shipowner must practice diligence in the selection of the vessel's design and construction; an efficient and capable shipyard; and a classification society. Once in control of the vessel he must assure also the enrollment of a qualified master and crew; furnish proper equipment and supplies; pass all the statutory inspections required, etc. Many of those requirements are part of the objective aspect of seaworthiness governed by the conventions adopted by the IMO.[337] All these tasks are obviously very difficult, if not impossible, to be performed just before the ship's departure. Therefore, it would be expected that the shipowner acts diligently in the whole preparation of the voyage, much in advance of it, to assure that the ship he offers for the contract, can positively accomplish it. This, in our opinion, would be a better interpretation of the expression "before". Right at the beginning of the voyage the only thing that the shipowner can do is to

---

[331]Tetley, *Marine Cargo Claims*, Vol. 1, 893.

[332]Ibid., 894.

[333]Clarke, 234, at 235: "The common law case establishing the notion of river stage is weak, but the common law rules show a remarkable capacity for survival and the common law solution, accepted by Carver, will probably prevail."

[334]Ibid., 235.

[335]Tetley, *Marine Cargo Claims*, Vol. 1, 894. *See Orient Ins. Co. v. United S.S. Co.* 1961 A.M.C. 1228 (S.D.N.Y. 1961).

[336]Clarke, 126.

[337]SOLAS, STCW, etc.

check again all those aspects and to confirm they are in working order. If they are not, he must repair or replace whatever is necessary.

In this regard, Clarke reports the opinions of two authors addressing this question. Zaphirou points that the obligation "is performed by making in one exhaustive operation the ship seaworthy, in a wide sense, before she receives the cargo and before she is about to sail."[338] Götz, for his part, states that such a period does not refer to any time before the beginning of loading, because the Convention does not mention any express rights or duties previous to that period.[339] Certainly, the Convention does not clearly state any express duty before that time, but it is implied. Clarke criticizes Götz's opinion asserting that under such a construction, the carrier would not have any duty to prepare the holds, which are supposed to be ready and "cargo worthy" to receive the cargo.[340] Making the holds cargo worthy is a task that must be obviously performed before the beginning of the loading of the goods. The same is also inferred from the cases dealing with the human worthiness condition of the vessel.[341] As seen before, courts require the shipowners to practice due diligence in hiring the master and crew by practicing proper interviews and inspection of the mariners' documents. This task is, or at least in most of the cases, performed far before the loading of the cargo. Similarly, the acquisition, installations and test of new equipment require to be done much in advance to the period stated in the provision. The carrier must, therefore, provide a ship at the beginning of loading "on which due diligence has been exercised and continues to be exercised;" and this demands that due diligence "begins to be exercised at the earliest moment at which any part of the ship should last have been surveyed."[342] Hence, due diligence in preparing, the objective aspects, and some relative aspects as well, of the seaworthiness condition of the vessel could not be expected to be performed in this short period right before the commencement of the voyage. Rather, prior to the voyage, significant time must be taken to complete each task with due diligence.

3.2.2.4.2 No Doctrine of the Stages Under HR

The English law construed the doctrine of the stages.[343] This doctrine states that where the ship is in route and calls at an intermediate port, "substantial and actual intervention by the owner or its agents, as distinguished from the master and crew, will revive the duty of seaworthiness, so that the ship must be seaworthy at any

[338] Clarke, 233, *citing* Zaphirou 1963 J.B.L. 221, 225.

[339] Ibid., 233, *citing* Götz 38 Can. B.R. 96, 98.

[340] Ibid., 233–34.

[341] *See The Makedonia* [1962] 1 Lloyd's Rep 316; *Hong Kong Fir Shipping Co. v. Kawasaki Kisen Kaisha* [1962] 2 QB 26.

[342] Clarke, 234.

[343] *See Thin v. Richards*, [1892] 2 QB 141 (C.A.) and *The Vortigen*, [1899] Prob. 140 (C.A.).

particular stage of the voyage."[344] Under the Hague Rules, this doctrine has no application.[345] This shows the detachment of English courts from their own traditional maritime doctrines in the interpretation of this Convention.[346] The voyage starts independently for each cargo, and is construed as the period covered by the bill of lading, regardless of any call in intermediate ports.[347] The carrier must practice due diligence only at the port where the cargo is loaded. In a typical voyage of any liner, the ship stops at many ports to load new cargo. Tetley explains that for a specific cargo the voyage will start in the specific port where it is loaded. The carrier must practice due diligence in regards to such cargo in the specific port where it is loaded, not in the subsequent intermediate ports where the vessel stops to take on new cargo. If the ship comes into port unseaworthy either from the previous port or due to some defect, which may have arose during the previous voyage, the carrier is held liable for not making a diligent inspection and the necessary reparations before loading the new cargo.[348] Therefore, if a seaworthiness defect arises while the vessel is at sea, the shipowner will not be liable for damages caused by such a defect to the cargo previously loaded. He will be liable only in respect to the damages caused to the cargo loaded after the same appeared, if due diligence was not practiced to make the respective reparation. An exception is made if a defect occurs immediately after the commencement of the voyage which would then raise the presumption that it is given to a want of due diligence.[349]

3.2.2.5 How to Practice Due Diligence to Make the Ship Seaworthy?

3.2.2.5.1 The Previous Standard

Determining whether or not due diligence has been in fact exercised is the main challenge when measuring whether the obligation of seaworthiness has been met. As a reference, the previous absolute nature of the obligation required that carriers: "not merely they should do their best to make the ship fit, but that the ship should

[344]Schoenbaum, 894. *See United States v. American Trading Co. (The Glymont)*, 66 F.2d. 617 (2nd. Cir. 1933); *The Steel Navigator*, 23 F.2d. 590 (2nd Cir. 1928) (dicta); and, *May v. Hamburg-Amerikanische Ppacketfahrt Aktiengesellschaft (The Isis)*, 290 U.S. 333, 54 S.Ct. 162, 78 L.Ed. 348 (1933).

[345]Girvin, 427, *citing Leesh River Tea Co. Ltd. V. British India Steam Navigation Co. Ltd.* [1967] 2 QB 250 (CA); *The Makedonia, see supra* note 230, at 160.

[346]As Lord Macmillian sets in *Stag Line LTD v. Foscolo, Mango and Co. Ltd., See supra* note 8, at 350.

[347]Ibid., 427.

[348]Tetley, *Marine Cargo Claims*, Vol. 1, 896. *The Fjord* [1999] 1 Lloyd's Rep. 307.

[349]Girvin, 427, *citing Phillips Petroleum Co. v. Cabaneli Naviera S.A. (The Theodegmon)* [1990] 1 Lloyd's Rep. 52.

really be fit."[350] The strictness of the rule is commented on in *Macffaden v. Blue Star Line* (1905), where Judge Channell quoted a passage of Carvers on Carriage by Sea:

> […] must have that degree of fitness which an ordinary careful and prudent owner would require his vessel to have at the commencement of her voyage having regard to all the probable circumstances of it. To that extent the shipowner, as we have seen, undertakes absolutely that she is fit, and ignorance is no excuse. If the defect existed, the question to be put is, would a prudent owner have required that it should be made good before sending his ship to sea had he known of it? If he would, the ship was not seaworthy within the meaning of the undertaking.[351]

As the ignorance of the defect was not an excuse, the shipowner was liable even for latent defects not discoverable by a diligent inspection. This was the consequence of the absolute obligation to provide a seaworthy ship. Such defects are found frequently in the construction of the ship and are not always easily discoverable to make the necessary reparation on time. The strict nature of this rule was doubtless a heavy burden on the carrier.

3.2.2.5.2 The Current Standar+d

The current system demands that the carrier provide a vessel with optimal hull, machinery, equipment, cargo capacity and personnel, to perform the contract. As pointed out above, it is clear that some of these elements are normally performed much earlier than the beginning of the voyage. A prudent shipowner will choose a reputable shipyard to design, build and equip his vessel. The same would be expected in selecting a classification society to make the inspections and issue the required statutory certificates. He must hire, as well, a reliable ship manning agency to select and recruit the proper crew. But the specific duty consists in that during the period "before and the beginning of the voyage", the carriers must perform a thorough inspection to confirm that all these elements are sufficient, in good condition so as to be working properly during the voyage. In other words, he must ensure there are no defects, or potential defects, or malfunctions that may hinder the safe navigation of the ship and the appropriate care of the cargo carried. Of course, once a defect is found, being diligent means making the proper repairs or replacements if necessary, with his own personnel, or when needed, by hiring a trustworthy shipyard or subcontractor specialized in the object needing repair. This

[350]Statement of Lord Blackburn in *Steel v. State Steamship Co.*, (1877) 3 App. Cas. 72, 86, as quoted by Wilson, 9.

[351]*See McFadden v. Blue Star Line* (1905) 1 K.B. 697, 706.

includes even trivial defects that have a remote chance to compromise the safety of the vessel and of the cargo.[352]

Now, these inspections are required within the parameter of reasonableness. It is said that the question to be asked is not what kind of examination should have been made, but what kind of examination could have been made.[353] The carrier, as a reasonably prudent person, is not expected to make the inspection using methods that are out of the normal practices of the industry during the relevant time; nor any that may cause an unreasonable delay at the port of departure.[354] It was evidenced in *The Amstelslot* (1962)[355]:

> The nature of the examination which a reasonable prudent man, skilled in the behavior of reduction gears, would make involves striking a balance between the likelihood of there being a fatigue crack detectable by way of the methods suggested by the plaintiffs and the delay, expense and possible risk involved by the use of such methods.[356]

The standard of the inspection requires a balance that must observe the economic implications and the risks to be prevented.[357] Notwithstanding, he must observe also the "seriousness of the consequences" in not practicing due diligence which may result in defects, if not discovered timely or not repaired promptly, may cause greater damage.[358] Therefore, he has to pay special attention to those specific and fundamental parts.

---

[352]*See Athenian Tankers Management S.A. v. Pyrena Shipping (The Arianna).* [1987] 2 Lloyd's Rep. 376, at 389: "…a prudent owner might well require even a trivial defect to be made good before sending his vessel to sea if, even in a remote contingency, that defect might jeopardize the safety of the vessel or its cargo, upon the basis that every defect, however small, that might do so, must, as a matter of prudence, be corrected before the vessel put to sea."

[353]Clarke, 224, *citing* Diplock L.J.

[354]*See Brown v. Nitrate Producers S.S. Co.* (1937) 58 Ll.L.R. 188, at 191 Lord Porter says: "The only question is whether by 'latent' it means that you have to use every possible method to discover whether (the defect) exists, or whether you must use reasonable methods. I cannot myself believe that in every case it is obligatory upon a ship's officer on the commencement of a voyage to go and tap every rivet to find if it has a defect or not. If that were so, ships would be held up in port for a very long time while the rivets were being tapped…" as quoted by Clark at 225, *citing also Cransfield Bros. v. Tatem S.N.Co.* (1939) 64 Ll.L.R. 264, 275; *Guan Bee v. Palembang Shipping Co.* [1969] 1 May L.J. 90, 91; *M.D.C. v. N.V.Z.M. Beursstraat* [1962] 1 Lloyd's Rep. 180; *The President Monroe* [1972] 1 Lloyd's Rep. 385, 386 (U.S. District Ct).

[355][1962] 2 Lloyd's Rep. 336.

[356]Ibid., 346–347, as quoted by Clark at 225.

[357]Clarke, at 225: "In *The Australia Star* [(1940) 67 Ll.L.R. 110, 118], it was said that the removal of insulation from a refrigerated hold before every voyage to seek oil leakage from adjacent tanks would be impracticable on account of the expense involved."

[358]Ibid., 228.

3.2.2.5.3 Type of Defects

There are two different categories of defects relating to unseaworthiness, distinguished by the function of the defective part within the ship. Clarke, regarding the method of proof of the unseaworthiness condition, describes the difference between those defects of a traditional maritime nature; and those inherent to more modern machinery and equipment.[359] The first, defined as the "traditional maritime breakdowns", are more easily detected by a reasonable and customary inspection and can then be repaired without much difficulty. But the second category, so-called "functional breakdowns", relates to defects in newer and more sophisticated equipment and machinery. They demand more expertise in determining whether or not they are in proper working order "before and at the beginning of the voyage". Clarke reports this distinction based on the analysis of Dunfield J. in the case *De Carvalho v. Kent Line* (1951)[360]:

> There are two classes of failure at sea with which we may be called upon to deal. Suppose a plate gives out, or a pipe; it is found to have been rusty or out of repair. Suppose a boiler leaks or explodes: the end plate or the tubing of some other part is found to be worn out. Suppose a wooden ship gives out; the sternpost is found to be rotten … these are things which could have been seen, examined and tried before sailing, and they should have been. In effect, the common sense position that the case really proves itself against the shipowner. If the plate is bad, it proves that he did not examine the ship properly or he would have found it. And time was when such simple positions covered the field, but they do so no longer. There is today another type or case, as here, the functional breakdown. A complicated internal combustion engine … fails … a subtle radar installation goes wrong by its own internal delicacy of constitution … nobody can say it is obvious that these machines must have been out of order when the ship sailed; the case does not prove itself like those others. So, somebody has to prove it.[361]

A more strenuous level of due diligence may be required in instruments which have a higher level of technological complexity, and because of that, require a further level of inspection. Defects in such devices or machinery may remain latent, as they require highly skilled personnel on board. The presumption based in the rule of *res ipsa loquitur* is challenged by this case.[362] The conclusion that the mere existence of the unseaworthiness condition proves immediately that due diligence was not practiced may apply regarding those basic conditions or characteristics, but may be too strict for newer and more technologically complex appliances. Hence, the courts may perhaps deem this as a matter of demanding perfection on the ship.

Notwithstanding, it must be taken into account the aforementioned evolving character of the concept of seaworthiness. Devices or systems that in the first years of its application are seen as extremely complex, become, over time, common in navigation, where the functions are then well-known and any defect in them would

---

[359] Ibid., 179.

[360] 32 M.P.R. 282.

[361] Ibid., 305–306, as quoted by Clarke, 179.

[362] Clarke, 179.

be categorized as a "traditional maritime breakdown". A good example is contained in the transcribed decision. It was issued in the middle of the last century, when the two technologies referred to therein, particularly radar, were still relatively new and not yet widely used. Nowadays the use of radars is commonplace; in fact, it has reached such a level of application in the industry, that its use on board is currently statutorily required by the SOLAS Convention.[363] It would be inexcusable not to assure through a diligent inspection that it is in optimal condition and properly working for the voyage. In addition, many of the currently new technologies have reached such a level of perfection that the probabilities of failure are far lower than in the past.

The problem of constraining or limiting the carrier's activity in the fulfillment of the duty to the mere inspection of traditional parts of the ships, where defects are more easily found, may have the effect of reducing his interest in the acquisition, application and correct use and inspection of new devices that improve the vessel's operation. Certainly, discovering a break in the hull might be easier than a defect in the electric system or in the software of a weather forecasting device functioning with satellite signals. Inspecting vessel hulls is doubtless for the carrier, routine. It is not the same for newer technological appliances. Therefore, courts have not always agreed to limit the carriers' obligation to the area of professional knowledge or to traditional maritime issues.[364] Such a limitation would be risky, as it discourages the proper inspection of equipment that greatly improves the vessel's conditions and cargo protection. Furthermore, defects in newer technological devices are not always difficult to find, if proper knowledge exists among the crew members and if proper inspections and tests take place. It obviously depends on the specific characteristics of the device.

3.2.2.5.4 Professional Inspections

Nowadays, for the purpose of confirming the optimal condition of the ship, given its complexity and the large scale of the business, there is a generalized practice on the part of shipowners to hire classification societies to develop inspections on their vessels. Such inspections may be undertaken in two separate phases. First, during the entire process of the ships construction; to assure that the shipyard fulfills the design stated in the drawings. Secondly, for the regular inspections statutorily required, or simply for the carriers own knowledge and assurance. By statute, and particularly attending to the international conventions, vessels require inspections carried out by the flag states as well as by the port state control. Flag states countries have among their essential prerequisites in granting their flag, the existence of a certificate of the ship's conditions issued by a classification society. In addition,

---

[363] *See supra* note 274.

[364] Clarke, 202, *citing The Muncaster Castle* [1961] A.C. 807 and *The Lady Gwendolen* [1965] 1 Lloyd's Rep. 335.

these states usually hire classification societies to perform, on their behalf, the periodical inspections statutorily required. Because these inspectors are more specialized on the subject, their expertise allows them to perform a more thorough examination, and certainly, should be better at discovering defects in the traditional maritime issues, as well as potential failures in new technological devices.

The result of those inspections must be documented in certificates demanded by statute. These certificates have been filed in courts as evidence of the vessel seaworthiness and the practice of due diligence, but were not always accepted as so. Indeed, as evidence of the practice of due diligence they may present some problems. The inspections may not correspond to the moment previous to and at the commencement of the voyage, or they may have been done negligently.[365] In addition, they normally certify in any case the optimal condition of the objective aspects of seaworthiness. It means the fulfillment of statutory requirements, but rarely, if ever, would they certify the optimal seaworthy conditions for the specific cargo and the particular voyage. Therefore, these certificates lack of full evidentiary value.

There are cases, however, where such certificates have been admitted as evidence of the seaworthy condition. This happens when the defective part of the vessel was inspected without discovering the defect, and that survey, according to the court, would have been the only method required to a diligent carrier for such a purpose.[366] In such a case, the seaworthy condition is not met due to latent defect, and yet the certificate is held up as conclusive evidence of the fulfillment of the duty. The certificates, however, would be conclusive under English law, if the survey was recent; the nature and extension of the survey must be evidenced; and, the part of the ship that was later defective must have been adequately covered.[367] The certificate will show in any case the carriers diligence in the periodical testing and maintenance of his ships, previous to the survey.[368]

3.2.2.5.5 The Presumption of Unseaworthiness

It is true that defects in more modern devices may pass unobserved and later appear as latent defects. For sure the carrier would be released from liability, but not for any cause that might be presented as such. In *The Torenia* (1983),[369] the court held the carrier liable because the cause for lack of seaworthiness was not latent, it was discoverable through the practice of due diligence.[370]

---

[365]Ibid., 217.

[366]Ibid., 217, *citing*: *The Assunzione* [1956] 2 Lloyd's Rep. 468, 487; *Cranfield Bros. v. Tatem S. N.Co.* (1939) 64 Ll.L.R. 264, 267.

[367]Ibid., 219–220.

[368]Ibid., 220.

[369][1983] 2 Lloyd's Rep. 210, 230.

[370]Tetley, *Marine Cargo Claims*, Vol. 1, 880.

But the question is, how to determine when a specific defect is latent or not? Tetley approaches the problem concluding that if due diligence has been properly exercised; there is no reason for the vessel to be unseaworthy.[371] The subject was analyzed in English courts creating the presumption of want of due diligence if the vessel is in fact unseaworthy.[372] It was further explained in the *Fjor Wind* (2000). The judge argued that "Putting the matter in simple terms, a ship should not be unseaworthy if proper care is taken".[373] The unseaworthiness in and of itself, leads to the presumption that due diligence was not properly practiced. When the cause of the damage is unknown, and it is presumable that it was due to a latent defect, it will be difficult for the carrier to convincingly prove that he practiced due diligence. In such cases, the courts apply presumptions and frequently conclude that the vessel was unseaworthy because of want of due diligence.[374] The presumption applies also when there is no precise details of the unseaworthiness condition, but there is an inference of the condition or defect because the carrier cannot prove that he practiced due diligence. The courts in these cases may hold carriers liable, even when the unseaworthiness condition is not clear.[375] The same criterion in applied by American courts.[376] Nevertheless, it must be acknowledged that such presumption is easier to argue when evaluating defects in traditional maritime matters.[377] Here is where the distinction made by Dunfield J., and transcribed above, seems to be useful.[378]

3.2.3 The Care of the Cargo

The second duty assigned to carriers by the HR, is established in Article 3(2) which states the obligation of due care of the cargo in the following terms:

> Subject to the provision of Article 4, the carrier shall properly and carefully load, handle, stow, carry, keep, care for, and discharge the goods carried.

---

[371]Ibid., 878.

[372]Ibid., 878. *See Scrutton on Charterparties* 14 Ed. 1839, at 110: "In most cases of the vessel is unseaworthy due diligence cannot have been used by the owner, his servant or agents; if due diligence has been used the vessel will in fact be seaworthy. The circumstances in which the dilemma does not arise (e.g. a defect causing unseaworthiness but of so latent a nature that due diligence could not have discovered it) are not likely to occur often." Also cited in *The Muncaster Castle* [1961] A.C. 807, at 873.

[373][2000] 2 Lloyd's Rep. 191, 205.

[374]Tetley, *Marine Cargo Claims*, Vol. 1, 901. *See Consolidated & Barge Co.* v. *Marcona Conveyor Corp.*, 716 F. 2d 1077, 1985 A.M.C. 117 (5th Cir. 1983).

[375]Ibid., 901, *citing The Theodegmon* [1990] 1 Lloyd's Rep. 52, 77.

[376]See *U.S. v. Ultramar Shipping supra* note 255 at 897: "Any doubt as to the seaworthiness of the vessel 'must be resolved against the ship owner and in favor of the shipper'", *citing* also *The Southwark see supra* note 123 at 6 or 16.

[377]Clarke, 201–202.

[378]Ibid., 201.

This duty is of central importance as well as the obligation of due diligence in making the ship seaworthy.[379] This is a statutorily and conceptually separate obligation from the duty of due diligence in providing a seaworthy ship, yet both are certainly interrelated.[380] If the cargo is damaged or there is a shortage in cargo, in the absence of any unseaworthy condition or any defense according to Article 4 (2), a breach of the obligation of due care would become, the central argument of the claimant´s case.[381] It implies, for example, the reasonable inspections of the goods for insect infestation or other problems that might develop during the period they are under carrier's custody.[382] But assigning liability for non-accomplishment of this obligation often presents a difficulty when determining exactly the cause of the damage. There is sometimes a thin line between damages caused by the want of cargo worthiness or stowage worthiness, errors in navigation or nautical fault and the lack of due care. The courts have dealt with the issue presenting some guidelines for such determination.

3.2.3.1 Precedents Under the Harter Act

The duty of due care in the context of carriage of goods by sea under bills of lading appears for the first time in the Harter Act, with no precedent in the common law as an explicit obligation.[383] The H/H-VR and consequently the American Carriage of Goods by Sea Act in Section 3(2) reproduces the obligation of care, established in the Sections 1 and 2 of the Harter Act. Hence, the American courts, when deciding cases under the Harter Act, presented the first analyses of this duty. Such analyses are equally valid for the interpretation of the American COGSA,[384] and similarly for the H/H-VR.

This is a separate duty of the carrier, and given the wording of the rule, it is frequently analyzed together with the carrier immunities.[385] One of the first decisions of the Supreme Court of the United States analyzing this obligation appears in the case *James Knott v. Botany Worsted Mills* (1900).[386] The case related to a cargo of wool damaged by drainage of wet sugar. The question was whether the cause of damage was due to negligence during the loading, stowage, or custody of goods, or faults or errors in navigation or in the management of the vessel. The decision was

---

[379]Girvin, 429.

[380]Schoenbaum, 897.

[381]Girvin, 429.

[382]Schoenbaum, 899.

[383]Ziegler, 145.

[384]Gilmore and Black, JR., 148–149. As the differences between the Harter Act and COGSA are mostly verbal and stylistic, American courts have frequently used decisions made under Harter act as reference when deciding cases under COGSA.

[385]Schoenbaum and Yiannopoulos, 388.

[386]179 U.S. 69, 21 S. Ct. 30, 45 L.Ed. 90.

based on the reason for the damage itself. The vessel was seaworthy and managed properly. Although there was some defect in its trim, it was the result of the changes in the cargo loading that made the vessel tilt forward. The cargo of wool, stowed forward, got wet from the sugar drainage that flowed in the cargo's direction. It was the sole cause of damage, which resulted from the negligence in the stowage of the cargo on the ship.

A similar situation can hold the carrier liable when the cargo is contaminated by another cargo. In *The Persiana* (1911),[387] cargo was damaged by whale oil leaking into the bilges, where it was intentionally accumulated to be saved and delivered. The Court decided that it was not an act of management of the vessel, but an act directly relating to the management of the cargo. Hence, the carrier was liable for negligently allowing a situation to occur that caused damage to other cargo.[388]

Liability for damages may also be found during the unloading of the cargo due to any negligence in such a process on the part of the carrier's employees or the employees of any independent agency hired by him to perform this operation.[389] In *The Germanic* (1905),[390] the Supreme Court of the United States held the carrier liable for damaging the cargo in a process of negligent discharge. The carrier proceeded to unload the cargo while simultaneously loading coal for the ship's bunker, as she was supposed to sail again shortly after her arrival. The coal loading produced a list to starboard that subsequently, sank the vessel. The question again was if the accident was caused by an error in management of the vessel or want of due care. The Court concluded that, considering the primary nature and objectives of the acts that caused the loss, a hurried and imprudent unloading caused the accident. Thus, the damage was attributable to negligence during the unloading.[391]

Similarly, in the case *Andean Trading Co. v. Pacific Steam Co.* (1920),[392] hatches negligently left open, allowing the entrance of seawater during heavy weather, was alleged to be an error in management of the ship. The court took into account that as the ventilators were removed, the only way to ventilate the cargo was through the opening of the hatches in fine weather. Such an act was regarded as relating "primarily, and indeed exclusively" to the care of the cargo, not to an act of the management of the ship. The court, citing the support of the precedent in the decision stated in *The Germanic*, held the carrier liable.[393]

[387]185 Fed. 396 (2nd Cir. 1911).

[388]Longley, 88.

[389]Ibid., 89.

[390]196 U.S. 589, 25 S.Ct. 317, 49 L.Ed. 610 (1905).

[391]Longley, 90.

[392]263 Fed. 559 (2nd Cir. 1920).

[393]Longley, 87. Similar considerations about ventilation were held in *The Jean Bart* 197 Fed. 1002 (D. Cal. 1911); *The Edith* (1926) 10 F.2d 684, 1926 A.M.C. 281 (2nd Cir.); and, the *W. T. Locket & Co. v. Cunard S.A. Co.* (1927) 21 F.2d 191, 1927 A.M.C. 1057 (E.D.N.Y.).

In *The Vallescura* (1934),[394] a court held that if there is a want of the duty of care and, at the same time, an exculpatory cause that makes it impossible to separate out the real cause of the cargo loss, the carrier is liable for the whole loss unless he can prove the portion attributable to the excepted peril.[395]

3.2.3.2 The Obligation Under the Hague/Hague-Visby Rules

3.2.3.2.1 Not an Overriding Obligation

English law has construed the duty for due care as equivalent to "*reasonable care*".[396] The fact that the provision is subject to Article 4, which relates to the carrier defenses and limitation of liability, indicates that, in contrast to the rule set in Article 3(1), it does not have an overriding effect.[397] While it is clear that under English law this obligation is not an overriding duty, in American Law there has been no such distinction made.[398]

3.2.3.2.2 Period of Time of the Obligation

Another contrast with rule 3(1) is that while this must be met during a specific timeframe, namely before and at the beginning of the voyage, the duty of care is an on-going obligation throughout the entire voyage.[399] The practice of this obligation is not limited to the timeframe "from tackle to tackle". The carrier must take care of the cargo during the entire process from loading to discharge of the goods; essentially the entire period of time the goods are under his custody.[400] The only exception to this would be, of course, if the shipper or the consignee assumes such operations.

3.2.3.2.3 Meaning of the Words "Properly and Carefully"

The words "properly and carefully" suggest initially a higher degree of attention and care. An analysis of this provision was made in *Albacora v. Westcott and Laurance Line* (1966).[401] In this case, the cargo interest argued that the word

---

[394] *Schnell & Co. v. S.S. Vallescura*, 293 U.S. 296, 1934 A.M.C. 1573 (1934).

[395] Tetley, *Marine Cargo Claims*, Vol. 1, 903.

[396] Wilson, 191.

[397] Girvin, 430.

[398] Margetson, 160.

[399] Schoenbaum, 898.

[400] Ibid., 897.

[401] [1966] 2 Lloyd's Rep. 53, 1966 S.L.T. 253.

"properly" of Article 3(2), meant to provide the special conditions that the cargo requires. This cargo, according to the plaintiff, required refrigeration, but the carrier ignored that and the shipper neither informed about such special requirement. The Court concluded that the obligation is to adopt a sound system, according to the knowledge of the carrier about the nature of the goods.[402] Lord Reid acknowledges that the word properly means "something slight different from carefully."[403] Properly means, he added, "in accordance with a sound system and that may mean rather more than carrying the goods carefully."[404] If the carrier had no means to know that the cargo required such a special treatment, there is no reason to expect he will provide so. The carrier was not given the special instructions about refrigeration to the cargo, but the cargo consisted of wet salted ling fillets, which should suggest to a professional carrier that refrigeration is required. However, this would depend on the common practice at that time to carry such cargo. The lack of knowledge of the carrier released him from liability. There is then an obligation from the part of the shippers to inform of the special conditions that their cargo requires. That knowledge will indicate what the cargo "properly" requires and what the sound system is for this cargo to arrive in good condition at destination. But the question was then what a sound system is. Lord Pearce, in the same case, addressed this issue:

> The word 'properly' presumably adds something to the word 'carefully'. A sound system does not mean a system suited to all the weakness and idiosyncrasies of a particular cargo, but a sound system under all the circumstances in relation to the general practice of carriage of goods by sea. It is tantamount, I think, to efficiency. To accept the pursuer's contention would be to import into the Hague Rules a revolutionary departure from the scheme of the common law.[405]

This construction of the word 'properly' has been applied and approved in many cases.[406] As this provision was based on the Harter Act, the cases decided under it showed a higher demand for cargo protection. We must remember that the Convention conceded to reduce substantially the high standard of the absolute undertaking. To compensate that, it seems that the lawmakers added these adverbs to the rule as an attempt to compel the shipowner to act more efficiently and provide a proper care of the goods and to assure the result of the contract. This was as a way, perhaps, to balance the risks distribution a bit more favorably towards cargo owners.

---

[402]Ibid., 254.

[403]Ibid., 254.

[404]Ibid., 254. *See Renton v. Palmyra Trading Corporation of Panama* [1957] A.C. 149, 166; *The Eurasian Dream, supra note* 797.

[405]Ibid., 256–257.

[406]Girvin, 431, *citing Great China Metal Industries Co. Ltd. v Malaysian International Shipping Corp. Bhd* (*The Bunga Seroja*) [1999] 1 Lloyd's Rep 512 (HCA), at 143; *Caltex Refining Co. Pty. V. BHP Transport Ltd.* (*The Iron Gippsland*) [1994] 1 Lloyd's Rep. 335, 337; *Parsons Corp. v. CV Scheepvaatonderneming* (*The Happy Ranger*) [2006] EWHC 122, [2006] 1 Lloyd's Rep. 649.

Under American Law, there is not an obligation on the part of the carrier to inspect the cargo once it is received, but there is also a duty on the shipper to inform the carrier of any special requirements for the cargo.[407] If the carrier is aware of such requirements and is in the capacity to carry them out, but does not properly fulfill them, he will be held liable. One example is the case of cargo requiring to be carried at a specific temperature. The carrier must adhere to the appropriate temperature, as well as provide proper ventilation in order to protect cargo from excessive moisture.[408]

Therefore, the proper and carefully care of the cargo will depend on the following factors: (1) the type of the cargo itself; (2) the information provided by the shipper about the cargo and the special conditions that it requires; (3) the knowledge of the carrier on the cargo, even if special features are not expressly reported by the shipper, but the carrier is expected to know; and, (4) the measures commonly taken by a diligent professional carrier as regular practice at the moment of the carriage (sound system) for the load, handling, stowage, care and discharge of that cargo. It does not require extraordinary measures, unless, of course, those unusual actions are contractually agreed.

3.2.3.2.4 Load, Handle, Stow, Carry, Keep, Care for and Discharge the Goods

The first and simple reading of this part of the rule suggests that the carrier himself must perform all of these duties. Notwithstanding, the operations of loading and discharging are frequently left to other persons different from the crew or carrier's personnel such as port operators. In the words of Lord Devlin, this rule "is not to define the scope of the contract but the terms on which that service is to be performed."[409] English courts have said that this rule is a mere description of how such duties must be performed, but, that it does not mean that such duties must be undertaken exclusively by the carriers or that they are to be liable for such operations even when they are not undertaken by them.[410] This position has been criticized as inconsistent with the objective of the rules and has not been adopted by other States.[411] Under English law the carrier is allowed to contract out this

---

[407] *See Tenneco Resins, Inc. v. Davy Int'l, A.G.*, 881 F.2d 211, 1990 A.M.C. 402 (5th Cir. 1989); *O'Connell Machinery Co. v. M/V Americana*, 226 F.2d 1130, 1986 A.M.C. 2822 (2nd Cir. 1986); *Lekas & Drivas v. Goulandris*, 306 F.2d 426, 1962 A.M.C. 2366 (2nd Cir. 1962).

[408] *See Intercontinental Trading Co. Inc. v- M/V Zenith Sun*, 684 F. Supp. 861, 864 (E.D.Pa. 1988); *The Poleric* (*Bank Line v. Porter*), 25 F.2d 843, 1928 A.M.C. 761 (4th Cir. 1928); *James Knott v. Botany Worsted Mills*, 179 U.S. 69, 21 S.Ct. 30, 45 L.Ed. 90 (1900). A similar view is held under English law, *See* The "*Mekhanik Evgrafov*" and "*Ivan Derbenec*", [1987] 2 Lloyd's Rep. 634.

[409] *Pyrene Co. Ltd. v. Scindia Navigation Co. Ltd.* [1954] 2 QB 402, 418.

[410] Baatz, Michael Tsimplis "Obligation of the Carrier", 37. *Citing Jindal Iron Steel Co Ltd and others v. Islamic Soldarity Shipping Co Jordan Inc.* (*The Jordan II*) [2005] 1 Lloyds Rep 57; *Balli Trading Ltd v. Afalona Shipping Co. Ltd.* (*The Coral*) [1993] 1 Lloyd's Rep 1.

[411] Ibid., 37.

obligation, and then held liable only if and when he himself undertakes to perform them. In other words, when the carrier agrees to undertake such a process, they must fulfill it "carefully and properly".[412]

It is common to find nowadays the FIO, FIOS or FIOST clauses[413] inserted in the bills of lading. This means that the cargo owners undertake the performance of some or all the obligations described in Article 3(2), releasing the carrier from any responsibility for such operations.[414] In the United States, such clauses have been recognized in the *Sigri Carbon Corp. v. Lykes Bros. S.S. Co. Inc.* (1987).[415] There the carrier was relieved of liability for faulty stowage made by the stevedores contracted by the shipper. Notwithstanding, the jurisprudence has not been conclusive, and other opinions have been handed down in other cases.[416] Neither this opinion is shared equally in all jurisdictions.[417]

Once the carrier agrees to undertake the loading, unloading, stowage and trimming processes, he is held responsible for the performance of the master, his servants, stevedores, agents, and subcontractors under his delegation. In this sense, this obligation is as personal as the obligation of due diligence in making the ship seaworthy.[418] Both duties can be performed by third parties contracted by the carrier, but the liability arising from inadequate, insufficient, or nonperformance of such duties continues to be held by the carrier.[419]

Another problem of the Convention regarding this obligation relates to the possible overlapping of this provision with the exception cause of negligence in the management of the ship that cause damage or loss of the cargo. Acts of negligence or default on the part of the master or crewmembers in the management of the ship are held as an exculpatory cause set in Article 4(2)(a). The determination of whether a specific act of careless conduct conforms to a "lack of proper care of the cargo" or

---

[412]Girvin, 431.

[413]FIO: free in and out; FIOS: free in and out, stowed; FIOST free in and out, stowed and trimmed.

[414]*See Jindal Iron & Steel Co. v. Islamic Solidarity Shipping Co. see supra* note 410. Also in *The Arawa* [1977] 2 Lloyd's Rep. 416, 424—5; *GH Renton & Co. Ltd. v. Palmyra Trading Corporation of Panama* [1957] AC 149.

[415]655 F. Supp.1435, 1988 A.M.C. 1787 (W.D.Ky. 1987), at 1438: "…the inclusion of an FIOS term in the bill of lading should not be disregarded as inconsistent with the COGSA so long as it is understood that the term in no way relieves the carrier of responsibility for its own acts or for the acts of other under its control".

[416]Schoenbaum, 896, note 2. *See Metals & Mineral Corp. v. M/N Arktis Sky*, 978 F 2d. 47, 1993 A.M.C. 509 (2nd Cir.1992), and *Atlas Assur. Co. v. Harper, Robinson Shipping Co.*, 508 F.2d 1381, 1975 A.M.C. 2358, 2689 (9th Cir. 1975).

[417]Girvin, 431.

[418]Ibid., 432.

[419]Schoenbaum, 897. *See Nichimen Company v. M/V Farland*, 462 F. 2d. 319, 1972 A.M.C. 1573 (2nd Cir. 1972).

"error in the management of the ship" in assigning liability to carriers has been shown to be fraught with conflict as evidenced by the unpredictable decisions in the United States as well as in other countries.[420]

3.3 Conclusion

The shortcomings of the Convention had to be, and have been, largely met by courts' decisions. The ambiguity and scarcity of the rules obliged the parties to irremediably submit their controversies to the judicial systems or to arbitration, mostly, as pointed out, in the United States and England. The common law courts have made large analysis and vast construction of the carriers' duties set out in the Convention, which, doubtless, has provided better clarity. Certainly, the English courts, in the interpretation of the Convention and considering to its international character, tried to deviate from some aspects of their traditional maritime law, but they kept almost the same meanings of the basic concepts of due diligence, seaworthiness, among others. American courts, on their part, continued partially the same constructions made under the Harter Act of those terms. The courts of both jurisdictions have described the duty of practicing due diligence in making a ship seaworthy as an overriding obligation of a prudent shipowner to exercise reasonable diligence in making the ship reasonably seaworthy under the relative circumstances. For the second duty of properly and carefully load, handle, stow, carry, care for and discharge the cargo, it is required a "sound system under all circumstances".

Reasonableness then, becomes a key element to assess the performance of these duties. It is precisely what forces the parties to submit their disputes to the Courts, leaving frequently little place for direct negotiation and dispute settlements. What is reasonable under the circumstances? This is the question that requires, in the end, that reasonable judges decide what is reasonable in every case. To assess this element of reasonability, the courts have remarked on an important aspect. The ship's seaworthiness has a relative, and especially, an evolutionary character. These characteristics had already been recognized prior to the adoption of the HR. At the beginning of this century, judges still recognize that the seaworthiness condition must be assessed attending to "the standards and practices of the industry at the relevant time."[421] These relative and the evolutionary characteristics of the concept make necessary the observation and, frequently, the application of new technological advances. The vessel's seaworthiness is a state that is constantly upgraded and improved, thanks to the continuing experience, the advance of knowledge, and the discovery of new techniques of shipping. Therefore, it has been stated that the

[420]Honnold, 98–99. *See* Schoembaum, Admiralty and Maritime Law (1987), 237; 2A Benedict on Admiralty § 11 (7th Ed. 1992); Tetley, *Marine Cargo Claims*, Vol. 1, 398–399, 404–406, 408–410.

[421]*The Eurasian Dream* [2002] 1 Lloyd's Rep. 719.

ship must be kept “up to date”. Yet, that does not mean that the ship has to be perfect or to be a totally “accident free vessel”. There is a middle point between these two extremes that is decisive in the assignation of liability. The same reasoning applies for the “sound system” required in the management of the cargo.

To facilitate the analysis and assignation of liability, the courts held the useful fragmentation of the concept of seaworthiness. Such fragmentation in different essential components of the ship was not new. As it was seen in the previous part, it existed already in the laws of Rhodes and the Consulate of the Sea. This division into specific aspects enables the analysis on the observation of technologies not only regarding ships’ design, but particularly in three very important subjects: cargo worthiness, equipment worthiness, and human worthiness. These are the aspects that are mostly impacted by technological change.

Therefore, the H/H-VR case-law shows that, in practicing due diligence and caring the cargo, it must be observed the state of the knowledge and the technologies available at the time it must be practiced. What kind of technologies may be reasonably applied? What knowledge regarding technology must have the master and the crew at their disposal? What is a sound system today? The case law of the jurisdictions under analysis has provided some hints on this regards. But an answer to these questions is still not clear. Having established the general legal framework governing the contracts of carriage of goods by sea, and the interpretation and construction made by the common law courts of the Rules, we try to approach the problem and provide an answer in the next part.

Chapter 4
The New Technologies Applied in Maritime Transportation

Since the last decades of the 20th century, and most notably in the present century, technology has shown signs of unprecedented development. The maritime transport industry has been strongly benefited by the introduction of new electronic devices, radio and satelital communication, the implementation of innovative methods of loading, stowage and unloading of cargo, among others. This has facilitated both, the navigation and the care of the cargo, making it more effective and secure. In the same way, new designs of more sophisticated, larger and efficient ships have allowed the transport of bigger amounts of cargo, in less time, at a lower cost and in a much safer way. Such a reality has evidenced and confirmed that the concept of seaworthiness of the ship is changing and that it evolves according to the times and the improvements that the scientific development it allows to implement. This affects both, the hull of the vessel and her equipment, as well as the knowledge that her captain and crew must have of these new instruments. From the case law providing analysis on technologies that burst onto the scene in the last century, where their implementation, or lack of it, affected the occurrence of damage to or loss of the cargo, three groups of technologies appear, whose application, in the technological era, is mandatory regarding the obligation to exercise due diligence, established in the Hague/Hague-Visby.

4.1 The Technological Era

4.1.1 The Development of Shipping Technology

Technology is the knowledge fortified by appropriate machinery, tools and "know how".[1] It is a consequence of human curiosity, the use of intellect applied to the solution of problems, the fulfilment of human needs and the facilitation of life. Maritime navigation is in itself a consequence of technology. The need for trans-

[1]T. J. Schoenbaum and A. N. Yiannopoulos, *Admiralty and maritime law: Cases and materials*, Contemporary legal education series (Charlottesville, Va: Michie Co. 1984), 14.

V.H. Chacón, *The Due Diligence in Maritime Transportation in the Technological Era*, Springer Series on Naval Architecture, Marine Engineering, Shipbuilding and Shipping 5, DOI 10.1007/978-3-319-66002-8_4

portation through rivers, lakes and sea drove people since ancient times to think, taste, discover, apply, develop and constantly improve different modes of aquatic navigation. Since the last two centuries, the achievements in naval and shipping technologies have substantially improved the industry. As a consequence of the industrial revolution, in the early stage of the 19th century, the Hudson River witnessed the navigation of the *Clermont*, the first steamship successfully engaged in commercial operations.[2] The steamship technology was initially developed in the United States for navigation in rivers and inland waters.[3] Later, the British improved the technology and applied it to ocean going vessels, starting in 1836 with the first effective steamship service connecting Europe with North America.[4] In 1845, another "landmark of the whole history of shipping" took place, with the introduction of the *Great Britain*, "the first iron-hulled, double-bottomed, screw propelled and steam power Atlantic liner".[5] During the second part of the 19th century, there occurred a second industrial revolution and the world faced an expansion in the intercontinental exchange of goods. One of the technological achievements of this period was the adaptation of refrigeration systems to vessels. This increased the volume of trade of perishable goods on a much larger scale.

Another outstanding and more significant development in shipping technology occurred after the Second World War. Schoenbaum and Yianopoulos listed six different categories in which technology impacted the marine and shipping issues. Four of them relate directly to the shipping industry:

- First, the advent of larger ships called "superships" such as tankers with much larger cargo capacity;
- Second, the introduction of new types of vessel such as the big container ships and new means of cargo handling in ports;
- Third, the use of new devices that allow the determination of the precise location of a ship and other structures on the oceans; and,
- Fourth, the use of sonar and satellite technology which enabled ships, through precise maps and nautical charts, to know the exact positions and extension of geological structures under the sea, the direction of ocean currents, as well as having permanent communication between the vessels on the sea and their offices at the shore.[6]

---

[2]E. Gold, *Maritime transport: The evolution of international marine policy and shipping law* (Lexington, Mass: Lexington Books 1981), 90. It was in 1807. *See* McDowell, C.E. and Gibbs, H. M. *Ocean Transportation*, New York: McGraw-Hill 1954, 21.

[3]Ibid., 90–91, *citing* McDowell, p. 28, and David Howarth, *Sovereign of the Seas* (New York: Atheneum 1974), 319ff.

[4]Ibid.

[5]Ibid., 91.

[6]Schoenbaum and Yiannopoulos, 14–15. These authors also pointed out, as other areas of where technology has an impact, the deep-diving submarines and other vessels that enable to explore the ocean depths and to drill and to locate very large platforms in deep water which facilitates the extraction of resources and the discovery of new ocean uses.

Larger and more sophisticated steel-built engine vessels with higher speed were now made safer and faster than the previous wooden made variety.[7] In the same way, the introduction of radars and similar devices, together with modern charts, reduced the risks of groundings and collisions with other vessels or icebergs.[8] The improvements in computerized radios and satellite enable carriers to be in permanent contact with their vessels and agents.[9] In addition to these aspects, the invention of new devices and systems and the discovery of safer methods of stowage enabled carriers to provide a better care and preservation of cargo. This has been especially important for the carriage of perishable cargo, which commonly requires refrigeration and special conditions of humidity, ventilation, etc.

This technological revolution was accompanied by the functional and financial development of the industry. It paved the way for corporate structuring of large scale and interconnected units of finance and production, initially in the western world's economy.[10] In the last decades, this model has been followed in a more active way by the Asian and Latin-American economies. The times when the common carriers were only one person with one vessel and there were a series of separated intermediaries who participated in transport chain gave way, in the 20th century, to large organizations owning or chartering entire fleets of vessels and pursuing a more integrated transport.[11]

4.1.2 The Development of Containerization

Another very important aspect of the changes occurring after the Second World War was the scale effects of increased trade volumes and the introduction of containerization.[12] The latter reduced significantly the ocean shipping costs and the time of loading and unloading at ports.[13] This, in addition to the use of larger and faster ships, has improved the quality and the transportation time of ocean liners over the past 30 years.[14] By 2011, the total amount of container traded summed 151

---

[7]H. Karan, *The carrier's liability under international maritime conventions: The Hague, Hague-Visby, and Hamburg rules* (Lewiston, NY [u.a.]: E. Mellen Press 2004), 89.

[8]Ibid., 89.

[9]Ibid., 89.

[10]L. Gorton, *The concept of the common carrier in Anglo-American law*, Scandinavian university books (Gothenburg: Läromedelafërl. (Akad.-förl.) 1971), Vol. 43, 22.

[11]Ibid., 22.

[12]D. Hummels, 'Transportation Cost and International Trade in the Second Era of Globalization' (Summer 2007) 21, The Journal of Economic Perspectives, 131–154, 140.

[13]Ibid., 150: "...evidence of this effect in aggregate data was originally overshadowed by the dramatic increases in inputs costs in the 1970s... it was only when crude oil prices began to drop in the mid-1980s that ocean shipping costs really began to fall".

[14]Ibid., 150.

million TEUs, with some 1.4 billion tons of dry cargo.[15] At the beginning of the year 2012, the world fleet reached a total tonnage of 1534 million dwt., with some 104,305 seagoing commercial ships in service.[16] Container ships represented the 12.9 percent of the world fleet, but their impact on world trade is by far higher, as they make up 53% of the seaborne trade in dollar terms.[17] On average, each dwt of container ships carries 27 times more seaborne trade than a dwt of dry bulk carriers.[18] The tendency in building more technologically advanced and larger container ships continues.[19] Most of the new delivered container vessels were gearless, and by 2011, they were 34% larger than those delivered the year before.[20] This segment of ships also leads the age distribution of the merchant fleet, with an average age per dwt of below 9 years, with almost 64 percent of the fleet younger than 10 years old.[21] For refrigerated cargo, container ships account for approximately 60% of reefer cargo, and new deliveries include larger reefer capacities.[22] It is estimated that by 2014, 74% of the perishable reefer goods will be transported by container ships, which provide 95% of the entire reefer market cargo capacity.[23] The container lines report the increasing introduction of new regular services for the transportation of perishable goods between the largest centers of production to the largest markets of Europe and North America.[24]

[15]UNCTAD, '*Review of maritime transport 2012*', 19.

[16]Ibid., 34.

[17]*World Shipping Council*. http://www.worldshipping.org/ Accessed 04.11.213; UNCTAD, '*Review of maritime transport 2012*', 35, 40. The largest tonnage of container ships, about 37 per cent, belong to German shipowners, followed by Japan (8.8 dwt) and Greece (6.8 dwt). Consequently, Germany has the largest share of global seaborne trade carried. Though Germany owns the largest container ship fleet, these vessels are mostly chartered to big cargo liners of other countries. In fact, the main cargo liners, at the 2012, are the Danish Maersk Lines, the Swiss MSC and the French CMA CGM, which together operate 30 per cent of the global container cargo capacity.

[18]UNCTAD, '*Review of maritime transport 2012*', 35.

[19]Ibid., 89. By 2012, Maersk Lines ordered 20 vessels with capacity for more than 18,000 TEU, known now as the super-post-Panamax, ultra-large container ships (ULCSs). These new dimensions demand also port adaptions, for example, shore-side gantry cranes, which must be larger (72 meters) and higher (52 meters).

[20]Ibid., 35.

[21]Ibid., 39.

[22]Ibid., 36.

[23]Ibid., 58. *See* Drewry (2011), http://www.baridmaritime.com/index.php?option=com_content&view=article_&id=1116:charter-rates-thaw-amid-shifting-reefer-market&catid=66:container&Itemid=57. Accessed 20.02.2011.

[24]Ibid., 58.

4.1.3 *New Risks in Maritime Navigation*

During the last decades, apart from the traditional risks that threaten maritime navigation, the international carriage of goods has been particularly threatened by some new phenomenon or new forms of old threats. The climate change and piracy have demanded special actions on the part of governments, but more directly from carriers. The climate change has had an impact in coastal states, ports and navigation in general. Directly in port, the fluctuations in water level are responsible for port inundation or submersion, increased runoff and siltation that require continual dredging.[25] The deterioration of the physical conditions of port structures such as the docks or the integrity of the pavements hinders port access and the operation of cargo loading/unloading. It affects, among others, the cargo capacity, sailing and loading schedules, storage and warehousing, etc.[26] On ship, voyages are threatened by unusual rains, storms, changes in marine currents and sea levels, etc. The melding of the polar ice caps may also cause changes in routes of navigation which sooner or later should be reflected in the ocean charts. These new challenges, being noticed and reported with certain anticipation, may not always be reputed as force majeure or act of God. As the carrier must foresee possible risks, the vessel must be prepared to face them, and actions must be taken to prevent such events to cause damage of or loss of the cargo.

Another threat that has reappeared more recently and with more force is piracy. Over the last 10 years, piracy has become an increasing threat in the West Indian Ocean, the Gulf of Aden and in the Gulf of Guinea. It has caused a rise in the insurance premiums of vessel sailing in that area.[27] Carriers must incur in additional costs for rerouting, security equipment and fuel costs related to navigation in high speed to avoid the harmful effects of piracy.[28] These costs for the shipping industry have been estimated between $3.4 billion and $8.7 billion in 2010.[29]

These new threats demand a higher level of care from the carriers. They must now pay special attention to them and take actions regarding both the ship's seaworthiness, as well as the care of the cargo.

---

[25]Ibid., 23. *See* UNCTAD (2011), Ad Hoc Expert Meeting on Climate Change Impacts and Adaptation: A Challenge for Global Ports. Information Note. 29 September. UNCTAD/DTL/TLB/2011/2.

[26]Ibid., 23.

[27]Ibid., 64.

[28]Ibid., 64. "Up to 2011, no ships has been successfully hijacket that operated at 18 knots or higher" *See* One Earth Future Foundation (2011). The Economic Costs of Somali Piracy. http://.www.oceansbeyondpiracy.org/sites/default/file/economic_cost_of_piracy_2011.pdf. Accessed 12.06.2012.

[29]Ibid., 64.

4.2 The Application of New Technologies in the Carriage of Goods by Sea

4.2.1 Introduction

Given the importance of the international carriage of goods by sea, the new challenges it faces and the failure in the attempts to modernize the governing legislation, technology appears as a valuable tool to prevent, avoid or at least to reduce cargo damage and losses. We have mentioned some of the shortcomings of the H/H-VR regimes. As Tetley points out, a proper code or a well drafted statute "should be able to encompass and regulate new advances of science and civilization unknown at the time the law was adopted."[30] This Convention does not fulfill this requisite. It is, among other reasons, because the technological development that have taken place after the 1920s when these rules were discussed and redacted has been so extraordinary, that its redactors could not have foreseen them. One example is the development of radio communications. The exception of liability for error in navigation or negligent seamanship was based on the impossibility of shipowners to have control of the vessel once it had departed from port. Now, radio and satellite communication enable them to be in permanent contact with the master and crew while at sea or in foreign ports. Certainly, the carriers are not yet capable of total control of the master and crew's activities, but even partial control has been enough to raise discussions on the validity of this exception of liability.[31] This is evident in the abolition of this immunity in the Hamburg and Rotterdam rules. In addition to radio communication, there are a vast number of new technologies that facilitate carriers to have more control on their ships. Undoubtedly, there will be many more to come, which obliges the law to be broad and flexible in order to adapt to these new realities created by technological development.[32]

In the absence of an updated legislation addressing this issue, the judicial enforcement of new technologies, as part of the obligation of practicing due diligence in making the ship seaworthy and caring for the cargo, may provide some answers. To find out what technologies should be required as part of the seaworthiness condition of the vessel, we have to look at the constructions made by the courts on the subject. New technologies have been subject to the court's analysis in maritime claims. They have analyzed, adapted and applied the wording of the HR to the new situations created by technology. Courts have raised the standard of care by adapting new methods to combat fire, installation of modern scientific navigation equipment on board the vessel, such as radio communications, radars, etc., as part of the obligation of practicing due diligence in making the ship seaworthy.[33] As other examples, courts have also adapted these rules to the use of light aboard

[30]W. Tetley, 'Interpretation and Construction of the Hague, Hague/Visby and Hamburg Rules' (2004) 10, The Journal of International Maritime Law, 30–70, 63–64.

[31]Karan, 89–90.

[32]Ibid., 90.

[33]Tetley, *Interpretation and Construction...*, 63–64.

ships, known also as LASH Barges[34]; or the determination of what a package is in a "Ro-Ro" carriage[35]; or the carriage of containers on decks.[36]

Therefore, to answer the question of what technologies can be reasonably required to apply as part of the carrier's statutory duties under the studied regime, we must primarily look at the case-law on the topic. The obligation of practicing due diligence, in the way it has been construed by the courts, offers some flexibility when requiring the application of some new technologies in order to reduce some common risks of the maritime adventure. It is particularly possible when considering the relative and, especially, the evolving character of the seaworthiness condition.

The problem lies in the exact determination of which technologies might be fairly and reasonably required as part of the obligation. There is not a formula or an exact criterion that allows judges to declare that the application of a specific technology can be reasonably expected on the part of the carriers for a specific case. It is clear that not every new technology can be required. It would be unfair to demand that carriers apply technological devices that are not well known or that may unreasonably increase their operating costs and affect their commercial competitiveness. The general criterion stated by the courts and exposed in the previous part, is that the vessel must be "up to date" in its seaworthiness condition, but that does not mean perfection or that it has to be an "accident –free" vessel. Yet this range is ambiguous. Determining which technologies fall within that range is the objective of this part.

4.2.2 The Evolving Nature of the Seaworthiness Condition

4.2.2.1 The Condition of the Vessel and its Equipment

In the previous part, we briefly pointed out the evolutionary character of the seaworthiness condition. Since the advances reported in means of maritime transport after the first industrial revolution, judges pointed out that the seaworthiness condition was an evolving one. The first judicial analyses of this characteristic are

[34]See *Wirth Ltd. V. S/S Acadia Forest*, 537 F.2d 1272, 1976 A.M.C. 2178 (C.A.LA. 1976), at 1276: "Our principal task in this case is to determine what Congress would have thought about a subject about which it never thought or could have thought and one about which we have never thought nor any other Court has thought. Technology has created a maritime transportation system unlike any which was in existence in 1936 when Congress enacted COGSA."

[35]*See Commonwealth Petrochemicals, Inc. V. S/S Puerto Rico*, 455 F. Supp. 310, 313, 1978 A.M. C. 2153 (D.C. Md. 1978), *rev'd on other grounds,* 607 F2d. 322, (4th Cir.1979).

[36]*See Du Pont de Nemours Inter, S.A. v. S.S. Mormacvega,* 493 F.2d 97, 102 1974 A.M.C. 67, (C. N.N.Y. 1974); *Electro-Tec Corp v. S/S DART Atlantica*, 598 F. Supp. 929 (D. Md. 1984), at 934: "Finally, I note that this Court, like the Second Circuit in Mormacvega, *supra*, is willing to consider the effect on COGSA and maritime law of technological innovation and changing vessel design."; Tetley, *Interpretation and Construction…*, 63–64.

found within the scope of marine insurance, but such considerations are similarly valid for the carriage of goods by sea. Perhaps the oldest case pointing to this aspect was *Tidmarsh v. Washington Fire & Marine Ins. Co.* (1827).[37] Judge Story recognized three factors that gradually raised the standard of seaworthiness within the last 30 years prior to his decision:

1. The perfection of knowledge of shipbuilding;
2. More experience of maritime risks; and,
3. Higher degrees of skills in navigation.[38]

This increase of the standard, continues Story, required, for example, that some equipment and the practice of carrying additional spare sails on the vessel were now deemed as essential, although previously not required.[39] By the second part of the 19th century, in *The Titania* (1883),[40] Judge Brown extensively quotes the judgment made by Judge Story 55 years before in *Tidmarsh v. Washington,* but adding that:

> The question of seaworthiness, therefore, as regards the implied warranty in favor of the insurer or of the shipper of goods, is to be determined with reference to the customs and usages of the port or country from which the vessel sails, the existing state of knowledge and experience, and the judgment of prudent and competent persons versed in such matters.[41]

Thus, since the 19th century, and long before the enactment of the Hague Rules, American Jurisprudence had recognized and established that the seaworthiness condition of the vessels had to be adapted to the new developments of knowledge and experience in the field, and subject to the progress of science and technology. Once the Harter Act was into force, the Supreme Court of the United States continued recognizing and assigning liability in cargo claims, attending to the evolving character of the seaworthiness condition. We referred previously to the case *Martin v. Southwark* (1903),[42] a landmark case regarding this issue. Justice Day supported his decision quoting the definitions of seaworthiness provided by Bouvier, where this author states the impossibility to assess the seaworthiness condition according to objective rules, because such a condition is constantly modified by the discovery and application of technical improvements. The exact quotation of Justice Day was:

> And the same author further says: 'it can never be settled by positive rules of law how far this obligation of seaworthiness extends in any particular case, for the reason that improvement and changes in the means and modes of navigation frequently require new

[37]4 Mason 439; 23 F. Cas. 1197. (C.C. of Mass.).

[38]Ibid., 1198. See complete quote *supra* at page 128.

[39]Ibid., 1198. However, for this case, such recognition did not held the ship liable for lack of seaworthiness, as Judge Story held that the condition must be judged according to the practices of the port of the country where the vessel belongs, not of that where the insurance policy was issued.

[40]19 F. 101 (D.C.N.Y. 1883).

[41]Ibid., 107.

[42]191 U.S. 1, 24 S.Ct. 1, 48 L.Ed. 65.

improvements, or new forms of old ones; and these though not necessary at first, become so when there is an established usage that all ships of a certain quality, or those to be sent on certain voyages or used for certain purposes, shall have them.'[43]

Indeed, the fact that making a ship seaworthy involves more factual issues than legal ones limit the enactment and application of positive rules to strictly determine what the extension of this obligation is.[44] This criterion was shared also in the case of T.J. *Hooper* (1931), where Coxe D. J. expressly mentioned that the advances of science and the equipment available, specifically referring to radio communication devices, makes seaworthiness an ever-changing concept:

> The standard of seaworthiness is not, however, dependent on statutory enactment, or condemned to inertia or rigidity, but changes 'with advancing knowledge, experience, and the changed appliances of navigation.' It is particularly affected by new devices of demonstrated worth, which have become recognized as regular equipment by common usage.[45]

English law, for its part, has taken a similar approach as evidenced in *The Lendoudis Evangelos* (2001),[46] where it was stated that "Seaworthiness must be judged by the standards and practices of the industry at the relevant time, at least so long as those standards and practices are reasonable."[47] The concept of reasonability on a specific issue is subject to an evolution too.[48] What was considered sufficient some decades ago, may not be currently sufficient. The relativity of this obligation greatly demands the consideration of the time when the duty is to be fulfilled. Here is where the evaluation of the impacts of new technologies is fundamental for an appropriate assessment of the duties set in the H/H-VR. The current access to new technological appliances as well as new and more efficient methods of inspection or stowage which prevent, avoid or reduce risks of damages and losses, makes the standard of seaworthiness progressively higher.[49] It may create liability for the carrier, when the technologies are deemed reasonable and yet not applied, or when they are on board, but are defective or are not properly used, as stated by some courts. Notwithstanding, such increase in the standard does not necessarily affect the condition of old ships.[50] New standards of seaworthiness will

---

[43]Ibid., 8–9.

[44]J. Bouvier, *Bouvier's Law Dictionary and Concise Encyclopedia*, 3 rev. (being the eigth Edition) by Francis Rawle, 2 vols. (Kansas City, Mo.: Vernon Law Book Company 1914), II, 3028.

[45]53 F.2d 107 (D.C.N.Y. 1931), 111; 1931 A.M.C. 1764.

[46][2001] 2 Lloyd's Rep. 304.

[47]Ibid., 306. Quoted also in *The Eurasian Dream* [2002] 1 Lloyd's Rep. 719, 736.

[48]*See Electro-Tec Corp. v. S/S Dart Atlantica* (1984), 598 F. Supp. 929, 934 (D. Md. 1984). The court stated on carriage of containers on deck that "what was unreasonable yesterday, may be reasonable today".

[49]W. Tetley, *Marine Cargo Claims*, 4th edn., 2 Vols. (Cowansville, [Toronto]: Blais; Thomson Carswell 2008), Vol. 1, 930.

[50]Ibid., 930.

affect only new ships or old ships which have been subject to major repairs and reconstruction.[51]

Therefore, seaworthiness, as expressed by the courts, depends on the improvements and changes in the means and modes of transport. New technologies have, without a doubt, a substantial impact on navigation. They enable shipowners to provide more technologically advanced vessels that can assure a more efficient service. Technology enables also ship inspectors to make more accurate surveys and improve the effectiveness in the job they undertake. Having access to the advances of science, carriers can perform with greater accuracy their contractual responsibilities of making the ship seaworthy while properly caring for the cargo, and thus substantially reducing the chances of accidents or the effects of other threats.

An important point, however, is that the existence of such new techniques does not necessarily results always in less breakdowns as the new vessels are equipped with machinery and equipment of a certain complexity that require a very high standard of operation, maintenance and care.[52] Thus, this technical evolution must be accompanied by the technical training of the master, officers and crew members who, in the end, will be in charge of operating the vessel and the technological equipment needed for the maritime adventure.

4.2.2.2 The Technical Knowledge of the Crew

The human worthiness of a ship is also evolutionary in its character. As explained, the master and crew must be duly qualified and trained with the competences to work on the ship. To operate the ship properly, the shipowner must make the master and crew aware of any particular features, uncommon designs, characteristics or special needs. That includes, of course, the technological devices installed on board. The master's lack of knowledge regarding an unusual construction of the ship that may require special attention and that was known by the shipowner makes the ship unseaworthy.[53] Judge Hand discussed this issue in *Compañia General de Tabacos de Filipinas v. United States (The Elkton)* (1931).[54] Certainly, a ship may be

---

[51]Ibid., 930, *citing Peter Paul Inc. v. M/S Christer Salen*, 152 F. Supp. 410, 1957 A.M.C. 2141 (S. D.N.Y. 1957).

[52]Ibid., 930–931.

[53]*See Standard Oil Company of New York v. Clan Lines Steamers, Ltd.* [1924] A.C. 100, 113, (1923) 17 Ll. L. Rep. 120, 122. The ship had an unusual construction that made her unstable in certain loading conditions when the ballast tanks were pumped out. The owner knew about it, but did not inform the master and a casualty occurred while the ballast tanks were pumped out, causing a loss of cargo. The case was decided under the Harter Act, which was incorporated as applicable law into the Charter Party.

[54]*The Elkton* 49 F.2d 700, 1931 A.M.C. 1040 (C.A. 2 1931). The vessel in that case did not need special attention on a technological issue, but on a particular condition that required special care on the part of the crew. A broken vent pipe let fuel to leak from the top tank into holds and contaminated a cargo of sugar.

seaworthy though not in all respects fit for the service when she leaves the port.[55] There are some issues that can be duly attended to during the voyage. But the carrier is responsible to make such issues known to the master and crew, so they can pay proper attention to them. If damages or losses result from a condition that is unattended to by the crew or due to their "ill handling of adequate equipment" due to lack of knowledge, the carrier is liable for "not carrying a crew properly advised as to her needs".[56] The failure to inform the master and crew of those special needs is held as want of due diligence. The carrier could be exonerated if the master and crew know that special attention is required on a particular issue and they do nothing, or proceed negligently when some actions are needed.[57] If it is the case, this situation falls within the exception of negligent seamanship or errors in navigation.

A carrier accepting cargo on an unseaworthy ship, whose condition may be rendered inoffensive through special care of the master and the crew, adds an additional risk on the shipper if the servants do not provide such special precautions during the voyage. The statute, in this case the Harter Act, does not impose such a risk on the shippers and the carrier must do "his best to remove all risks".[58] A shipowner cannot rely on the exoneration of liability for errors in navigation, when he offers a vessel that requires a higher level of care than usual in order to be seaworthy.

This reasoning clearly applies for technological devices on the ships as well. This aspect takes on more relevance, when the safety of the ship and the people on board depend on these new advances. The shipowner must inform the master and crew of all technologies and any special attention they require, confirming that the personnel he hires to operate his vessel has the technical knowledge to manage such equipment. This was evident with the introduction of radar on merchant ships. *The Lady Gwendolin* (1965)[59] particularly addressed this issue. A vessel properly equipped with radar had a collision because her master had no experience with it and was given no instructions on how to use it.[60] The court remarked that the main concern of shipowners was the safety of life at sea. For that purpose, it was recalled that within the obligation of making the ship seaworthy, properly manning the ship is included. This required the presence of capable master and officers duly instructed on how to manage the equipment. The shipowner must even alert the

---

[55]Ibid., 701. *citing*: *The Silvia*, 171 U.S. 462, 19 S.Ct. 7, 43 L.Ed. 241; *International Navigation Co. v. Farr*, 181 U.S. 218, 21 S.Ct. 591, 45 L.Ed. 830; *The Steel Navigator*, 23 F.(2d) 590 (C.C.A. 2); *The President Polk*, 43 F.(2d) 695 (C.C.A. 2); *The Manitoba* 104 F. 145(D.C.N.Y. 1900).

[56]Ibid., 701, *citing*: *The Schwan*, L.R. (1909) App. Cas. 450; *See Standard Oil Company of New York v. Clan Lines Steamers, Ltd., see supra* note 53.

[57]Ibid., at 701 Judge Hand provides some examples: "For instance, if the rudder be bent, the quartermaster may be instructed to make proper allowance in his helm; if the boilers be weak, the engineer, to keep a low head of steam."

[58]Ibid., 701.

[59][1965] 1 Lloyd's Rep. 335.

[60]Ibid.

master that such equipment does not allow them to sail at high speed.[61] Judges were especially concerned about the false sense of total reliance given over to radars that created a situation where master and crew did not pay sufficient attention when navigating their vessel. In some cases, it even encouraged masters to proceed in high speed during fog, as was the case here. When this collision occurred in 1961, there were already many cases covering this issue. The false sense of safety that radar gave the master was common knowledge in the shipowners' community, and thus could be attributed as a standard for the owner of the faulty vessel.[62]

As a prudent shipowner, knowing those facts, he had the obligation to make his master aware of this issue. Technology offers advantages for safe navigation and the care of the cargo, but it must always be remembered that safe navigation depends on the master, officers and crew on board the ship. It cannot be supposed that all these valuable inventions will function by themselves with no errors.[63] The competence and capacity of the master and crew when operating technological devices must be ensured by the shipowners.

4.2.3 The Classification of Technologies Used in Maritime Transportation

The shipping industry has constantly introduced new technologies in its operations. Technological appliances are implemented as a fundamental condition for the carriage of certain cargoes; as a way of precautionary measures for protection against common risks of navigation, or to improve the care of the cargo. Through continued use, some technologies have become a common practice in the industry. Still others have reached a level of importance that has become mandatory through national statutes, international regulation or standards set by the classification societies. In some other cases, carriers offer special services based on technological discoveries that make possible the carriage of particular cargoes or provide better care of traditional ones. In these cases, the implementation of these technologies is contractually agreed upon with shippers, who normally assume an additional payment for the same. Another group of technologies do not reach a mandatory enforcement nor are they contractually agreed to, but shipowners continue applying

[61] Ibid., 335–336. At trial it was evidenced that his master was used to navigate at excessive speed in fog and that the shipowner knew about this "addiction to speed", which they should have prevented. The court said that there is an obligation to provide particular vigilance to the radar from the part of shipowners.

[62] Ibid., 339. Even the court pointed out that such a problem was of common knowledge since before 1953 when the vessel was fitted with radar and made her maiden voyage.

[63] *See The Fogo* [1967] 2 Lloyds Rep. 208, at 221 Judge Cairns of the Admiralty Division said that automatic steering was a "most valuable invention if properly used. It can lead to disaster when it is left to look after itself while vigilance is relaxed. It is on men that safety at sea depends and they cannot make a greater mistake that to suppose that machines can do all their work for them."

them as common practice of the industry. Carriers implement the latter unilaterally to make more efficient the service they offer. Thus, it is possible to approach our analysis through a classification of these technologies according to the source from where they are applied. First, we distinguish technologies applied unilaterally by carriers. Second, technologies offered by the shipowners to shipper for the special care of the cargo and are contractually agreed upon. And third, the technologies required by statutes, regulation or standards of classification societies. We will address them in this same order.

4.2.3.1 Technologies Applied Unilaterally by Carriers

As mentioned, the shipping industry has beencontinuously applying new scientific advances and new technological devices to improve its operations. New vessel design, more advanced equipment, and new methods of inspections and stowage are constantly tested and applied. They seek to make processes safer, faster, while reducing costs. Once a new technology is repeatedly used over a period of time, it becomes a common practice or a standard of the industry. Sometimes, it is included in regulations and becomes statutorily mandatory, as we will see later. But, in other cases, the use of a technology remains as a common practice of the industry and is never included in statutes, or only much later. When a new technology becomes a custom within the industry, its application is expected. These are then equally required as part of the seaworthiness condition of the ship.

This is exactly the group of technologies that offers some problems for the determination of liability for damages or losses caused by its non-application or its defectiveness. The problem stems from trying to determine when the use of a technology has become a common practice. On common practices in navigation, Judge Hand in *Anglo-Saxon Petroleum Co. v. United States* (1955)[64] stated that: "…in order to give any custom the force of a law there must be evidence that it is a definite, uniform and known practice."[65] In addition, he pointed that even if it is a generally followed practice, it must be also considered how convenient it is or "how far it would be of much value".[66]

Certainly, many of the technologies that in the past were held as a common practice are now part of one of national or international regulations. This is the case for example with the compass, radio communications, radars, routing systems, etc. But there are always new technologies coming into the market that offer more efficiency and eliminate or substantially reduce the possibility of occurrence of some risks. When can these new technologies be required to be applied as part of

---

[64]222 F.2d 75, 1955 A.M.C. 985 (C.A.2 1955).

[65]Ibid., 77 at f.n. the Judge cites: *Chicago, M & St. P. Ry v. Lindeman,* 143 F. 946, 949 (8th Cir. 1906); *McClellan v. Pennsylvania R. Co.*, 62 F.2d 61, 63, (2nd Cir. 1932); *Bagwell v. Susman,* 165 F.2d 412 (6th Cir. 1947); *Albert v. R.P. Farnsworth & Co.,* 176 F.2d 198, 201 (5th Cir. 1949).

[66]Ibid.

the H/H-VR duties? The case law relating to these technologies, in some cases when they were not statutorily ordered, offers some analysis on the application of technologies that may shed some light for the current technologies and those that are to come in the future.

4.2.3.1.1 Radio Communications

Since the discovery of the use of electric waves to transmit wireless messages,[67] a whole industry of wireless world-wide communications, direction finding, radars, etc. was developed.[68] One of the first users of the Marconi's technology was Lloyd's, which installed his system in all of its signal stations on the coast to receive and transmit wireless communications since 1898.[69] The ship *Kaiser Wilhelm der Grosse,* owned by the German Company Norddeutscher Lloyd, was the first merchant vessel equipped with a Marconi's wireless telegraphy system around 1900.[70] By 1903, some liners equipped 19 of their vessels with same system.[71] Ten years later, by June 1913 there were 686 transatlantic ships using Marconi's wireless telegraph stations, and some additional ones using other routes.[72] During the First World War more improvements were achieved, such as the increase of the range of coverage and wireless telephonic communication between ships.[73] The English Merchant Shipping Act (Wireless Telegraphy)of 1919 ordered that every ship carrying 200 passengers or more hadto carry three Radio Officers for all voyages exceeding 48 hours from port to port.[74] The use of radio wireless proved to be so convenient and became so extended that in 1923 legislation was introduced in England making it compulsory for foreign-going vessels with more than ten lifeboats to equip at least one with wireless telegraphy.[75]

---

[67]H. E. Hancock, Wireless at Sea: The First Fifty Years (London: Percy Lund, Humphries & Company, Limited 1950), p. 1. In 1895, Guglielmo Marconi discovered the possibility to transmit and direct electric waves over distances. Notwithstanding, though other scientists predicted, proved the existence of and experimented with electro-magnetic waves previous to Marconi, he was who develop its first practical application of the theories on the matter.

[68]Ibid., xi. The first wireless message in history was sent across the water in May 1897. Ibid. p. 10.

[69]Ibid., 15.

[70]Ibid., 28. In November 1900 the Belgian Royal Mail Steam Packet, *Princesse Clementine* was equipped so; and in 1901, the *Lake Champlain* became the first British merchant vessel also equipped with same system.

[71]Ibid., 43 For example Hamburg-Amerika Line, Norddeuscher Lloyd, Cunard Company, Red Star Line, Compagnie Générale Transatlantique.

[72]Ibid., 78.

[73]Ibid., 109.

[74]Ibid., 127. If the voyage was longer than 8 hoursbut less than 48 hours, the act still required the presence of two radio officers. It was perhaps in respond to the *Titanic* accident. Another vessel was only twenty miles away but her radio officer was off duty at that moment and did not get the *Titanic's* message.

[75]Ibid., 131.

The issue of the 30th of October of 1941 of the shipping journal *Fairplay* referred to the long-distance wireless as "common" in the industry.[76]

(1) **The Case Law**

Though the use of radio communications on board ships rapidly became an extensive practice and a mandatory obligation for some types of vessel, its lack of implementation on vessels not covered by such legislation was also argued as want of due diligence.

An American landmark case on the implementation of new technologies and seaworthiness related concretely to the use of radio communications. The *T.J. Hopper* (1931)[77] related the loss of two barges loaded with a cargo of coal that sank while being towed by the tugs *T.J. Hopper* and the *Montrose*, during a sudden storm in the coast of New Jersey in March 1928. Four other tugs navigating in the same area at the same time decided during the afternoon or in the evening of the previous day of the storm, to go for refuge to the nearest breakwater. The master of three of them made that decision because these tugs were equipped with radio apparatus on board. That enabled them to receive weather reports and be up to date on the bad weather conditions to come. Instead, the two tugs subject to this case kept going on its way to its destination. None of them were provided with similar equipment to receive weather forecasts. The captains of both tugs had personal radios, but both of them were inoperative. The judge said that, if while tugs are not insurers of the cargo; their operators were under an obligation to use reasonable care and skill in the service they were hired for. That obliges them to pay attention to the barometer readings and the weather indications.[78]

District Judge Coxe recognized that there was not statutory regulation requiring the installation of radio on tugs. However, he pointed that seaworthiness was not subject to rigid standards or statutory laws, and is affected by the advances of science and by new devices of demonstrated worth that have been recognized as common usage.[79] It was consigned in the trial through testimonials, that the use of radio was a well extended practice, including the point that 90% of the coast tugs operating in that coast were so equipped.[80] Therefore, he concluded "…the use of the radio was shown to be so extensive as to amount almost to a universal practice

[76]Ibid., 15. The *Fairplay* journal further added: "The regular introduction of wireless had been of great advantage to underwriters, because a vessel could now indicate her position, and, if in distress, assistance could be sent to her. In pre-wireless days a vessel braking down at sea might be at the mercy of the elements for a considerable period before by chance she was sighted and there was no doubt that many vessels were lost which would have been saved if they could have indicated their position."

[77]53 F.2d 107, 1931 A.M.C. 1764 (D.C.N.Y. 1931), *aff'd* 60 F.2d 737 (2nd Cir. 1932), cert. den'd; *Eastern Transportation Company v. Northern Barge Corporation*, 287 U.S. 662, 53 S.Ct. 220, 77 L.Ed. 571.

[78]Ibid., 109. *See The Hercules* 73 F. 225(C.C.A.); *Maryland Transp. Co. v. Dempsey* 279 F. 94(C. C.A.); *The Victoria* 79 F. 122(D.C.).

[79]Ibid., 111, *citing TheTitania* and *TheSouthwark*, *see* supranotes 40 and 42 respectively.

[80]Ibid., at 111: "Captain Powell, master of the Menominee, who was a witness for the tugs, testified that prior to March, 1928, his tug, and all other seagoing tugs of his company, were

in the navigation of coastwise tugs along the coast"[81] If the tugs *T.J. Hopper* and the *Montrose* had been equipped with appropriate and duly operative radio communication devices on board, it was clear, as acknowledged by the same masters of both tugs, they would have gone to the breakwater as well as the other four tugs had done, and the loss would have been avoided.[82] The tugs were held then unseaworthy for lack of radio communications on board.

This decision was subject to further analysis and affirmed in a second instance.[83] Judge Hand added additional arguments supporting the use of new technological devices as part of the seaworthiness condition of the vessel. Contrary to the statement made by the judge in the first instance, he deemed it not fair to affirm that there was a generalized practice of carriers to equip their tug with radio receiver sets. The radio sets, he observed, were brought on board mostly by the masters or crew, not by their owners, nor were they supervised by them. However, he noted that "an adequate receiving set suitable for a coastwise tug can now be got at small cost and is reasonably reliable if kept up; obviously it is a source of great protection to their tows."[84] Considering the difficulties met in the operations performed by tug towing of heavy loaded barges, he held radio sets as a valuable tool that provides protection against some risks that cannot be known in any other way.[85] Therefore, creating distance from the common usage criterion as the sole argument for assessing the seaworthiness condition of a ship, he bluntly explained the issue as follows:

> There are, no doubt, cases where courts seem to make the general practice of the calling the standard of proper diligence; we have indeed given some currency to the notion ourselves. Indeed in most cases reasonable prudence is in fact common prudence; but strictly it is never its measure; a whole calling may have unduly lagged in the adoption of new and available devices. It never may set its own tests, however persuasive be its usages. Courts must in the end say what is required; there are precautions so imperative that even their universal disregard will not excuse their omission.[86]

(Footnote 80 continued)

equipped by the owner with efficient radio sets, and that he regarded a radio as part 'of the necessary equipment' of every reasonably well-equipped tug in the coastwise service."

[81]Ibid., 111.

[82]Ibid., 111.

[83]*The T.J. Hooper*, 60 F.2d 737, 740(2nd Cir. 1932), 1932 A.M.C. 1169, cert. den'd; *Eastern Transportation Company v. Northern Barge Corporation*, 287 U.S. 662, 53 S.Ct. 220, 77 L.Ed. 571.

[84]Ibid., 739.

[85]Ibid., at 740. He further said: "Such a set is the ears of the tug to catch the spoken word, just as the master's binoculars are her eyes to see a storm signal ashore. Whatever may be said as to other vessels, tugs towing heavy coal laden barges, strung out for half a mile, have little power to maneuver, and do not, as this case proves, expose themselves to weather which would not turn back stauncher craft. They can have at hand protection against dangers of which they can learn in no other way."

[86]Ibid., 740.

Common usage of the industry was the customary criterion to assess seaworthiness. Imperative precautions come as another criterion in demanding the use of new technological appliances. Although there was not a generalized custom to equip ships with receiving sets, it was still necessary to have information of common risk in navigation such as severe weather. Radio was the only means of access to that timely information and preventing the possible disasters it could cause. Another element taken into account was the cost of radios devices. Radios were available for insignificant costs. As evidenced in the trial, many of them were brought into tugs by the master. It did not require a large investment to have them on board. Hence, in addition to its imperative need, its cost efficiency should have also been considered.

However, the criterion exposed in the aforementioned case was not consistently maintained in the American jurisprudence. Twenty years after that case, when the use of radio was much more extensively in use, a District court ruled that the mere absence of a radio on a tug did not make her unseaworthy. It was stated in *Hendry Corporation v. Aircraft Rescue Vessels* (1953).[87] The Judge cited, in support of his conclusion, another decision issued by the same Judge Hand just three months before the *T.J. Hopper*. In the *Pacific Fir,*[88] the discussion revolved around the number of ventilators a ship should have. Judge Hand stated that a shipper cannot expect the "latest designs" in ships as mandatory for seaworthiness, rather, that old well-built ships in good conditions may also be seaworthy.[89] This citation ignores the fact that a radio is not part of the latest design, but equipment, which may be installed on ships after their construction. This equipment was also not the latest equipment, but in use in navigation for at least four decades. Requiring the use of radio by the middle of the 20th century was not unreasonable, taking into account its low cost, availability, and that in the case of tugs performing inland or coastal carriage, would not have required very sophisticated radio equipment. Carriers were well aware of the usefulness of radios. Even a shipowner had previously argued to have on board Radio Aids to Navigation as evidence of having practiced due diligence in providing "ample and correct navigation data".[90]

[87]113 F. Supp. 198, 201 (D.C.La.1953). The court, however, did not make major analysis on the use of radio because there was not causal link between the losses and the lack of this device. District Judge Wright observed first, that there were not storm warnings received by other tugs the day of the accident. Second, the vessel was constructed far before the use of radio became in use onboard ships. And, third, that there was no evidence that the failure to have a radio contributed to the stranding.

[88]*See Spang Chalfant & Co. v. Dimon S.S. Corporation* 57 F.2d 965, 1932 A.M.C. 738, (C.A. 2, 1932).

[89]Ibid., 967: "The Pacific Fir was one of many fabricated vessels built during the war at Hog Island after a single design. At that time and until not long before the voyage in question it was customary to build ships with only two ventilators, though latterly these have been doubled, no doubt an improvement. But two ventilators may be enough; shippers are not entitled to the latest design; ships, well built in their time, may still carry cargo unless they become so clearly out of fashion as to be an anachronism. The shipper bargains for no more than usual carriage."

[90]*See The Maria* 91 F.2d 819, 1937 A.M.C. 934, (C.A. 4 1937). It was a case of a ship's stranding occurred in 1932. The shipowner alleged bad seamanship as the reason for the stranding.

With Judge Hand's decision, the courts had the chance to provide some references or some minimum standards of due diligence regarding the application of technologies. That could have also fulfilled some of the gaps left by the HR or COGSA regarding the use of newer appliances and partially compensated the imbalances of the H/H-VR. Instead, they abdicated such opportunity. In later decisions, the courts sent back to shipowners the power to rule or set standards of seaworthiness.[91] If seaworthiness is completely left to what the industry regularly does, or what they commonly and regularly use or apply, that may restrain the progress of the condition of seaworthiness and discourage the implementation of technological advances that serve as imperative precautions.

4.2.3.1.2 The Compass

The compass is perhaps one of the oldest and most useful technological instruments applied in navigation. Though the compass was known in the West after the 9th century, it was not yet universally available and not always accurate.[92] It is said to have been discovered and used in maritime navigation in England after the 12th century.[93] At the beginning of the 19th century, the compasses' condition was still not optimal, because, with the slightly magnetized needle, accuracy became harder to achieve through the increasing use of iron on ships.[94] Seamen paid little attention to this issue and regarded many of the disasters resulting from this lack of accuracy as acts of God.[95]

(1) **The Case Law**

By the middle of the 19th century, the use of the compass was already considered part of the seaworthiness condition of the vessel. Once of the first American cases on the subject was the *Bazin v. The Liverpool & Philadelphia Steam-Ship Co.* (1857).[96] Here a cargo of perfumes shipped out from The Havre, France to be delivered in Philadelphia was partially lost and the rest severely damaged due to the shipwreck of the carrying vessel. It was a brand new iron vessel, "well built, well rigged, well manned and in charge of a master of reputed skill and experience".[97] During the ship's first voyage it wrecked in a fog, but in calm weather and in a very well-known cape

---

[91] *See Nuzzo v. Rederi*, 304 F.2d 506, 510 (2nd Cir. 1962), where the Judge Hand's opinion was widely criticized.

[92] G. J. Mangone, *United States admiralty law* (The Hague, Boston: Kluwer Law International, 1997), 12. A rudimentary version was the only instrument available for navigation during a large part of the middle ages.

[93] E. P. Wheeler, *The modern law of carriers: The limitation of the common-law liability of common carriers under the law merchant, statutes and special contracts.* (New York: Baker, Voorhis & Co. 1890), 5.

[94] A. E. Fanning, *Steady as she goes: A history of the Compass Department of the Admiralty* (London: H.M.S.O; National Maritime Museum [distributor] 1986), xi.

[95] Ibid., xi.

[96] 3 Wall. Jr. 229, 5 Am. Law. Reg. 459.

[97] Ibid., 459.

area. The judge concluded that in such a situation the cause of the wreck was a defective compass. The assumption was made that an iron vessel may have caused the needle to deflect, though there was no evidence of that. The fact was that the vessel was sailing between 30 or 40 miles out from the proper route. The bill of lading included a clause releasing the carrier from "accidents from machinery, boilers, steam, or any other accidents of the seas, rivers, and steam navigation, of whatever nature or kind so ever". This clause was not recognized because the wreck was not the result of an accident of the sea, but a consequence of a defective compass. The defect in this device made the vessels unseaworthy. Grier, C.J., expressly stated:

> If the compass on the new iron vessel was not sufficiently protected to traverse correctly, the vessel was as little seaworthy as if she had no compass and this should have been carefully ascertained before she started on her voyage. If there was no fault in the compass, then it is very evident that the officer who is thirty or forty miles wrong in his calculation, and driving through a thick fog with a full head of steam, and first discovers his true position by running on an island, a cape, or a continent, has neither the skill nor the prudence to be entrusted with such a command and for want of such an officer the vessel is not seaworthy.[98]

The vessel's defective compass was not the only reason for unseaworthiness. She was also not well manned. Both situations made the vessel unseaworthy. The judge did not enter into any more consideration on the lack of testing the compass. The sole fact that it was not working properly on the ship, make her immediately unseaworthy.

The same criterion was reaffirmed in *Richelieu & O. Nav. CO. v. Boston Marine Ins. Co.* (1890).[99] It was not a cargo claim, but a claim on an insurance policy for losses resulting from the stranding of the vessel. The judge ruled that regardless whether or not the stranding occurred during a heavy fog that prevented the mate from seeing an island, "the predominating and efficient cause was the negligence or unseaworthiness, and must be regarded as the proximate cause, under the circumstances."[100] The unseaworthiness was because of a defective compass. The court recalled that seaworthiness was a warranty, and whether the unseaworthiness condition was known or unknown it was immaterial in assigning liability to the shipowner.[101] Again, the fact that the vessel ran ashore 17 miles out of her course raised the presumption that the compass was defective. It was so confirmed by the testimonials of the master and crew members, who declared they were aware of such defect.[102] In the decision it was expressly stated:

---

[98]Ibid., 467.

[99]136 U.S. 408, 10 S.Ct. 934, 34 L.Ed. 398.

[100]Ibid., 426 or 939. The Judge cites: *Waters v. Insurance Co.*, 11 Pet. 213; *Insurance Co., v. Transportation Co.*, 12 Wall. 194, 199.

[101]Ibid. 429 or 940. *The court cites: Work v. Leathers*, 97 U. S. 379 (18t8); *The Glenfruin*, 10 Prob. Div. 103; Union *Insurance Co. Philadelphia v. Smith*, 124 U. S. 405, 8 Sup. Ct. Rep. 534 (1888).

[102]Ibid., 428–430 or 940.

> In order to be seaworthy, the steamer must have been supplied with a good and reliable compass or compasses, which must have been kept in proper repair and condition for the safe navigation of all waters described in the policy.[103]

By the middle of the 20th century, the use of the compass on board was already included in some national regulations as mandatory for some types of vessels. However, though it was not mandatory for tugboats, a court considered it necessary for its seaworthiness. In *The Claribel* (1965),[104] the US Court of Appeals of the 5th Circuit held the defendants liable for an accident caused by the lack of a properly calibrated compass on the tug. Since the installation of the compass on the tug back in 1956, it had not been checked nor calibrated, even when its accuracy would have been affected also by two factors credited in the process.[105] In addition, not the captain, nor the pilot had the minimum knowledge on how to correct a defective compass or how to compensate the deviation regarding the factors that had an influence on its accuracy.[106] Such failure in the compass, together with the lack of knowledge of the master and pilot and their inability to correct or compensate the deviation, were considered the proximate causes of the accident.[107] It was questioned if an accurate compass and knowledgeable crew regarding this apparatus were required as part of the tug's seaworthiness. The defendants argued that the Coast Guard regulations did not demand the appointment of licensed navigators as master for inland tug operations.[108] Masters with the understanding on the basics of compass compensation were an exception, not the rule.[109] The court acknowledged that compass repair and calibrations were, indeed, a casual practice.[110] However, the court paid special attention to the standard of care required in the lake where the accident occurred. It was noted that the lake

---

[103]Ibid., 429 or 940.

[104]*Greater New Orleans Expressway Commission v. Tug Claribel*, 222 F. Supp. 521, 524, 1964 A.M.C. 967. (D. C. La. 1963), *aff'd sub nom Coleman v. Jahncke Service Inc.*, 341 F.2d 956, (C.A. La. 1965), 1965 A.M.C. 535. (5th Cir.), at 957 the decision describes the casualty: "Early on the morning of January 27, 1960, in heavy fog, a barge in tow of Jahncke's tug Claribel hit a support piling of the Lake Pontchartrain Causeway, 3.4 miles south of the north draw. Two spans of the causeway collapsed. A Winn-Dixie trailer-truck, which happened to be crossing the causeway at the point of impact, was damaged and lost some of its freight. The Greater New Orleans Expressway Commission, owner and operator of the causeway, filed a libel in rem against the Claribel, and a libel in personam against Jahncke."

[105]Ibid., 958. First, short before the accident, electric arc welding work were done on the vessel, which normally affects the compass in a steel-hulled vessel. Secondly, the size, position and number of barges being towed have also an influence on the compass.

[106]341 F.2d 956, 958, 1965 A.M.C. 535 (5th Cir.).

[107]Ibid., 958. The decision of first instance explained better this issue at 222 F. Supp. 521, 524, 1964 A.M.C. 967 (D. C. La. 1963). The captain and pilot had never been instructed in how to use a compass, nor were licensed by the Coast Guard.

[108]Ibid., 958.

[109]Ibid., 958.

[110]Ibid., 958.

presented some other inconveniences,[111] which demanded, as confirmed by the experts in the process, that any tug navigating there must be equipped with proper compass and qualified crew. It was concluded that the tug was unseaworthy because the area of navigation demanded a higher level of care, which required the application of a compass. The particular characteristics of the lake made this technological device of imperative application, despite it not being in common practice, because "usage is not, and should not be, an inevitable standard".[112]

4.2.3.1.3 Navigation Charts

During the 19th century, the British naval vessels surveyed oceans and coasts of the world and issued navigation charts and sailing directions that were unique or better than any other in existence.[113] Navigation charts became an essential instrument in navigation. To carry them out of date was a cause for unseaworthiness. In *The Maria* (1937),[114] faulty charts caused the stranding of the vessel. A new set of charts reporting changes in the lightship and the buoy were made known by the United States. Official publications reporting such changes were available in the United States, England and Italy, as well as corrected charts, pilot books and daily memoranda being available at the ports this ship was calling on prior to her stranding.[115] The shipowners alleged that the information on board the ship

[111]Ibid., 959: "But the evidence also establishes that visibility is often not good. Lake Pontchartrain is large and shallow, subject to sudden weather changes, treacherous squalls, and frequent fog. Commercial traffic is heavy, and the barges that ply the lake are big and unwieldy."

[112]Ibid., 959. It quoted part of the statement made by Judge Hand in the *T.J. Hooper*, *see supra* note 77, at 740. The Judge also cited: *Texas & Pacific Railway Co. v. Behymer*, 1903, 189 U.S. 468, 470, 23 S.Ct. 622, 47 L.Ed. 905; *Pure Oil Co. v. Snipes*, 293 F.2d 60, 71 (5th Cir. 1961); *Kane v. Branch Motor Express Co.*, 290 F.2d 503, 507 (2nd Cir. 1961); *June T., Inc. v. King*, 290 F.2d 404, 406 (5th Cir. 1961); *Schlichter v. Port Arthur Towing Company*, 288 F.2d 801, 804 (5th Cir. 1961) (dictum); *Santomarco v. United States*, 277 F.2d 255, 257 (2nd Cir. 1960), cert. den'd, *American Stevedores Inc. v. United States*, 364 U.S. 823, 81 S.Ct. 59, 5 L.Ed.2d 52 (1960); *Troupe v. Chicago D. & G. Bay Transit Co.*, 1956, 234 F.2d 253, 260 (2nd Cir.); Prosser, Torts, 3rd ed. p. 170.

[113]Gold, 124. Navigation charts are not a new issue. Since between 1771 and 1775, Captain James Cook charted vast tracts of the then unknown oceanic and coastal areas especially in the South and North Pacific. It gave place later, in the nineteenth century, to the real birth of the science of hydrography. *See* J.C. Beaglehole, *The Life of Captain James Cook* (Stanford: Stanford University Press 1974); Allan Villiers, *Captain James Cook* (New York: Scribner 1967); Susan Schlee, *The Edge of an Unfamiliar World: A History of Oceanography* (New York: Dutton 1973). See also Howarth, *Sovereign of the Seas*, p. 325.

[114]91 F.2d 819, 1937 A.M.C. 934 (C.A. 4 1937).

[115]Ibid., at 821: "There were branch offices of the United States Hydrographic Office in New Orleans and Galveston, in all of which corrected charts, light lists, pilots books, daily memoranda, and notices were available for examination by all mariners, domestic and foreign. Both the Hydrographic Office and the Coast and Geodetic Survey had agents in New Orleans and Galveston for the sale of their respective publications."

provided by such navigation charts, light lists, etc., were an issue of the navigation of the ship and not related to its seaworthiness.[116] Then, the owners attempted to construct an exception of error in the management of the ship to exclude their liability. But the court found that the vessel was navigated using the out of date charts and navigational data provided by the shipowner. The court categorically rejected such an argument and held the carrier liable for lack of due diligence. It was stated that "charts, light lists, and similar navigation data are essential equipment for the safe navigation of a ship, that she is unseaworthy without them, and it is the duty of her owner to supply them."[117] All those documents are necessary sources of information for the navigator and the shipowner must provide them. Though such a function is often delegated to ship officers, the carrier remains liable if that is not diligently done. What the carrier is delegating is an issue related to the vessel's seaworthiness. A failure in performing this task by a crew aware of that condition constitutes unseaworthiness and cannot be transformed into bad seamanship.

This was perhaps the first case stating that navigational charts are part of the vessel's seaworthiness. In a previous case,[118] the shipowner was released of liability because, though the charts were not updated, there were on board navigational supplements and documents that showed the changes making it possible to correct the charts. The shipowner practiced due diligence as he had provided "adequate equipment or *sources of information*". The stranding was due to the failure of the master to correct the charts with the information that was available on board. Then, the conclusion reached in *The Maria* followed the reasoning of this case. The cases opened the path for the analysis of electronic charts. The SOLAS Convention accepts the use of electronic charts on board.[119]

4.2.3.1.4 The Radar

The Radio Detection and Ranging, better known as Radar, has been one of the most useful technological discoveries to improve safety in navigation, especially for the prevention of collision. After the Second World War, the shipping industry started considering the use of radars for commercial navigation. The shipping press of that time reported that to become commercially used a "first-class servicing

[116]Ibid. The shipowner defended this theory based on two cases: *The Jason* (1912), 162 F. 56 (D. C.); Id., 178 F. 414 (C.C.A.); Id., 225 U.S. 32, 32, S.Ct. 560, 56 L.Ed. 969, and *The Murrell* (1911), 200 F. 826 (D.C.); Id., 195 F. 483 (C.C.A.). Judge Soper analyzed both cases concluding that none of them supported such at theory because they did not address the specific issue of navigation charts as part of the essential equipment of the vessel.

[117]Ibid., 824.

[118]*See The Steel Scientist,* reported as *Daisy Philippine Underwear Co. v. United States Steel Products Co.,* 11 F. Supp. 175(D.C.), and in *United States Steel Products Co. v. American & Foreign Insurance Co.* 82 F.2d. 752 (C.C.A.).

[119]Chapter V, Regulation 19.2.1.4.

organization" was needed.[120] It was also necessary to assure its reliability, ease of maintenance, and that the use on board could be handled by deck officers instead of a skilled radio officer.[121] Although at that time such equipment was expensive, the savings that large vessels could realize in arriving on time at their destination, when the difference of a few hours could result in economic losses, making the use of radars was highly profitable.[122] The economic benefit was, of course, more evident in the prevention of collisions or strandings. The application of a device used and developed during war times, available now for commercial navigation during peacetime depended in its capacity to fulfil some requirements. It had to be: (1) reliable, (2) simple to operate, (3) easy to maintain, and (4) cost efficient.[123] It had to provide a continuous and outstanding service for a minimum care, attention and cost.[124] Its application in navigation was the subject of discussion in the International Conference on Safety of Life at Sea of London in 1948. It was largely agreed that this device had become one of the most helpful pieces of equipment for navigation. The Conference issued a recommendation establishing certain minimum performance standards such as range and bearing discrimination, urging Governments to promote its application with the recommended standards.[125] However, its implementation on ships remained just a recommendation, not compulsory equipment. The technology became compulsory some decades later. The analysis of this device in the case law has been initially and mostly within the scope of collision cases. Later, its failure or lack of it on board the vessel was judged as a want of seaworthiness.

(1) **Collision Cases**

The effectiveness and usefulness of radars in maritime navigation was analyzed, perhaps for the first time, by an American court in *The Medford* (1946).[126] In 1945, a steamship of the US government navigating in a very high speed through a fog-bank collided with a fishing vessel causing it to sink. The steamship was equipped with radar, which was in perfect condition and two radar men were on board. However, the radar was not used. The accident could have been prevented if this device, considered already at the time the "best anti-collision device yet perfected", had been put into operation.[127] The court held this failure as the sole cause of the collision, holding the steamship totally liable for the accident. The remarkable issue of the decision was the awareness of the court regarding the value of

[120]Hancock, 196.

[121]Ibid., 196.

[122]Ibid., 197. Reported by *Syren and Shipping* on July 3, 1946.

[123]Ibid., 197.

[124]Ibid., 197.

[125]Ibid., 196.

[126]65 F. Supp. 622 (D.C.N.Y. 1946).

[127]Ibid., 624. The Judge cited the 'Electronic Navigational Aid' document published by the United States Coast Guard—U.S. Government Printing Office 1945.

radars as the best device to prevent collisions. The court even used severe language to describe this failure:

> The failure of the Barry to use her radar is the most serious and sinister aspect of these causes. The perfection of that device is thought to have invoked a new concept of the responsibilities attaching to vessels so equipped, touching their handling and operation in or near a fog-bound area [...] I cannot so read that opinion as to find refuge for the Barry in its ample folds, for the stipulated proof here is that the offending ship could have informed herself of the presence and track of the Medford in abundant time to have avoided by a wide margin any danger whatever of striking her. Under such circumstances, it is impossible to yield to the argument for the Barry, that her conduct is to be condoned to any extent, in view of her failure to employ the very device which was installed to prevent a collision, and to operate which she carried two men having special rating in the U.S. Navy to attest their qualifications, and who had no duty on the ship other than to operate the radar unit.[128]

One year later, the district court of New York addressed this issue in another collision case. In *The Hindoo* (1947),[129] the criterion of the former case was reaffirmed. This time, it was a collision between two merchant vessels that occurred in the Caribbean Sea in 1944. The *Australian Star* was equipped with radar, but her master did not pay enough attention to its indication nor request more reports after the radar operator reported two objects. The court said that this vessel, having radar, could observe the other vessel and determine her direction and speed with great precision.[130] Knowing that, the *Australian Star* should have reduced her speed or deviate her course. That would have avoided the collision, but this vessel did nothing. The court concluded that the negligent use of her radar contributed to the casualty.[131] It was clarified that it was not the court's intention to penalize the *Australian Star* for being equipped with radar, but to hold her responsible because "she chose to remain blind when she had the means to see".[132] *The Hindoo*, on her part, was a vessel owned by the United States. She was not equipped with such a device. Although her navigation was held also imprudent, it was not attributed to her lack of having radar on board, which would also have helped to prevent the collision. The court stated that prudent navigation demands the use of all the means available for such a purpose:

---

[128]Ibid., 626–627.

[129]74 F. Supp 145, (D.C.N.Y. 1947), 1947 A.M.C. 1630, affirmed on this point in *United States v. The Australian Star*, 172 F.2d 472 (2nd Cir. 1949), certiorari denied, 1949, 338 U.S. 823, 70 S.Ct. 69, 94 L.Ed. 499.

[130]The court of appeal however clarified that "...radar can show only the distance and bearing of a vessel, not her heading at any moment. By successive radar readings the general direction in which the vessel has moved between them can be plotted, but her course will be discovered more promptly and more accurately by observing her navigation lights." 172 F.2d at 478 f.n. 2.

[131]Ibid., 148: "By means of the intelligence radar supplied she could have navigated safely with respect to the Hindoo without relying on the surmises... Had the master made more intelligent use of this radar he would have known at 9:24 p.m. that he was almost certainly on a collision course and would have taken precautionary measures."

[132]Ibid., 149.

> Prudent navigation involves taking advantage of all the safety devices at hand. Insofar as it is the judicial function to fit scientific discoveries into the framework of laws not tailored to their measure, the function should be carried out with an eye to the general purposes of the law, and to desirable social ends.[133]

To reach this conclusion, Judge Rifkind observed a provision of the International Rules for Navigation at Sea establishing an obligation of the vessel to take any precaution required by special circumstances.[134] That provision did not state that the use of radar on board was mandatory. Due to the absence of an express statutory regulation on this issue, the judge invoked the interest of the public as a criterion to assign liability on the *Australian Star* for not using her radar. The prevention of collisions and marine casualties is obviously a desirable social end. Given the lack of exact regulation on the topic, general principles of law and social ends must be attended to when deciding on the implementation of new technologies not ordered by the law, but whose application can prevent marine accidents. Of course, such application will depend also on other factors, such as the technology's cost, as we will see later.

In these two cases, however, equipping the vessel with radar was not argued to be part of her seaworthiness. This argument was presented in another collision case. In *Anglo-Saxon Petroleum Co., Ltd., of London, England v. United States* (1950),[135] the district court of Massachusetts categorically said, and with no further explanations, that the failure to carry a radar did not make the vessel unseaworthy.[136] This collision occurred in 1942, when the use of radars was still under experiment and could not be expected to be part of the seaworthiness.

In another collision case occurring in 1951, one of the colliding vessels was equipped with radar. Despite having detected the other vessel and being able to plot its course and speed with enough time to avoid the collision, the master did nothing. For this, among other reasons, it was held at fault. It was decided in *Polarus Steamship Co. v. The T/S Sandejord* (1956).[137] Judge Medina pointed a possible problem that may arise when applying some new technologies. He said that radar, when not properly used, may increase the possibilities of collisions because it "gave

---

[133]Ibid., 149.

[134]Ibid., 149: "Article 29 of the International Rules for Navigation at Sea, 33 U.S.C.A. § 121 provides: 'Art. 29 Nothing in these rules shall exonerate any vessel, or the owner or master or crew thereof, from the consequences of any neglect to carry lights or signals, or of any neglect to keep a proper lookout, or of the neglect of any precaution which may be required by the ordinary practice of seamen, or by the special circumstances of the case.'"

[135]88 F. Supp 158, 1950 A.M.C. 631, (D.C. Mass. 1950). It was a collision occurred in 1942 between the tanker *Davila*, owned by the plaintiff and the destroyer *U.S.S. Wilkes* escorting at that moment a cruiser during the Second World War.

[136]Ibid., 160. The case had the particularity that the destroyer escorting a cruiser was indeed equipped with an air search type of radar, intended for detection of aircraft approach, but which was not suitable for navigational purposes. The court found no fault in the action taken by the destroyer to avoid the collision.

[137]236 F.2d 270 (C.A. 2 1956), 1956 A.M.C. 1779.

a false sense of security".[138] He made similar warning in *Afran Transport Co. v. The Bergechief* (1960).[139] The US court of appeals of the second circuit decided this case assigning liability for lack of proper use of radar. This time, Judge Medina remarked that the mere use of radar is not a justification for not observing the other rules of navigation.[140] The court even rested upon the faulty vessel the burden of proving that her failure in using the radars did not contribute to the collision.[141] The court recognized that by 1955, when this collision occurred, there was not yet any regulation requiring ships to carry radars, neither was it required by any judicial decision. It recalled the decision of the *Anglo-Saxon Petroleum v. U.S.*,[142] establishing no lack of seaworthiness for not having radars. The court recognized, however, that the value of such a decision was not as strong as it was before, because certain conditions had changed. The use of radar on board ships was more extended. The court invoked the *T.J. Hopper*, where, despite that there was no statute requiring the use of a radio receiving apparatus, it was held as an imperative precaution and then as part of the vessel's seaworthiness. However, as this issue was not a subject of discussion in this case, Judge Medina limited his ruling and went on to say that "a rule requiring radar, subject to some limitation and qualification, will sooner or later be formulated",[143] but declined to establish such a rule. He continued with the traditional requirement that if a vessel has radar, it must to be used properly.[144] Notwithstanding, the case created the burden for the vessel to prove that her failure to use the radar did not contributed to the collision.

The use of radar on vessels as a rule of seaworthiness was almost judicially recognized in *Farrell Lines, Inc. v. S.S. Birkenstein* (1962).[145] The court cited Judge Hand's analysis in the *T. J. Hooper* as well as Judge Medina's "prophecy", to justify the imposition of a standard of conduct that had not yet been set by statue nor by a general practice, but that "was evident that the practice ought to become so".[146] But as no fault could have been found on the part of the tug involved, nor lack of radar was proved to be the cause of the accident, the judicial rule requiring

[138]Ibid., 271.

[139]274 F.2d 469, 1960 A.M.C. 1380 (C.A.2 1960).

[140]Ibid. at 472: "Radar is an additional safeguard, and a failure to use it may constitute negligence, as we shall see, but a master who relies on radar alone and disregards any or all other precautions and requirements, statutory or otherwise, does so at his own risk."

[141]Ibid., 473.

[142]*See supra* note 135.

[143]Ibid., 474.

[144]Ibid., 474.

[145]207 F. Supp. 500, 1963 A.M.C. 1846 (D.C.N.Y. 1962).

[146]Ibid., 510. He cited also *Texas & Pacific Ry. Co. v. Behymer*, 189 U.S. 468, 470, 23 S.Ct. 622, 47 L.Ed. 905 (1903), at 470 or 623: "What usually is done may be evidence of what ought to be done, but what ought to be done is fixed by a standard of reasonable prudence, whether it usually is complied with it or not."

radars was not set. Nor can a "second eye", as the radar was compared to, be demanded just to be useful in the abnormal conditions reported in this case. Judge Friendly also considered that in obliging a tug to have radar on board involved the inclusion of an extra crew member to manage the radar. Therefore, it continued the rule that prudent navigation means to take advantage of all the safety devices on board. If there is radar on the ship, it has to be properly used, but it was not demanded to have it as part of the vessel's equipment.[147]

English courts on their part, followed a similar approach. Judge Willmer, of the Admiralty Division, issued four decisions, between 1955 and 1956, that underlined the position of English law regarding radars. The functionality of this device was initially compared to "a potential silent look-out."[148] In *The Chusan* (1955),[149] the court reaffirmed that it is the "seaman's duty, in the exercise of reasonable care, to take full advantage of any equipment with which the vessel is equipped."[150] The cases made clear that the accuracy and reliability of radars were not in doubt. This decision went even further regarding this certainty. The judge said that by 1953, when this collision took place, it was already expected that a large passenger ship, such as the *Chusan*, was equipped with a radar set.[151] If having radar was expected, it suggests that this was seen already as a common practice, at least for big ships. As a modern steamship, radar may not be the only type of equipment one would expect the vessel to be supplied with, said the judge.[152] This vessel was assigned part of liability in the collision for its lack of proper use of this appliance. Similarly to the warnings issued by Judge Medina in the American cases, Judge Willmer also pointed out the possibility of extreme reliance of masters and officers on radars. He clarified that making mandatory the proper use of radars, and keeping a continuous watch on them does not mean that a vessel is allowed to sail at an excessive speed in a fog.[153] Equipping the ship with radar was considered a "great advantage over other vessels which are not similarly equipped", said the same judge in *The Nora* (1956).[154] As this equipment provides a higher degree of information and enables the vessel to make better

---

[147]The rule was also held in *Orient Steam Nav. Co. v. U.S.*, 231 F. Supp. 469, 474 (D.C.Cal. 1964), 1964 A.M.C. 2163, where the plaintiff's vessel, *The Oriana*, was even equipped with a "Deca true motion radar", recognized by 1962 as the "highest achievement in the field of radar", but it was not in use previous to the collision.

[148]*See The Esso Plymouth* [1955] 1 Lloyd's Rep. 429, 437. It was about a collision occurred in 1952. Judge Willmer said that if the radar had been switched on timely and the other vessels were noticed, *The Esso Plymouth* would have taken some steps to reduce the risk of collision.

[149][1956] 1 W.L.R. 72 (Probate, Etc., Division).

[150]Ibid., 76.

[151]Ibid., 74. The same Judge Willmer in *The Esso Plymouth* [1955] 1 Lloyd's Rep. 429, 437, decided some months before *The Chusan* and related to a collision occurred in 1952, held that if the radar had been switched on timely and the other vessels were noticed, *The Esso Plymouth* would have taken some steps to reduce the risk of collision.

[152]Ibid., 76.

[153]Ibid., 75.

[154][1956] 1 Lloyd's Rep. 617, 626.

decisions in avoiding collisions, that gives vessels to less of an excuse for committing a mistake, and that must account for the assignation of liability.[155] He remarked, as well as he did in *The Chusan*, that this equipment must be "intelligent and reasonably" used.[156] This navigation aid creates a greater duty to take the best advantage arising from having it on board, as argued in *The Sedgepool* (1956).[157] A vessel not having it may claim to receive a lenient treatment for not having the "fortune" to access the additional information provided by such equipment.[158]

(2) **Cargo Claims**

Radar as part of the vessel's seaworthiness came to be analyzed in a cargo claim in *The President of India v. West Coast Steamship Co., (The Portland Trader)* (1962).[159] The district court of Oregon, and later confirmed by the court of appeals in 1964,[160] held that a vessel crossing the Pacific Ocean was not unseaworthy for not being equipped with radar and loran.[161] District Judge Kilkenny recognized that such devices were valuable aids to navigation and that if they were on the ship, the grounding would have been avoided. Besides that, he stated that the standard required by this obligation is not an "accident-free ship, nor an obligation to provide a ship or gear which might withstand all conceivable hazards."[162] He comes to the conclusion that to demand this technology as essential part of the vessel's equipment it has to be an established by a general usage:

> The overwhelming weight of the evidence in the case is that there is no worldwide or American practice or custom with reference to the use of radar or loran as aids to navigation...Here, we have no well established usage. Some companies use radar or loran; others do not. The evidence shows that none of the hundreds of Liberty ships built during the war originally carried either loran or radar.

---

155 Ibid., at 625–626, the Judge even applied maxim which says: "For unto whomsoever much is given, of him shall be much required".

156 Ibid., 626.

157 [1956] 2 Lloyd's Rep. 668, 679.

158 Ibid.

159 213 F. Supp. 352, 1963 A.M.C. 649 (D.C. Or. 1962), The SS Portland Trader was a 4455 tons net vessel carrying under a Charter Party 9800 long tons of bulk wheat from Vancouver to Calcutta, came in contact with a reef in January of 1960. The master worked for 9 years with the same company, but it was his second voyage as a master and the first one in the Sulu Sea. The Sulu Sea, where the stranding occurred, was blanketed by loran signals, but, though the captain had received training as radar observer, there were on the ship neither radar nor loran. The only equipment was a radio direction finder but without RDF coverage to blanket the area. Both, the master and the shipowner had knowledge about these facts. The reputed American Bureau of Shipping had classified the vessel with the outstanding note of A-1, but on its certificate, there was no reference to electronic aids to navigation, neither indication whether the ships had radar or loran.

160 H. R. Baer, Admiralty Law of the Supreme Court, 3rd edn. (Charlottesville, Va: Michie Co. 1979), 501.

161 Ibid., 501.

162 213 F. Supp., 356.

> Some have been so equipped in recent years. Most of those engaged in the tramp trade, such as the TRADER, were not so equipped.[163]

This reasoning made clear that this technology could only be required as part of the seaworthiness when they are commonly and extensively used. The mere helpfulness in avoiding hazards so dangerous as collisions was not enough to demand its application. It must then be determined whether or not there is such a common practice. Certainly, since the end of Second World War, the use of radar was increasingly implemented in maritime commercial navigation. As the collision cases showed, by the middle of the 1950s, many vessels were already equipped with radar and that a big modern vessel was expected to be so. The extension of the use of radar was evidence in this case in the fact that the master of this vessel had received training on reading this device.[164] Judge Kilkenny, however, recognized the evolving character of the seaworthiness condition. While pointing out that Judge Hand's decision in the *T. J. Hopper* had been subject to controversy in his circuit and even rejected in posterior cases,[165] he agreed that the standard of seaworthiness changes with the knowledge, experience and the appliances of navigation. However, he subjected such evolution to the common use of new scientific advances. The use of radar and loran in tramps steamers "under the facts and circumstances here in question, is still in a period of transition and at the time of this occurrence was not so essential to navigation that its absence would give rise to a finding of unseaworthiness."[166] He echoed the visionary statement of Judge Medina, adding that he "would be one of the first to recognize that in the not too distant future the absence of such navigational aids on such ships might well make them unseaworthy."[167]

These decisions, both American and English, while they clearly recognized radar's reliability, accuracy and convenience in preventing collisions, did not order the use of it as part of the vessel seaworthiness. Judges promoted the correct, prudent and intelligent use of this device as a measure of maximization of the advantages that it provides. However, it was still not held as custom in navigation or as an imperative precaution that would require it as part of the vessel seaworthiness. Observing these cases where the radar was analyzed, it seems that modern devices are not strictly held as part of the vessel seaworthiness, and its non-application does not constitute unseaworthiness, unless when they are on board, they are defective.[168] This approach, in a certain way, even discouraged the implementation of radar on board. If radar was so efficient in preventing catastrophic casualties resulting from collision, it would be applicable to the criterion of

---

[163]Ibid., 356–357.

[164]Ibid., 354.

[165]Ibid., at 357: "The precise meaning and overall judicial effect of Judge Hand's decision in Hooper is in serious dispute in the Circuit of its origin. *Nuzzo v. Rederi*, 304 F.2d 506, 510 (2nd Cir. 1962), and dissenting opinion of Judge Clark, 511. Be that as it may, I believe in Judge Hand's truism that the courts must in the end fix the standards of what is or is not a seaworthy ship."

[166]Ibid., 357.

[167]Ibid., 358. *See also* Baer, 501.

[168]Tetley, *Marine Cargo Claims,* Vol. 1, 915.

Judge Hand on the imperative that does not require it to be a custom to be enforced. In the case law found, there is not enough explanation for not requiring its application on board as part of the ship's seaworthiness. It must be assumed, comparing radio sets to radar, that the cost of both systems played an important role in deciding its reasonableness in navigation.

4.2.3.1.5 Newer and Safer Methods

Some new modalities in the stowage, ships' inspections, and carriage of the cargo that improved the operations are expected to be applied by shipowners. However, new methods that render more efficiency in operations cannot be immediately required as part of the obligation. For the enforcement of these new modalities applies also the rule of regular use in the industry. This was evidenced in the *The Venice Maru* (1943).[169] A fire of a highly flammable cargo, sardine meal, was attributed to an improper stowage. Improper stowage makes the vessel unseaworthy. The sardine meal was shipped in 1934 using rice ventilators in the holds to prevent excessive heating. This method was already largely accepted as appropriate for the carriage of heat sensitive cargoes, but it was not totally safe. By that time, a newer method of stowage had been tried and proved to be more efficient to prevent excessive heating and fire. This method, however, did not become the common standard until 1935. The court held that the carrier had used the common methods of stowage. Therefore, he was not at fault for not applying a newer system that was not yet of general use. It could not be expected that the shipowner apply the new method, although it was a better way of stowage, the year prior to it becoming the standard.

Other cases relating to more efficient methods of stowage also argued against the opinion expressed in the *T.J. Hopper*. It was in the same Second Circuit in *Nuzzo v. Rederi* (1962).[170] Judge Hincks questioned the arguments of Judge Hand, based on two decisions of the Supreme Court issued after the *T.J. Hopper* that reaffirmed the criterion of the common usage as the standard for seaworthiness. It was in *Boudoin v. Lykes Bros. S.S. Co., Inc.* (1955),[171] and *Mitchell v. Trawler Racer, Inc.* (1960).[172] None of them related exactly to the use of new technologies on ships, but affirmed that the standard of seaworthiness was not a one of perfection. It was claimed that the stowage was incorrect, and that by that time there existed a safer method. The court held that the vessel does not need to be prepared to face any single hazard that may be encountered in the maritime adventure. Such a statement

---

[169] *See Consumers Import Co., v. Kawasaki Kisen Kabushiki Kaisha,* 133 F.2d 781, 1943 A.M.C. 277 (C.A.2 1943).

[170] 304 F.2d 506, 510 (2nd Cir. 1962). It was a claim for personal injuries suffered by a longshoreman when unloading a cargo of lumber from a ship, due to an alleged unseaworthiness caused by improper stowage of the cargo onboard.

[171] 348 U.S. 336, 75 S. Ct. 382, 99 L.Ed. 354 (1955).

[172] 362 U.S. 539, 80 S.Ct. 926, 4 L.Ed.2d 941 (1960).

seems reasonable at a first look and we may agree. But the shipowners should not ignore that the new scientific advances allow them to detect and foresee some of the common risks of navigation, as well as any new risks, and must take reasonable actions against them. In this case, the no need for perfection criterion was applied to deny claims where unsafe stowage cause personal injuries of stevedores. Judge Clark dissented in this opinion[173] remarking that the holding of this case might introduce concepts of negligence in the law of seaworthiness, which in cases of personal injury, was still an absolute obligation. He criticized that injured persons must be the ones who bear the consequences of the vessel's unseaworthiness.

New methods of carriage have required the modification of some judicial rules on carriers' liability. Carriage under a clean bill of lading was traditionally interpreted as cargo carried below deck.[174] Carriage on deck without agreement with the shipper, or when it was not a customary port to trade common practice for the specific cargo, was regarded as an unjustifiable deviation that made the carrier liable as an insurer.[175] With the implementation of containers, carrying them on deck of ships specially designed for this type of carriage was held as a reasonable deviation.[176] The newer ship's design did not expose the cargo on deck to greater risk than those stowed under deck, nor when it was carried in other type of ships. It was affirmed in *Du Pont de Nemours Inter, S.A. v. S.S. Mormacvega* (1974).[177] The court found that the ship was specially reconstructed for the purpose of carrying cargo safely on deck, and concluded that "technological innovation and vessel design may justify stowage other than below deck".[178] The same criterion was adopted in *Electro-Tec Corp. v. S/S Dart*

[173]*See supra* note 170, at 511.

[174]*See St. Johns Corp. v. Companhia Geral Commercial Do Rio de Janeiro,* 263 U.S. 119, 124, 44 S.Ct. 30, 31, 68 L.Ed. 201 (1923); *Encyclopaedia Britannica, Inc. v. S/S Hong Kong Producer*, 422 F.2d 7, 14 (2nd Cir. 1969), cert. denied, 397 U.S. 964, 90 S.Ct. 998, 25 L.Ed.2d 255 (1970); *Calmaquip Engineering West Hemisphere Corp.*, v. West Coast Carriers, Ltd., 650 F.2d 633, 638–639 (5th Cir.1981).

[175]Such deviation nullified the exception of liability as well was the limitation of liability set in the bill of lading. *See Niles-Bement-Pond Co. v. Kampkiesa Kieselskabet Balto*, 282 Fed. 235 (2nd Cir. 1922); *S.S. Willdomino v. Citro Chemical Co.*, 272 U.S. 718 (1927); *Jones v. The Flying Clipper*, 116 F. Supp. 386, 388 (S.D.N.Y.1953); *Encyclopaedia Britannica v. S.S. Hong Kong Producer, see supra* note 174, at 16–17; *Ingersoll Milling Mach. Co. v. M/V BODENA*, 829 F.2d 293, 299 (2nd Cir.1987); *Sequros Banvenez. S.A. v. S/S Oliver Drescher*, 761 F.2d 855, 859, 1985 A.M.C. 2168, 2173 (2nd Cir.1985); *Constructores Tecnicos S. de R.L. v. Sea–Land Serv., Inc.*, 945 F.2d 841, 845, 1992 A.M.C. 1284, 1290 (5th Cir. 1991).

[176]According to the Section 4(4) of COGSA: "…any reasonable deviation shall not be deemed to be an infringement or breach of this chapter or of the contract of carriage, and the carrier shall not be liable for any loss or damage resulting therefrom…".

[177]493 F.2d 97, 102 1974 A.M.C. 67, (C.A.N.Y. 1974).

[178]Ibid., 102. The decision cites: *The Salvore*, 60 F.2d 683, 685 (2nd Cir.), cert. denied, 287 U.S. 653 (1932); *Lagerloef Trading Co. v. United States*, 43 F.2d 871, 872 (S.D.N.Y.1930); *The William Crane*, 50 Fed. 444 (D.Md.1889); The Neptune, 17 Fed.Cas. 1343, 1344 (No. 10,118) (C. C.S.D.N.Y.1868); *See Encyclopaedia Britannica Inc. v. Hong Kong Producer, see supra* note 1313, at 18 n. 12 (2nd Cir. 1969); *Standard Electra S.A. v. Hamburg Sudamerikanischer*, 375 F.2d 943, 945 (2nd Cir.), cert. denied, 389 U.S. 831 (1967).

Atlantica (1984),[179] adding that "[w]hat was unreasonable yesterday may be reasonable today".[180] Indeed, by 1998 it was expressly recognized by an American Court that since 1991, after 35 year of the launch of the first carriage using containers,[181] "on-deck stowage of containers on containerships was a well established custom of the trade in ocean transportation on a world-wide basis".[182] That meant this type of carriage was not considered an unreasonable deviation. It must be observed, however, that such common use of containers was already in use far before 1991. The technology developed for carrying containers on open decks of vessels specially designed for that purpose, made it safer, reducing the risk of falling overboard. It proved to be more efficient and became a common practice in international trade that could not be categorized as a deviation anymore.

New methods of stowage have not been the only improvement in the shipping industry. New technological methods that might help to reveal more precisely the condition of the ship before sailing have been a subject of discussion.[183] Under English law, a more advanced method of inspection of the vessel's hull was alleged by claimants to be part of the obligation of practicing due diligence. It was analyzed in *Union of India v. NV Reedrij Amsterdam (The Amstelstot)* (1963).[184] A system called Magnaflux offered a more accurate method to inspect the hull and detect cracks.[185] The process may have helped to discover the cracks in the vessel that were not discoverable through a visual inspection. Such cracks made the ship unseaworthy, causing later claims for damages. The allegation was however rejected by the first judge and was not discussed again in the appeal. It was decided that the proper sort of examination was the mere visual inspection. Lord Reid expressly pointed that:

> [...] that test is never applied to this kind of machinery unless there is some special reason for applying it, and there was no such a reason here... for it would be plainly impracticable to make elaborate scientific test for every defect which could possibly be present in any part of the machinery.[186]

---

179 598 F. Supp. 929 (D. Md. 1984).

180 Ibid., 934.

181 *See supra* note290 and accompanying text.

182 *See Konica Business Machines, Inc. v. Vessel "Sea-Land Consumer"* 153 F3d. 1076, 1078 (C. A. 9 (Cal.) 1998), 1998 A.M.C. 2705, 98 Cal. Daily Op. Serv. 6955, 98 Daily Journal D.A.R. 9595, citing the judgment of the first instances district court 1996 WL 468770; *Insurance Co. of North America v. Blue Star (North America), Ltd.* Not reported in F. Supp., 1997 WL 345235 (S. D.N.Y. 1997), 1997 A.M.C. 2434.

183 Tetley, *Marine Cargo Claims,* Vol 1, 930, note 269 cites *The Australian Star* (1940), 67 Ll. Rep. 110. Defects in the rivets and butts in the fuel oil tanks were not discovered before the beginning of the voyage, causing later a leakage. Due diligence was practiced according to the court, because it had to be judged according to the shipowners knowledge at the relevant moment.

184 [1963] 2 Lloyd's Rep 223 (HL).

185 The test consisted, as explained by the same Lord Devlin, of spreading oil with iron particles on the vessel's gear and then charged with electricity that produces the iron particles to be aligned at any crack.

186 [1963] 2 Lloyd's Rep. 223, 230–231. As cited by M. A. Clarke, *Aspects of the Hague Rules: A comparative study in English and French law* (The Hague: Nijhoff 1976), 225.

This test seemed to be highly advanced and its practice on a ship seemed complicated. That certainly would be out of the scope of reasonableness. According to case law, the application of new methods that prove to be safer and more efficient will be enforceable only if they are a common practice in the industry. However, a first failure in the application or functioning of the method or a first breakdown of a particular issue does not mean immediately a want of due diligence. After the first failure, the carriers must pay special attention to the particular issue.[187]

4.2.3.1.6 Other Equipment

Another invention of Marconi applied to maritime navigation was the direction-finding device. This equipment enabled the master to receive the essential bearing for determining location and direction when those astronomical observations were prevented due to climatological elements such as fog, clouds or heavy rains.[188] Because of its cost, shipowners were initially reluctant to apply it to their merchant vessel, but its value in saving time and money was rapidly recognized and many shipping companies adopted the technology by 1923.[189] During this time, it proved its convenience in saving lives, cargo and ships in cases of casualties and above all, the saving of time during heavy fog, evidencing, therefore, its cost efficiency.[190] The British Chamber of Shipping recommended to its members in 1927 to install this technology in all their ships.[191] In 1929 it became mandatory for British ships, as consequence of the Safety of Life at Sea Conference of that year, that ordered all passenger ships of 5000 tons gross and upwards, to bear a directions-finding apparatus.[192] Then in the Convention for the Safety of Life at Sea of June of 1948, it was made compulsory for ships of 1600 tons gross.[193]

This technology was used by the *Sealand Voyager*[194] in November of 1988. The ship was routed by Navitech, "a vessel routing service, which made recommendations

[187]Tetley, *Marine Cargo Claims,* Vol. 1, 931. Tetley cites the Canadian case *Health Steel Mines Ltd. v. The Erwin Schroeder* [1970] Ex.C.R. 426, where for the carriage of wet cooper concentrates was necessary the use of separators in the ship's holds, but that was not known. After this, it became a custom for the carriage of such cargo.

[188]Hancock, 116. It was known as the Marconi-Bellini-Tosi system. The first *Mauretania* was the first merchant vessel using a wireless device for directions and position finding installed experimentally in 1912 in transatlantic voyages. This vessel was owned by the British Cunard Line. It became more popular after the First World War for its contribution in winning the war.

[189]Ibid., 116, 119. For example Cunard, Canadian Pacific Ocean Services, Elders and Fyffers, the P. and O. Royal Mail, Red Star, and White Star lines.

[190]Ibid., 124.

[191]Ibid., 116–117. Even in 1924, Mr. J Herbert Scrutton, a well-known shipowner said: "The direction finder strikes me as the most extraordinary advance in navigation that one has ever heard of. I am all out for putting these things on board a ship. There is no question whatever about their usefulness".

[192]Ibid., 124.

[193]Ibid., 124.

[194]*Eastman Kodak Company v. S.S. Sealand Voyager* 1991 A.M.C. 2356.

prior to sailing giving advice on the route to take according to predicted weather."[195] The cargo damage was attributed to a containerunseaworthiness caused by a want of due diligence before the voyage. There were strong winds during the voyage yet they were below the 9 of the Beaufort scale which was not considered storm weather capable of affecting the ship as long as it was well prepared for normal weather. The use of the information provided by this technology was not analyzed by the court.

A fathometer is a device used to measure the depth of the water beneath the ships. It belongs to the group of Echo Sounding Devices. In *The Chickasaw* (1969),[196] previously mentioned above, the judge of the first instance held the carrier liable for having an inoperable Fathometer. The master was made aware of the defect, but he did nothing to repair it. As mentioned before, the Court of Appeals of the Ninth Circuit affirmed the first instance decision, not only due to the defective Fathometer, but also due to the failure to have on board an up-to-date correction card or chart for the radio direction finder apparatus, made mandatory by statue to have such a device in proper order.[197] However, in another case, the lack of a Fathometer did not render the vessel unseaworthy because on board the vessel were other instruments that allowed fixing the position of a vessel involved in lightering.[198]

The use of weather forecast systems has been held as necessary to assure that a port is safe. In *Islander Shipping Enterprises, S.A. v. Empresa Maritima del Estado, S.A.* (*The Khian Sea*) (1979),[199] Lord Denning MR decided a case where the warranty of nominating a safe port was in discussion, argued that if a port is to be held safe, it must be provided with this system.

4.2.3.2 Technologies Contractually Agreed

Some measures of special care for particular cargoes are frequently offered by the carriers as part of their services. As mentioned, some cargoes required specific care and handling that can be only provided by the application of specific technological appliances. In these cases, as technologies are strictly necessary for the nature of the contract, its failure in providing it or any defect in the technology that results in damage, holds carriers liable. One of the first and best examples of this type of technology is the refrigeration apparatus. This is obviously not a new technology anymore, as it has been in use for over a century. But its initial application in maritime transportation and the first courts' analyses deciding cargo claims originated from

[195] Ibid., 2358. This ship received reports of Navitech periodically during the voyage and daily weather reports by facsimile.

[196] 265 F. Supp. 595, 1966 A.M.C. 2219 (S.D. Cal. 1966), upheld in 414 F.2d 724, 1969 A.M.C. 1682, [1970] 1 Lloyd's Rep. 437 (9th Cir. 1969).

[197] At the time of the stranding it was set under 47 U.S.C. § 351 (a) (2), (1964) and 47 C.F.R. § 8.517 (a), at the time of the decision it was 47 C.F.R. § 83.459 (1968).

[198] *U.S. v. Ultramar Shipping*,685 F. Supp. 887, 1988 A.M.C. 527 (S.D.N.Y., 1987), *see* at 899–900 or 545.

[199] [1979] 1 Lloyd's Rep. 545.

failures in these systems, offer a source of reference that can be applied now, for example, deciding liability when reviewing any new technology such as the controlled atmosphere system for containers.

4.2.3.2.1 Refrigeration Systems

The first attempts for the carriage of refrigerated cargo were reported by the end of the 1860s and early 1870s, when the United States, Argentina and Australia tried to export meat to Europe using refrigeration systems, but with not much success.[200] The first entirely successful carriage in the world's history of frozen meat occurred in 1877.[201] During 1878 and 1879, some shipments of cooled meat, not frozen, were sent from the United States.[202] Another success was reported in 1880 by the *Strathleven*, carrying beef and mutton that were frozen on board, from Melbourne, Australia to London.[203] Similarly, in 1882 a New Zealand sailing ship, *The Dunedin*, equipped with insulated meat chambers, boilers and refrigeration machinery, sailed from Port Chalmers, New Zealand to London, carrying frozen mutton and lamb carcasses, arriving in perfect condition at their destination.[204] After these voyages the industry experienced a gradual growth in the carriage of meat and dairy products from Australia, New Zealand and Argentina.[205] In the first decades of use of refrigeration systems, some cargo claims arose from damage caused to goods that required refrigeration. Judges assigned liability based on the obligation of providing such apparatus in good condition as this special service was explicitly contracted.

---

[200]O. E. Anderson, JR., *Refrigeration in America: A History of a New Technology and its Impact* (United States of America: Princeton University Press 1953), 124; J. R. Stott, *Refrigeration at sea*, 2nd edn. (London: Applied science 1978), ix; J. T. a. R. J. Critchell, *A History of the Frozen Meat Trade: An Accound of the Development and Present Day Methods of Preparation, Transport, and Marketing of Frozen and Chilled Meats* (London: Constable & Company Ltd. 1912), 23, 27.

[201]Critchell, 28. It was a carriage from Buenos Aires to Marseilles on the *S.S. Paraguay*, equipped with an ammonia-compression machine. However, the real beginning of these exportations from South America came later.

[202]Critchell, 12–13, 26. The United States, however, is regarded as the pioneer of the exportation of chilled beef and the developer of the transport of meat under refrigeration. The Bell-Coleman Mechanical Refrigeration Co., developed a cooling and refrigeration system to be used on board ships.

[203]Critchell, 30–31.

[204]Stott, ix.; McLintock, A. Hare, J. O. Wilson, D. S. C, M. COM, A. L. A, C. Librarian, G. A. Library, Wellington and New Zealand Ministry for Culture and Heritage Te Manatu Taonga, *DUNEDIN—Te Ara Encyclopedia of New Zealand.* http://www.teara.govt.nz/en/1966/ships-famous/page-6 (18 November 2013); Critchell, 40.

[205]Anderson, JR., 124, 125. The machines used however, were not so efficient and the technology had to be improved, which drove to its replacement by some more efficient equipment after 1890; Stott., ix.

(1) **The Case Law**

In less than 15 years of extensive commercial use of refrigeration systems on ships, a defect in the apparatus that caused damages to the cargo was held as want of seaworthiness in *The Maori King* v. *Hughes* (1895).[206] The bill of lading for this carriage was headed "Refrigerator bill". The court found that, by agreeing to transport frozen meat, there was an implied warranty on the part of the carrier to provide refrigerating machinery, which had to be fit at the time of the shipment for the carriage of this kind of cargo.[207] The carrier had collected a higher freight for this special service. When the shipper agrees to pay an increased price, it is with the expectation that the carrier will provide that equipment in good working order, said the court.[208] The bill of lading contained a clause of exception of liability for damages caused by defects in the refrigeration machinery. The Court of Appeals made clear that such a clause could only be enforceable if the breakdown of the refrigeration apparatus occurred after the vessel sailed, but not if it occurred before. The Court remarked the existence of an implied warranty of seaworthiness at the beginning of the voyage. If the breakdown occurred before the commencement of the voyage, then it was inferred that the ship was not fit to carry frozen mutton. The refrigeration system was required because the nature of the contract so demanded.

The court, however, seems to have hesitated if a defect in the refrigerating systems constituted exactly a condition of unseaworthiness. The concept of seaworthiness was originally applied to the ship in itself. A refrigeration system was a totally new issue in shipping. However, by that time, it was already understood that seaworthiness meant also the capacity of the vessel to transport the specific cargo contracted. Lord Esher M.R., of the Court of Appeals, noted that when the judge of first instance referred to seaworthiness, it was not of the ship, but to the seaworthiness of her machinery, which must also be fit for the carriage.[209] The Court of Appeals affirmed the decision of first instance, recognizing the concept of seaworthiness as extending to the cargo capacity of the vessel. In consequence, the refrigeration capacity of the ship was part of her seaworthiness. Some years later, *Rowson* v. *Atlantic Transp. Co.* (1903),[210] came to clarify and include expressly within the concept seaworthiness, the capacity of the refrigeration machinery:

> Now, a vessel which has to carry cargo which can only be safely carried if its refrigerating machinery is in proper order is one which, at the present day, according to a series of

[206][1895] 2 QB 550. It was for the carriage of 4553 carcasses of "hard frozen mutton" from Melbourne, Australia, to London.

[207]Ibid., at 554: "An obligation, therefore, is to be implied from the bill of lading to have such machinery on board for the purpose of receiving the frozen meat; and the implication arises in the way in which all implications are made by law, and the only way in which they can be made, namely, that the Court can see that the implied obligation must have been in the contemplation and intention of both parties to the contract."

[208]Ibid., 554.

[209]Ibid., 555–556.

[210][1903] 1 K.B. 114.

> decisions, both in this country and in America, cannot properly be regarded as seaworthy unless it has that machinery in proper order. The term 'seaworthiness' is one which was originally, no doubt, used in days when refrigerating apparatus and other modern appliances for the safe carriage of cargo were unknown. In a sense it is obviously not a happy term to use, except with regard to that condition of the vessel which enables the owner to avoid exposure of the cargo to the perils of the sea.[211]

In the United States, it was similarly decided in the case *Martin v. Southwark* (1903).[212] Judge Day, of the Supreme Court of the United States,deciding the case under the Harter Act, cited as reference the twoaforementioned English decisions. The trade of meat had become bythat time an important part of the US foreign commerce. Shipownersstarted offering vessels equipped with refrigerating apparatus. Suchapparatus were under the sole supervision and care of the shipownersor his employees, without any control on the part of shippers. Thefreights were proportionally higher for this new service they offered. Thus, it was expected that such equipment was fit to receive the cargo contracted. In this case, in a previously superficial inspection, no defect was found in the apparatus. However, it broke down shortly after the vessel sailed for Liverpool and was repaired, but it broke down again and became incapable of reducing the temperature to the degree the cargo required. It caused the loss of a quantity of dressed beef. As the break down occurred within 1–3 hours after sailing, it raised the presumption that the vessel was not seaworthy at the time of sailing. The judge affirmed that for the particular characteristics of the cargo, it could not be transported on such a long voyage without that specific system. Therefore, considering that the vessel's seaworthiness must regard the special characteristics of the cargo to be transported, accordingly the malfunction of the refrigerating apparatus was held as a want of seaworthiness.[213]

Now, it must be kept in mind that the liability for defective refrigerating apparatus will depend in any case on the source or the character of the defect. The carrier would be exonerated if the malfunctioning of the device was the consequence of a latent defect. In *The Prussia* (1899),[214] for example, the apparatus was exhaustively tested beforeit was installed in the ship. Once in the ship, the system was againduly tested for over 24 hours prior to the sailing with. During that period, the machine worked properly and did so for over a month. Once a break down in the system arose, a cargo of dressed meat was injured due to lack of adequate temperature. A further examination of the machinery revealed that it had a latent defect from the manufacturer.[215] The court found that due diligence was duly practiced and the carrier provided a suitable vessel for the carriage of the specific cargo. The damage was due to latent defects in the

---

[211]As cited in *The SouthwarkSee supra* note 42, at 11.

[212]*See supra* note 42.

[213]Judge Day cited the following cases: *Stanton* v. *Richardson*, L. R. 7 C.P. 421; *Lyon* v. *Mells*, 5 East, 428 (1804); *The Thames*, 10 C.C.A. 232, 8 U.S.App. 580, 61 Fed. (F.) 1014 (C.A. 4 1894).

[214]93 F. 837, 35 C.C.A 625.

[215]Ibid. A leather washer was left into the apparatus probably inadvertently by the employer of the makers when putting it together. That could only be discovered by putting the apparatus apart and open by an engineer specially sent to inspect it.

apparatus, existing without knowledge or negligence of the carrier. However, the court seemed more comfortable in releasing the carrier from any liability, based on a clause of exoneration for latent defects included in the bill of lading that the court held as valid, under the Harter Act.[216]

It must be also considered that a breakdown in a ventilating system due to a faulty installation on the ship does not constitute a latent defect if such wrong installation can be detected by a reasonable inspection. In *Navigazione Libera Triestina v. Garcia & Maggini Co.* (1929),[217] some witnesses testified that the installation was madewith extreme care and following the prescriptions and regulations of the classification society Registro Italiano, which in turn, were exactly the same of those of the Lloyd's Register. However, the ship was found unseaworthy and that due diligence was not practiced. The court said that "the breakdown of the ventilating system was the result of a construction so faulty in the proportion, strength, and adjustment of its part that its insufficiency should have been discovered on inspection, and that there was absence of evidence of latent defects."[218]

We must observe that by the end of the 19th century and the beginning of the 20th, when most of these cases were decided, refrigeration systems were still new to the shipping industry. The courts made clear that once a technology is strictly necessary for the performance of the contract, because the related cargo depends on that, then that technology becomes part of the vessel seaworthiness. Refrigeration was required for the shipment of a cargo that, at that time, was also of certain importance for international trade. It was not an extra precautionary measure for the care of the cargo, but fundamentally necessary to assure the delivery of the meat in proper condition for human consumption. Therefore, the carrier must provide such apparatus and assure its proper functioning. However, these cases made clear that the carrier was released if the failure was caused by latent defect not discoverable through due diligence. Today, the assignation of liability for damages caused for defective reefer containers follows the same criterion.

4.2.3.2.2 Controlled Atmosphere Containers

A new technology that makes the carriage of perishable goods for long distances more efficient is the controlled atmosphere containers. Reefer ships and containers with modified or controlled atmosphere (CA) employ a technology that delays the time of ripening of fruits and vegetables. The metabolism of these products runs very fast

---

[216]It was in fact a negligence clause. The carrier set there a total release of responsibility for damages arising from defects or insufficiency in the refrigeration chambers or machinery, and even if caused by negligence, default, error in judgment on the master and other crew members or servants. The court held that, though this clause does not exempt the carrier for damages caused by his own negligence, it does when the damage arises from latent defect.

[217]30 F2d. 62 (C.A. 9th, 1929). It was a claim for damages to a cargo of bananas and pineapples caused by a breakdown in the refrigeration system.

[218]Ibid., 64.

since they are cut at their plantation to be transported long distances to the market where they will be sold. They are affected by the temperature, the level of oxygen they are exposed to and some other surrounding conditions.[219] Modified atmosphere (MA) was used first in reefer ships. It consisted of the application of a specific temperature while the level of oxygen was reduced.[220] But controlled atmosphere uses a more complex technology that enables some perishable products to have longer shelf life.[221] By 2006, 20 percent of the big reefer ships used the CA system, while 50 percent of the reefer ships run usually MA or CA systems. Around 3 to 4 million tons of bananas are transported with this system with an increasing tendency.[222]

This technology is being increasingly offered by cargo liners for the transport of perishable goods. It is especially useful for the carriage of perishable goods to the new Asian markets, which are experiencing fast growth and big demands for food. Exporting perishable goods from Latin America to the Asian markets may require this technology to assure a longer shelf life of the products. Without carriage in a controlled atmosphere container, the transport of some goods would be perhaps impossible, or at least not convenient nor profitable. The carrier´s failure in the preparedness of such systems before and at the beginning of the voyage, that then cause any malfunction during it, may raise liability for want of due diligence. The reasoning applied to cases on defective refrigeration may be extended to cases arising from defects in this technology.

4.2.3.3 Technologies Ordered by Statutes, International Regulations and Standards of Classification Societies

Through experience and the scientific development some minimum standards that allow safer navigation have been stated in statutory regulations. The opinion of Bouviers issued over a century ago, and reproduced in *The Southwark*,[223] may no longer hold up in court. Some aspects of the seaworthiness condition can be, and have been, established by positive laws. This is what we have explained as the objective

---

[219]K.-H. Hochhaus and H. Glandien, *Cool: Reefer-Technik mit Zukunft: Kühlschiffe; Markt, Transport und Perspektive* (Hamburg: Seehafen-Verl 2006, c 2006), 53. Though it was known since the time of the ancient Egypt and China that a reduction in the level of oxygen stops the respiration of the fruits and vegetables and its metabolism, this technique came to be commercially applied since the twentieth century.

[220]Ibid., 53. The reduction of the oxygen level normally occurs by placing the goods in the isolated ships holds as the fruits consume the surrounding oxygen, but such a process is quite long. To accelerate the process, the crew reduces or "cleans" the oxygen in the holds by adding nitrogen, which stays in the hold as the main component.

[221]Ibid., 53. In addition to the specific temperature and the control of the level of oxygen, the value of CO2 is also adjusted. It creates a regulated atmosphere that allows the reduction of loss of quality during the transport. As the levels of CO2 are higher, it acts as an inhibitor for the growth of fungi and bacteria which makes safer the transport of sensible cargo. For example, apples can be carried up to months conserving same fresh qualities.

[222]Ibid., 55.

[223]*See supra* note 42.

aspect of seaworthiness. Since the last century, public policy has been evidenced in the increasing and generalized governmental interest in regulating the ocean transportation. National and international regulations have been adopted to govern every aspect, from the financing of ships' construction until its sale as scrap.[224] At the present, there are regulations for some operative issues such as equipment, crewing, cargo handling, operations at ports, and environmental issues.[225] Such regulations cover not only an interest for safety, seafarer's health and liability issues, but also respond to national interests for security, prestige, competition, policy and trade protection.[226] Regulation established in national laws, international conventions, coast guard rules, or classification societies relating to the vessel's construction, equipment, manning, stowage methods, among other aspects, must be accomplished by the shipowner or they may be held liable for want of due diligence.[227]

Liability for not complying with statutory regulations related to navigation was already established in the 19th century in American law, particularly in collision cases. In *The Pennsylvania* (1873),[228] the Supreme Court of the United States ruled that the failure to comply with statutory regulations intended to prevent collision, creates a reasonable presumption that such failure "if not the sole cause, was at least a contributory cause of the disaster".[229] On that presumption, the court established that the shipowner must prove "not merely that her fault might not have been one of the causes or that it probably was not, but that it could not have been"[230] the cause of the incident. The rule is hard, but as the judge explained, it intended to enforce the obedience to the statutory regulations.[231]

This rule was initially and mostly applied for non-observance of rules of road and manning. Later, it was extended to cases for failure in providing equipment required by statute, as stated in *The Chickasaw* (1969).[232] In this case, the stranding of the vessel

---

[224]T. J. Schoenbaum, *Admiralty and maritime law*, Practitioner treatise series, 5th edn. ([St. Paul, MN]: Thomson/West 2011), 754.

[225]Ibid., 754.

[226]Ibid., 754.

[227]Ibid., 897, at f.n. 3 cites: *United States v. Ultramar Shipping Co.,see supra* note 198, at 898; *The Black Gull*, 250 F.2d 777, 1958 A.M.C. 277 (2nd Cir.); second appeal, 269 F.2d 68, 1960 A.M.C. 163 (2nd Cir.); *Columbus Co. v. Shore*, 276 F. 2d 93, 1961 A.M.C. 2557 (5th Cir. 1960) [refrigeration breakdown]; *State Marine Corp. of Delaware v. Producers Co-operative Packing Co.*, 310 F.2d 206 (9th Cir. 1962) [canned goods and dried fruits damp, stained, rusty, and moldy. *Elia Slazman Tobaco Co. v. S.S. Mormacwind*, 371 F.2d 537, 1967 A.M.C. 277 (2nd Cir. 1967) [Failure to care for the cargo of tobacco because of improper dunnage and ventilation]; Tetley, *Marine Cargo Claims,* Vol. 1, 899, *see The Kapitan Sakharov* [2000] 2 Lloyd's Rep. 255, 266 (C.A.).

[228]86 U.S. 125, 1998 A.M.C. 1506, 22 L.Ed. 148, 19 Wall. 125 (1873).

[229]Ibid., 136.

[230]Ibid., 136.

[231]Ibid., 136.

[232]*Waterman S. S. Corp. v. Gay Cottons*, 414 F.2d 724, 737, 20 A.L.R. Fed. 810, 1969 A.M.C. 1682, 13 Fed.R.Serv.2d 783414 F2d. 724, 737. (C.A. Cal. 1969.) To support this statement, Judge Duniway cited the following cases: *The Martello*, 153 U.S. 64, 74, 14 S. Ct. 723, 38 L.Ed. 637 (1894); *Richelieu & Nav. Co. v. Boston Marine Ins. Co. (The Spartan)*, 136 U.S. 408, 10 S.Ct.

caused its total loss and substantial loss of the cargo. The occurrence of the stranding was attributed to defective equipment on the ship. The ship was so badly equipped that the captain had no accurate source of information to determine her location. Apart from an inoperative Fathometer, the mechanical sounding device was taken away, the radar was broken, and the radio direction finder was defective with no recent compensation or deviation cards on board. By the time of the stranding, the US Code established that vessels leaving harbors or ports of the United Sates navigating in open seas must be equipped with an efficient radio direction finding apparatus, properly adjusted, in operating condition and approved by the Federal Communication Commission.[233] Applying the Pennsylvania rule, the court held the carrier liable for not having such technological devices in good working order.

Therefore, the lack of fulfillment of regulationsestablishing the application of certain technological equipment onthe vessel or newer and safer methods of stowage raise liability for the carrier if such a failure is the cause of the damage or proximately contributed to the damage or loss. Regulations establishing technological equipment are to be found in the national laws of some countries. But it has been in the committees of the International Maritime Organization (IMO), that such newer standards for safety of navigation have been more studied and then established through the adoption of some international conventions. Classification Societies and the International Organization for Standardization also offer some recommendations regarding navigation or safer stowage and care of the goods.

a) The Conventions of the InternationalMaritime Organization

(Footnote 232 continued)

934, 34 L.Ed. 398 (1890); *Pacific Tow Boat Co. v. States Marine Corp.* (*The Cotton State*), 276 F.2d 745, 749 (9th Cir. 1960); *Van Camp Sea Food Co. v. Di Leva*, 171 F.2d 454, 456 (9th Cir. 1948); *The Niagara*, 84 F. 902, 904–905 (2nd Cir. 1898); Wood v. *United States* (*The Wilson Victory*), 125 F. Supp. 42, 52–53 (S.D.N.Y. 1954); *The Iowa*, 34 F. Supp. 843, 848–849 (D. Or. 1940); *The Trave*, 55 F. 117, 119–120 (S.D.N.Y. 1893), rev'd on other grounds, 68 F. 390 (2nd Cir, 1895), cert. denied, 163 U.S. 692, 16 S.Ct. 1203, 41 L.Ed. 309 (1895); *The Bolivia*, 43 F. 173 (E.D.N.Y., 1890), modified on other grounds, 49 F. 169 (2nd Cir. 1891). See, *Coryell v. Phipps* (*The Seminole*), 317 U.S. 406, 409, 63 S.Ct. 291, 87 L.Ed. 363(1943); *China Union Lines, Ltd. v. A. O. Anderson & Co.* (*The Union Reliance*), 364 F.2d 769, 787(5th Cir., 1966); *Coleman v. Jahncke Service Inc.*, (*The Claribel*), 341 F.2d 956, 958 (5th Cir. 1965); *States S.S. Co. v. United States* (*The Pennsylvania*), 259 F.2d 458, 464 (9th Cir. 1958); *The Cleveco,*, 154 F.2d 605, at 614 (6th Cir., 1946); *The Severance*, 152 F.2d 916, 922 (4th Cir., 1945); *The E. Madison Hall*, 140 F.2d 589, 591(4th Cir., 1944); *Petition of Long* (*The Smith Voyager*), 293 F. Supp. 172, 178 (S.D.N.Y., 1968); *Navegacion Castro Riva, S. A. of Panama v. The M. S. Nordholm*, 178 F. Supp. 736, 741 (E.D.La., 1959); *Kutger v. United States*, 169 F. Supp. 104, 106 (N.D.Fla. 1958); *Kulack v. The Pearl Jack*, 79 F. Supp. 802, 806 (W.D.Mich. 1948).

[233] 47 U.S.C. § 351 (a) (2), (1964). The court also quoted the provision stated in 47 C.F.R. § 8517 (a) [at the time of the decision it was 47 C.F.R. § 83.459 (1968)], providing that the Radio Direction Finder must: "(4) be accurately calibrated for the purpose of taking bearings from which the true bearing and direction may be determined for actual use in radio location service and maritime radio navigation service."

The International Maritime Organization, since its creation in 1948, has adopted a series of international conventions establishing specials rules covering primarily safe navigation, the protections of the environment and the facilitation of commerce. It was explained that the relative character of the seaworthiness conditions, depends, among other factors on: (1) the type of the vessel; (2) the characteristics of the ports of loading and discharge; (3) the route needed for the voyage; (4) the season of the year; (5) the foreseeable weather condition; and of course; (6) the type of cargo and the method of stowage it requires. Similarly, we have seen the different components in which the concept of seaworthiness has been fragmented. The IMO has adopted some conventions and codes establishing minimum safety standards regarding some of the objective aspects of the ship's seaworthiness. These standards are precisely the result of the perfection of knowledge, the experience of marine risks, higher degrees of skill in navigation and scientific discoveries.[234]

Hence, there are currently international regulations relating to vessel's design, construction and equipment, manning and training of seafarers, safe container, port conditions and processes, methods for safe stowage, carriage of certain types of cargo, among others.[235] The most significant of them, regarding the technologi-calequipment and newer standards for the vessel, is the Convention forthe Safety of Life at Sea (SOLAS). It contains two codes that relatedirectly to the vessel's sea-worthiness: the ISM Code, the ISPS Codeand the CSS Code, all of them setting some standards regarding save navigation, port operation and proper stowage.

(1) **International Convention for the Safety of Life at Sea (SOLAS 1974)**

This Convention is recognized as the most important international instrument on safety of merchant ships. The first version was adopted as consequence of the Titanic incident. For its third version, the International Conference on Safety of Life at Sea held in London in 1948 decided to include as mandatory, that all vessels over 1600 tons gross involved in international voyages, be they passenger or cargo carrying vessels, must carry some technological devices such as direction finding apparatus, and must keep continuous radio watch.[236] Furthermore, cargo ships between 500 and 1600 tons gross have to be equipped with a radiotelegraph or a radio telephone installation, and with it, a respective radio officer or a qualified radio telephone operator.[237] The standard of the radio used was also increased, on ships over 1600 tons gross to include radios with a capacity of 150 nautical miles of minimum range.[238] Radars were not initially included as compulsory in this

[234]*See* Judge Story opinion in *Tidmarsh v. Washington Fire & Marine Ins. Co., supra note* 54.

[235]*See* International Convention on Load Lines (1966); Convention on the International Regulations for Preventing of Collisions at Sea (COLREGs 1972); International Convention for Safe Containers (CSC 1972); International Convention on Standards of Training, Certification and Watchkeeping for Seafarers (STCW 1978); International Convention on Maritime Search and Rescue (SAR 1979).

[236]Hancock, 204.

[237]Ibid., 204.

[238]Ibid., 204.

Convention; it was thought that the time had not yet come to make it a mandatory application.[239] This was established later as this Convention has been kept updated with certain regularity.

Two more versions of the Convention were adopted, one in 1960 and the last one in 1974. The latter introduced a mechanism for the permanent introduction of amendments for its update, currently known as SOLAS 1974 as amended. In its consolidated edition of 2009, the current Chapters IV and Chapter V, order the use of Radio communications and equipment for safe navigation respectively. Both chapters establish the obligation for shipowners to equip their ships with technological devices that assure safer navigation. Chapter V orders for example, and varying according to the tonnage of the ships, the use on board of ships routing,[240] ship reporting systems.[241] Among the most significant provisions regarding the technological equipment, the regulation 19.2.1 presents a list of equipment that must be on board every vessel:

> Regulation 19.
> 2. Shipborne navigational equipment and systems
> 2.1 All ships, irrespective of size, shall have:
> .1 a properly adjusted standard magnetic compass, or other means, independent of any power supply, to determine the ship's heading and display the reading at the main steering position;
> .2 a pelorus or compass bearing device, or other means, independent of any power supply, to take bearing over an arc of the horizon of 360°;
> .3 means of correcting heading and bearing to true at all times;
> .4 nautical charts and nautical publications to plan and display the ship's route for the intended voyage and to plot and monitor position throughout the voyage; an electronic chart display and information system (ECDIS) may be accepted as meeting the chart carriage requirements of this subparagraph;
> .5 back-up arrangements to meet the functional requirements of subparagraph .4, if this function is partly or fully fulfilled by electronic means;
> .6 a receiver for a global navigation satellite system or a terrestrial radionavigation system, or other means, suitable for use at all times throughout the intended voyage to establish and update the ship's position by automatic means;
> .7 if less than 150 gross tonnage and if practicable, a radar reflector, or other means, to enable detection by ships navigating by radar at both 9 and 3 GHz;
> .8 when the ship's bridge is totally enclosed and unless the Administration determines otherwise, a sound reception system, or other means, to enable, the officer in charge of the navigation watch to hear sound signals and determine their direction;
> .9 a telephone, o other means, to communicate heading information to the emergency steering position, if provided.

This is the basic equipment required by all ships. The regulation sets higher standards and more sophisticate equipment for bigger vessels, attending to its

[239] Ibid., 205.

[240] Regulation 10.

[241] Regulation 11.

tonnage. This regulation reflects the evolution of scientific advances in the means of navigation. Most of the technologies developed during the past century became of mandatory application. But the regulations stated in SOLAS are not limited to order the implementation of the technological equipment set there. The IMO has also issued a publication since 1988 entitled Performance Standards for Navigational Equipment.[242] This publication sets a series of basic performance standards that are regularly updated for all the radio communications and navigational equipment that must mandatorily be on board.[243]

On the carriage of cargoes, SOLAS does not established detailed standards for the care of cargo. It addresses in Chapter VI under the title "Carriage of cargoes" some very general provision on cargo information and stowage. It refers to the Code of Safe Practice for Cargo Stowage and Securing (CSS Code).[244] It also presents some provisions for bulk cargoes other than grain and for the carriage of grain. For the latter refer to the Code for the Safe Carriage of Grain in Bulk.[245] Chapter VII presents the standards for the carriage of dangerous goods.

Courts of the United States and England have established that not complying with the regulations contained in this Convention results in the vessel being unseaworthy. A failure in the application of the SOLAS rules relating to proper grain stowage rends a vessel unseaworthy. In *United States v. Ultramar Shipping Co* (1987),[246] though the SOLAS regulations did not exactly apply to the type of vessel involved, the judge held that the court had to determine if the inobservance of such standards was so "dangerous and unreasonable" as to be considered unacceptable.[247] The court held the vessel liable, stating that the improper stowage, not made in accordance of the SOLAS regulations, and the instability of the vessel contributed proximately to the loss.[248]

---

[242]"*General requirements and performance standards for shipborne radiocommunications and navigational equipment: Including related operational recommendations and guidance; incorporating the 2008 Amendments and all amendments adopted up to December 2010*", IMO publication, 6th. ed., 2011 ed. (London: IMO, 2011), iii.

[243]It includes standards for Magnetic and gyro compasses, heading control systems and automatic pilots, radar and ancillary equipment, electronic charts display and information systems, radio navigational equipment, radio direction-finding equipment, echo-sounding equipment, speed and distance indicators, rate in turn indicators, radar beacons, transponders and reflectors, satellite communications, among others.

[244]Chapter VI, Regulation 2.2.1. The CSS Code was adopted by IMO by resolution A.714(17).

[245]Chapter VI, Regulation 8. The International Grain Code was adopted by the Marine Safety Committee by resolution MSC.23(59).

[246]*See supra* note198, at 898.

[247]Ibid., 898. *See Bache v. Silver Line (The Silversandal),* 110 F.2d 60, 62 (2nd Cir.1940).

[248]Ibid., 900.

aa) The International Safety Management Code (ISM Code)

The International Safety Management Code (ISM Code) is the abbreviated name for the "International Management Code for Safe Operation of Ships and for Pollution Prevention". The code was adopted by the IMO on November 4, 1993, and came into force on the 1st of July of 1998. It was in response to a series of incidents involving roll-on/roll-off in the 1980s and the first years of the 1990s.[249] The IMO added this code in 1995 to the SOLAS Convention, as Chapter IX entitled "Management for the Safe Operation of Ships".[250] The Code develops a series of regulations establishing,among other issues, the obligation for shipping companies to developand maintain a Safety Management System (SMS). It applies to all oiltankers, chemical tankers, gas carriers, bulk freight and high speed craft of 500 gross register tons and over in international voyages.[251] All other cargo ships of 500 gross register tones are since the 1st of July 2002 subject to the regulation of this code, governing in consequence the majority of the merchant fleet of the world.[252]

The Code establishes a series of safety measures to be followed by the "Company", meaning not only the shipowner but extending the responsibility to managing companies, bareboat charters, and other organizations of persons who assume the responsibility for the operation of the ship.[253] The mandatory measures underline many aspects of the seaworthiness condition. The vessel must comply with a safety management system based in the existing mandatory rules and regulations. This SMS must take into account all the applicable codes, guidelines and standards established by the IMO, the national administrations, classification societies and maritime industry organization.[254] It addresses, for example, the human worthiness of the ship. The Company must assure that the master is properly qualified, conversant with the Company's SMS, and properly supported to perform his duties.[255] In the same way, the crew must be "qualified, certificated and medically fit, in accordance with national and international requirements."[256]

Among other issues, it addresses with special attention the performance of inspections. An audit process is required for the issuance, by the flag State, of a Document of Compliance (DOC) with the ISM Code.[257] This DOC must be kept on board.[258] Accordingly, a Safety Management Certificate (SMC) is issued once the ship has been inspected and it is confirmed that the management of the ship

---

[249]S. Girvin, *Carriage of Goods by Sea*, 2nd edn. (Oxford, New York: Oxford University Press, 2011), 393.

[250]Ibid., 393.

[251]Tetley, *Marine Cargo Claims,* Vol. 1, 939.

[252]Ibid., 939.

[253]Ibid., 939.

[254]Ibid., 942. *See* Articles 1.2.3 of the ISM Code.

[255]ISM Code Art. 6.1.

[256]ISM Code Art. 6.2.

[257]ISM Code Art. 13.4 and 13.5.

[258]Girvin, 393.

complies with the SMS.[259] The absence of such certificates onboard the vessel might be considered a cause of unseaworthiness and evidence of a want of due diligence.[260] It establishes an obligation to ensure that the vessel is maintained according to the pertinent regulations.[261] For such a purpose, the Company must ensure that inspections are made in an appropriate interval; make reports of the non-conformities and its possible causes; take proper corrective actions; and record all these activities.[262] It demands the existence of procedures of identification of technical systems and equipment whose failure might threaten the vessel's operation and safety. Similarly, the Company is obliged to take measures to promote the reliability of such equipment and its regular testing of stand-by arrangements and also of equipment or technical systems that are not in continuous use.[263]

All these provisions demand special attention of the shipowners or their agents. The non-compliance with these international standards is regarded as a want of due diligence in making the ship seaworthy. A failure in the inspection of technological equipment, that subsequently causes cargo damage or losses holds carriers liable for a want of due diligence. The impact of this code in the seaworthiness condition was analyzed in *The Eurasian Dream* (2002).[264] It was a claim for the total loss of a cargo due to a fire on board the ship. The vessel was held unseaworthy in three aspects: its lack of equipment, the master and crew's competence and the adequacy of the documents on board.[265] On the ship there were no records of inspections and examination of equipment such as fire extinguishers, gas tight doors and CO2 smothering equipment. Neither did the vessel have enough number of walkie-talkies and breathing apparatus sets on board. In addition, the master was not competent in this type of ship and the officers and crewmembers had no knowledge of safe operations for this type of ship.[266] To assign liability, the judge applied the standards established in the ISM Code. He affirmed that the seaworthiness condition of a vessel must be evaluated attending to the "standards and practices of the industry at the relevant time."[267] A large part of the current standards of the industry is precisely established in the ISM Code.[268]

Notwithstanding, such standards are not in and of themselves grounds for demanding perfection of the ship. In *The Torepo* (2002),[269] the vessel was accused

---

[259]Ibid., 393.

[260]Tetley, *Marine Cargo Claims,* Vol. 1, 943.

[261]ISM Code Art. 10.1.

[262]ISM Code Section 10.2.

[263]Ibid., 943; IMS Code art. 10.3.

[264][2002] 1 Lloyd's Rep. 719. (QBD (Comm. Ct.), Cresswell J.).

[265]P. Anderson, *ISM code: A practical guide to the legal and insurance implications*, Lloyd's practical shipping guides, 2nd edn. (London: LLP, 2005), 227.

[266]Ibid., 231.

[267][2002] 1 Lloyd's Rep. 719, at para. 127.

[268]Anderson, 233.

[269][2002] EWHC 1481 (Admlty).

of being unseaworthy for, among other reasons, not being equipped with adequate charts and because the echo sounder was defective. The court did not hold the vessel liable. It argued that the seaworthiness standard is not one of perfection; neither the courts nor the ISM demand perfection, but the best practices attending the particularities of the case.[270]

bb) The International Ship and Port Facility Security Code (ISPS)

The International Maritime Organization adopted this Code on December 12th, 2001, coming into force on July 1st, 2004. It was in response to the terrorist attack of September 11th, 2001. It was also incorporated into the SOLAS Convention as part of the Chapter XI-2. It applies to, among others, to cargo ships of 500 gross tonnage and upwards.[271] The "Company", meaning the same as for the ISM Code, must comply with new and more demanding standards of due diligence.[272] The obligations set in this code basically related to the safety of the vessel at sea, but more specifically at the port facilities. It sets some regulations to be accomplished by the port authorities as well. The code places again a strong emphasis in the vessel´s inspections. It orders a Ship Security Assessment (SSA) to be performed by persons with appropriate skills in such tasks or by a Recognized Security Organization (RSO).[273] It is required to obtain an International Ship Security Certificate (ISSC), issued by the contracting State, attesting to the compliance of the vessel with the ISPS regulations.[274] A copy of the Ship Security Plan (SSP) must be on board too. Its absence, as well as the absence of certificates, might be considered as a want of due diligence in making the ship seaworthy. In the same regard, the Company must ensure that the crew is properly trained for implementing the security measures on the vessel, if not, it may render the vessel unseaworthy.[275] On technological issues, there is an obligation, for example, to have security equipment on board that sends a signal to shore in case of any emergency or security threat.[276]

cc) Code for Safe Practice for Cargo Stowage and Securing (CSS Code)

As a response to accidents occurring due to improper stowage and inadequate securing arrangements on board ships, this code was adopted on 6th of November

[270]Anderson, 234. This vessel was 20 years old but was kept in good condition. The ship's stranding was attributed to mistakes and human errors, not to unseaworthiness.

[271]ISPS Code Section 3.1.1.2.

[272]Tetley, *Marine Cargo Claims,* Vol. 1, 946.

[273]ISPS Sections 8.2 and 8.3.

[274]Tetley, *Marine Cargo Claims,* Vol. 1, 948.

[275]Ibid., 948.

[276]SOLAS, Chap. XI-2, Regulation 6.

of 1991.[277] It was developed by the Sub-Committee on Containers and Cargoes upon instruction given by the Maritime Safety Committee.[278] Since then, it has been amended and updated to include some other related aspects. The advances of knowledge and technical skill in the operation of stowage of cargoes on ship are expressed in this code. Its main objective is to ensure the safety of life at sea through the prevention of accidents caused by faulty stowage. Its compliance, however, benefits the safe carriage of cargo as well. The Code presents a series of technical standards for safe stowage and securing of cargo such as containers on deck of ships not designed for carrying containers, portable tanks, wheel based cargoes, heavy cargo, coiled sheet steel, anchor chains, metal scrap in bulk, and some others. As the obligation of due diligence to make the ship seaworthy includes the proper stowage of the cargo on board, this code contains the current standards of the industry regarding this aspect.

(2) **International Convention for Safe Containers (CSC 1972)**

Since the 1950s, containers have been one of the most common and effective methods for the carriage of cargoes by sea. As a result, the IMO adopted in 1972 a Convention establishing some minimum safety standards, entered into force the 6th of September of 1977. Its adoption pursued two main objectives. The first was to ensure a high level of the safety of human life in the transport and handling of containers, and the second was to facilitate the international transport of containers by the creation of uniform rules of general applications.[279] This Convention does not apply for all type of containers, but only for those of a specific minimum size stated in the code and that have corner fittings.[280] The conventions basically establishes an obligation of the contracting parties to make inspections and test any new and existing containers to assure that they meet the basic technical standards set in the annexes.[281] The Administration must issue a Safety Approval Plate to be fixed on every approved container.[282] The owner of the container is responsible for maintaining its safe condition and must examine it when appropriate.[283] With this regulation, it is expected that a diligent carrier will provide containers that have a Safety Approval Plate, which supposes that the respective examinations have been performed upon it.

---

[277] *CSS code: Code of safe practice for cargo stowage and securing*, 3rd ed., 2011 ed. (London: International Maritime Organization, 2011), 1.

[278] Ibid., v.

[279] *International Convention for Safe Containers, 1972: CSC*, 5th edn. (London: IMO, 2012), v.

[280] Ibid., v. It excluded container used for the carriage by air.

[281] CSC Article IV.

[282] CSC Annex I, Chapter I, Regulation 1(a).

[283] CSC Annex I, Chapter I, Regulation 2.

b) The Standards of Classification Societies

As aforementioned, the ISM Code orders in the rule 1.2.3 the observation of the regulation established by the classification societies. However, the application of those standards as part of the vessel's seaworthiness was declared by judicial decision long before the enactment of this Code. Classification societies offer a source of some standards of seaworthiness that can be required as part of the obligation. This was demonstrated in the *Atlantic Transport Co. v. Rosenberg Bros. & Co.* (1929).[284] Cargo suffered damage due to sea water entering into the holds through the chain lockers during a storm. The American Bureau of Shipping issued a regulation in 1920 stating that all chain lockers abaft the collision bulkhead had to be water-tight. Similar regulation was later in 1925 published by the Lloyd's Register. The judge argued that for these organizations it was essential for the safety of the passengers, cargo and ships that chain lockers located abaft the collision bulkhead were kept water-tight. Making the chain locker water-tight was a reasonable method of protecting the cargo from water coming through the chain pipes. If such a regulation was the result of "the concensus [sic] of opinion of the shipping world, by reason of experience gained in shipping industry", then, the duty of the shipowner must be assessed taking into account this regulation.[285] The injuries to the cargo were not caused by the heavy weather, which might have exonerated the carrier from liability. The decision held that the injuries were the result of the shipowner having ignored the regulation prescribed by these "great organizations", as the judge called them.[286] The judge also considered the cost of adapting a permanent water-tight system to the chain locker. It could have cost between $300 and $400, and the expenses saved would have been much higher than that.[287] So, the cost was not so unreasonable in proportion to the damage it attempted to prevent, nor was there any evidence that accomplishing such a task was not possible. The shipowner neglected the reasonable precautions set out in the regulation causing the judge to arrive at the conclusion that due diligence was not practiced. The original construction of the vessel made her unseaworthy for the carriage of cargo in the forward hold at the time this carriage took place. After a new regulation of the classification societies was made known, stating the necessity of the water-tightness of a particular area of the vessel, the shipowner should have observed such regulation to grant the safe carriage.

However, the compliance with those regulations does not make the ship automatically seaworthy. The inspections and regulations made by classification societies following the standards provided in their manuals must be also performed with due diligence. In *Navigazione Libera Triestina v. Garcia & Maggini Co.* (1929),[288]

---

[284] 34 F.2d 843 (C.A. 9 1929), 1929 A.M.C. 1539.

[285] Ibid., 846.

[286] Ibid., 845.

[287] Ibid., 846.

[288] 30 F.2d 62 (C.A. 9 1929).

the carrier alleged in his favour that he had installed a ventilating system with the greatest care and according to the regulations established by Registro Italiano, which were the same used by the Lloyd's Registry.[289] Though both classification societies enjoyed at that time, and still do, a high reputation, the work done in the installation of such equipment was found faulty and the carrier was held liable. Judge Gilbert C.J. said that the diligence required "is diligence with respect to the vessel, not in obtaining certificates."[290]

c) The Standards Provided by the Other Organizations

The standards established by the IMO Conventions relate mostly to the safety of the vessel. They became part of the objective aspect of the ship's seaworthiness. But these international Conventions do not address directly issues relating to the care and protection of cargo. Only the Code for Safe Stowage (CSS) offers some guidelines that have more impact in securing the safe carriage of cargo. Also the International Grain Code and International Maritime Dangerous Goods Code (IMDG) present some standard for those specific types of goods. Setting those aside, there is not any other convention or code establishing safety standards for the protections of cargoes. The reason for the lack of international regulation on the subject may be found in the fact that it is not among the main objectives of the IMO. As established in their regulations, the IMO's primary interests are in the protection of the human life at sea, the protection of the environment and the facilitation of commerce.[291] Though the core of the shipping industry is the carriage of cargoes, this has not been subject to more attention. Another possible reason may be that discussing, determining and setting standards for cargo protections may be a too complicated task to undertake, considering the vast amount of different cargoes that are daily transported and the different circumstances involved in every shipment. A third reason is the reliance of the industry on insurance. Cargoes are usually double insured by the shipowners as well as by the same cargo owners. The effects of cargo damages or losses, though being negative not only for the parties but also for the countries' economy, are partially ameliorated by the compensation paid by the insurance companies.

Therefore, to determine the current standard of theindustry, courts have also taken into account the standards providedby private organizations such as the National Cargo Bureau as stated in the case of *U.S. v. Ultramar Shipping*.[292] Another possible source of standards recommending the use of technological advances for the protection of cargo is provided by the International Organization for Standardization. After the adoption of the ISPS Code, the ISO Technical

---

[289]Ibid., 63.

[290]Ibid., 64, *citing The Abbazia* 127 F. 495 (D.C.N.Y. 1904), 496.

[291]IMO | FAQs. http://www.imo.org/About/Pages/FAQs.aspx Accessed 08.05.2014.

[292]*See supra* note198.

Committee ISO/TC 8 published the ISO 20858:207, related to Ships and Marine Technologies—Maritime Port Facility Security Assessment and Security Plan Development.[293] Other contributions of this Organization are the ongoing development of the ISO 28000 on Security Management System for the Supply Chain and the ISO 28000:2007 on Specification for Security Management Systems for the Supply Chain. They both address the risks of the supply chain operations. The second one contains special guidelines on emergency preparedness, business continuity, sustainability, recovery, resilience and/or disaster management, whether related to terrorism, piracy, cargo theft, fraud and other security disruptions.[294]

4.3 Carriers Liability for Lack of Application of New Technologies

4.3.1 Introduction

To the question of what type of technologies can be enforced as part of the obligation set in H/H-VR, the case law revised in the previous section showed that four groups of technologies divided according the source or the reason of this application may result in liability when not applied:

1. Technologies ordered by mandatory regulation or standards of classifications societies;
2. Technologies which application is contractually agreed;
3. Technologies applied as a common practice;
4. Technologies applied as imperative precautions.

Once it is determined if a technology falls within one of these groups, liability can arise in three situations: (a) if it is not on board; (b) if it is on board but defective; and (c) if it is on board and in proper working condition, but the master, officer or crew lack the necessary knowledge for its correct use, operation or interpretation of the information it provides. Of course, liability will be assigned if there is a causal link between the failure and the damage or loss of cargo.

The assignation of liability for non-compliance with the application of technological devices or newer and safer methods set in the first two groups does not offer major problems. In the first case, it is a breach of the statutes, while in the second, it is a breach of the special conditions of the contract. Once the causal link is demonstrated between the non-application or its defectiveness or the lack of knowledge of the crew in its use, and the damage or loss claimed, a judge should not have major problems in holding carriers liable.

---

[293] UNCTAD, '*Review of Maritime Transport 2012*', 109–110.
[294] Ibid., 110.

Difficulties appear when the application of new scientific advances is claimed as common use or imperative precautions. These cases demand the establishment that a technology is currently in common usage or that it must be required as an imperative precaution. Both situations require evidence on the specific issue of the technology and, in consequence, parties would be obliged to go to court to determine the liability. Radio receiver sets and compass were both enforced as part of the ships seaworthiness in the *T.J. Hopper*[295] and in *The Claribel,*[296] respectively. Though these devices were not mandatory for tugs at that time, nor were applied as common practice, the particular circumstances persuaded judges to consider them as imperative precautions and assign liability for not having them.

But the same treatment was not given to radar. The accuracy and effectiveness of radar was sufficiently recognized in the prevention of casualties such as collisions. It was even shown that the expectation of judges was that big vessels were equipped with this device. Notwithstanding, during the first decades of commercial use of this technology, judges were reluctant to enforce it as part of the ship's seaworthiness. Courts demanded that it be a common practice or mandatory by statute, in order to demand it as part of the vessel's seaworthiness. But such an approach may have the effect of discouraging or delaying the use of technologies that have proven to be reliable and, given the risk it intends to prevent, it was also cost efficient. It raises then the question of how extensive the use of a technology must be to be considered a common practice. Judge Cox, in the first instance of the *T.J. Hopper*, took into account a witnesses' testimony asserting that 90% of the tugs in that area were equipped with radio sets.[297] A 90% rate of usage might be too high an expectation, making it impossible to require any technology, though while not yet reaching such a level of use, has shown its efficiency. The case law, however, does not offer an average of use extension that may be taken as a reference. That absence of reference will, therefore, demand more litigation for new technological issues.

The criterion of imperative precaution was not applied to radar. Although it was determined that certain navigational circumstances demanded a higher level of care, similar to those required for the compass, its implementation on board was not established as an imperative measure. In fact, the risks expected to be prevented by the proper use of compass were similar to those that could be avoided by the proper use of radar in many of the analyzed cases. During the 1950s and subsequent decades, when the international trade flourished after the Second World War, certain areas, maritime routes and ports saw heavy traffic. Radar was doubtless a very useful tool to prevent such catastrophic incidents as collisions or strandings, which cost not only the cargo but human life as well. The difference in the treatment given to these two technologies—radar and compass–, seems to be based on the complexity and cost of the technology. Keeping in mind that seaworthiness, as well as the duty of caring the cargo, must be measured according to a standard of reasonability; these elements have been taken into

[295] *See supra* note 77.

[296] *See supra* note 104.

[297] *See supra* note 77, at 111.

account. Certainly, reliability and accuracy of a technology is a requisite to demand its application. But its mere effectiveness has not been enough to require the application. The reasonability of the obligation must also take into account the economic implication of enforcing judicially the use of such technology. Courts and shipowners do not necessarily measure reasonable care in mathematical terms, but it cannot be denied that in their analysis considered the economic burden of demanding a technological precautionary measure.[298] Hence, another aspect that must be regarded is the cost efficiency of the technology.

4.3.2 The Cost Efficiency of New Technologies

The costs of applying a technology compared with the damages or losses that it intends to prevent must also be taken into account when assessing liability. Economic analysis of law has become ever more accepted and used instrument in measuring the consequences of the application of legal rules. Its functionality provides another valuable criterion to evaluate the convenience of enforcing the application of new technologies in maritime transportation.

4.3.2.1 The Consequences of the Minimum Liability Set in the H/H-VR

The liability rules, in general, serve more than only the purpose of forcing the negligent party to pay compensation to the injured party. In the specific case of the carriage of goods by sea, the purpose of the liability rules set in the H/H-VR is not only to require carriers to pay compensation to shippers for the damages or losses caused by their negligence. When the cargo is insured, as it mostly is, compensation from the liable party may even be irrelevant.[299] Liability also pursues the deterrence of negligence and the creation of incentives to take proper care.[300] If there were no threat of liability, shipowners would not have any incentive to prepare their ships and to take the necessary measures to accomplish the carriage safely and successfully. Such was the experience during the second part of the 19th century with the extended insertion of the negligence clauses into the bills of lading. No liability for shipowners resulted in many ships in poor condition being put out to for carriage which subsequently and frequently sank and were referred to as "coffin ships".[301] Lord Diplock commented on the consequences of total exclusion of liability for shipowners in the following terms:

---

[298]M. M. Billah, *Effects of insurance on maritime liability law: A legal and economic analysis* (Cham: Springer, 2014), 108.

[299]Ibid., 103.

[300]Ibid., 116.

[301]*See supra*chapter2, footnote 376.

> If, under the terms of the standard contract of carriage, the carriers were immune from all liability for loss or damage which could have been avoided by physical precautions taken while the goods are in his custody, he would have not commercial inducement to expend money on precautions to preserve the cargo from loss or damage which were not also required for the safety of the vessel, even if the cost were small in comparison with the resultant reduction in the risks of loss or damage.[302]

When liability is clearly allocated, the performance of the individuals subjected to such rule tends to improve, reducing damages and losses.[303] This is the main purpose of liability under the economic analysis.[304] A proper allocation of risks means allocating said risk to the party that can bear that risk for less cost.[305] Economic efficiency requires that the court holds liable for breach of contract the party that is in a better position to manage the risk.[306] A way to bear, prevent or at least reduce the risk is the implementation of precautions against it.[307] The process of getting the information to determine the appropriate precautionary measures carries with it also a cost. Moreover, the negotiation between carrier and shipper in deciding the application of precautions implies an additional cost. Hence, this is the function of establishing liability through statute.

The H/H-VR distributes the risks allocating to the carrier to bear the consequences of his negligence in the performance of the two duties that are obviously under his control. Keeping the vessel in proper condition of navigability requires information that may be only accessible to her owner. The same applies for the care of the cargo when it is under the carriers' custody.

The rest of the risks must be borne by the shipper, including negligent management of the ship by the carriers' servants, which the shipper has no control of. Risks related to the inherent condition or packagings of the goods are also allocated

---

[302]K. Diplock, 'Conventions and Morals—Limitation Clauses in International Maritime Conventions' (1970) 1, Journal of Maritime Law and Commerce, 525–536., 527.

[303]Tetley offers an interesting example of this phenomenon: "[…] as long a theft after discharge in the port of Montreal (and many other ports on the world) was not held to be the responsibility of the stevedore or terminal operator, thefts and losses continued unabated, insurances rates rose and cargo business was diverted elsewhere so that the port suffered a severe depression. When the responsibility of the stevedores and terminal operators was properly and clearly defined by the courts, the incidence of theft was greatly reduced and the port prospered". Tetley, *Marine Cargo Claims*, Vol. 1, 1829, f.n. 5. *quoting* Erling Selvig, "The Hamburg Rules, The Hague rules and Marine Insurance Practice" (1981) 12 J. Mar. L. & Com. 299 at p. 310 et seq.

[304]Billah, 103, *citing* Shavell, S. Foundation of Economic Analysis of Law. Harvard University Press, Cambridge. 267–269.

[305]R. Cooter and T. Ulen, *Law and economics*, 5th edn. (Harlow: Pearson Addison-Wesley, 2008), 7–8.

[306]Ibid., 8.

[307]Ibid., 8.

to the shipper.[308] The same for the rest of the exculpatory causes listed in the Article 4.2. All these causes refer to situations that are beyond the carrier′s control, and to enforce liability in those cases may oblige the shipowner to take excessive measures of care that may increase the operational costs.

The minimum liabilities set there do not create enough incentives, on the part of the carriers, to invest more in preventing damages and losses caused by negligence.[309] This disinterest is then evidenced in the most commoncauses of damages. For example, the exoneration of liability for bad seamanship is perhaps the reason of many casualties due to negligent navigation. Many of the cases related to the use of radar revealed bad seamanship. In addition, the maximum amount to which the liability is limited, as stated in the H/H-VR was relatively appropriate in 1924, when a package or unit of ordinary general cargo shipped in the conventional manner would rarely exceed such amount.[310] Today's cargo values far exceed the values during the first half of the 20th century.[311] So the low amount, in which such limitation is set, does not encourage carriers to take more precautionary measures. The minimum duties together with the limitation of liability have the effect of discouraging carriers from applying more efficient measures for the care of the cargo.

Indeed, it seems that carriers have invested moreoften in technologies to improve the performance of the ships, lower bunker consumption for its safety, than for the protection of cargo. Proper efficiency would demand on the part of the shipowner that he pursues a more integral benefit, for the ship herself as well as for the

---

[308]The court of the second circuit of the United States has clarified the criterion for this allocation on shipper in cases relating to liability for the carriage of dangerous goods in *Senator Linie GmbH v. Sunway Line, Inc.* 291 F.3d 145, 169, 2002 A.M.C. 1217, 1245–1246 (2nd Cir. 2002): "We conclude today that a strict-liability construction of §1304(6) will foster fairness and efficiency in the dealing of commercial maritime actors. In contrast to a carrier, which typically is in the position of taking aboard its vessel a large quantity and variety of cargoes, a shipper can be expected to have greater access to and familiarity with goods and their manufacturers before those goods are placed in maritime commerce. If an unwitting party must suffer, it should be the one that is in a better position to ascertain ahead of time the dangerous nature of shipped goods." Similarly, on the assignation of the burden of proof, the same court in *Caemint Food v. Lloyd's Brasileiro* 647 F. 2d 347, 354, 1981 A.M.C. 1801, 1812 (2nd Cir. 1981), stated: "It is fair to impose on the plaintiff the burden of showing the condition of the packaged goods on delivery because the shipper has superior access to information as to the condition of the goods when delivered to the carrier, just as the carrier has superior access to information as to what happened thereafter." As cited by Tetley, *Interpretation and Construction...,* 62, f.n. 224.

[309]J. O. Honnold, 'Ocean Carriers and Cargo; Clarity and Fairness—Hague or Hamburg?' (1993) 24, Journal of Maritime Law and Commerce, 75–109. at 107: "Thoughtful observers have concluded that the degree to which carrier are freed from responsibility for negligence reduces their incentive to make outlays that are needed for safe carriage", *citing* Kindred, at 613–614; O'Hare, Allocating Risk, 143–144, 152. Hellawel at p. 363–365; M. F. Sturley, 'Changing Liability Rules and Marine Insurance: Conflicting Epirical Arguments About Hague, Visby, and Hamburg in a Vacuum of Empirical Evidence' (1993) 24, Journal of Maritime Law and Commerce, 119–149.

[310]Diplock, 529, 530. Diplock also asserts that the amount of limitation of liability would be irrelevant in regards of the value of the goods so long as it is high enough as to act as an inducement to the carrier to take proper precaution to reduce the risks.

[311]Ibid., 530.

cargo.[312] The contract then, adds Lord Diplock, should impose enough liability as to create sufficient commercial inducement on the carriers to incent them to apply further precautionary measures.[313] This possibility is contemplated in the H/H-VR when it establishes the right of the parties to increase the carrier's liabilities, but in reality it rarely happens. The investment in precautionary measures would be justified by the reduction of risks of damage or loss, which in turn, should then have a direct effect on the insurance premium.[314] The obstacle for that is that the implementation of further precautionary measures involves an economic investment that shipowners are not always willing to pay.

4.3.2.2 The Cost of Taking Precautions

Rules establishing higher liabilities for carriers create the incentives on their part to apply new technologies as better precautionary measures. Taking additional precautionary measures represents an economic burden which usually increases the cost of transport. The decision to pay such costs will depend first, on how efficiently a technical device or newer method would reduce the risks and second, on how it might affect the profitability of the business. Such costs are then passed on to shippers through the freight and subsequently added to the final cost of the products being shipped.[315]However, carriers may choose not to apply technologies that produce such severe increases in freight costs as it may reduce their market competitiveness. One example of cost efficient technology was the direction-finding apparatus developed by Marconi. By 1923, it proved to be useful in the saving of life, ship and cargo. Additionally, it increasingly showed its high profitability as it enabled carriers "to effect economies in time, money and material" that fully justified its implementation.[316] It helped to save time especially during heavy fog when navigation in many cases was only possible because of this technology.[317] Therefore, to demand the application of a technology as part of the carrier's duties requires the analysis of the cost efficiency of the implementation of such a technology.

---

[312]Billah, 109.

[313]Diplock, 527. However, as such contracts are mostly performed under a bill of lading contained in a pro-forma document which clauses are pre-elaborated by the carriers, shippers have few possibilities, if any, to demand the carrier to assume higher liabilities.

[314]Ibid., at 527: "In competitive freight and insurance markets, which are fostered by the use of standard contracts of carriage, the cost of the precaution will be reflected in the charge for freight, but will be more than compensated for by the reduction in risk which will be reflected in the insurance premium."

[315]Ibid., 526.

[316]Hancock, 119. "The saving of only one tide in some cases saved the entire cost of the apparatus for the year."

[317]Ibid., 119.

Taking too few precautions may result in excessive damages, while applying too many precautions would increase the operating costs and make it unprofitable.[318] The expenditures for such devices would be obviously unproductive if their amount exceeded the value of the damage or losses over time.[319] The economic aim of any law in this respect should encourage the carrier to take such precautions as long as they are economically productive.[320] Judges have considered this element when assessing the reasonability of the application of new advances as part of the seaworthiness conditions.

A particular technological precaution will be applied if it results in a reduction of possible damage or losses that are greater than the cost of the technology to begin with.[321] This is the reasoning expressed by Judge Hand in 1947, known now as the "Hand Formula".[322] In algebraic terms, he stated that the burden of appropriate precautionary measures (B) should be less than the possibility of the damage (P) multiplied by the gravity of the injury (L). To assess the convenience of a precautionary measure under this formula implies another challenge. The carrier must have the information about the probability of occurrence of the damage and the cost of the damage or loss itself. The value of the technology is normally available, which will include not only the cost of the device, but also the cost of installation on board or in the containers, the cost of the operation and management of the device and its maintenance, which includes the training of the crew or hiring skilled personal to operate it. Similar elements have to be considered when applying new methods of stowage or inspections. In case of technological devices, it might not be implemented for a single shipment, so its cost must be divided by the number of voyages or shipments or period of time during which the device can be used with normal functionality.

Gathering information about the other two variables may offer some more difficulties. The value of the cargo may be regularly known, though with no total certainty. The prices of goods may vary with time because of the market fluctuations. The probabilities that a risk occurs and causes cargo loss are mostly unknown, or at least, not always possible to be determined with precision in advance.[323] The lack of proper information makes this approach uncertain, and the process of collecting the data and making the mathematic calculation represents a cost as well. The Hand formula is doubtless a helpful tool, but courts cannot rely solely on this approach. It may provide an approximation of the value (or lack thereof) of taking on a specific technological precautionary measure, but not with total accuracy.

---

[318] D. S. Peck, 'Economic Analysis of the Allocation of Liability for Cargo Damage: The Case for the Carrier, or Is It?' (1998–1999) 26, Transportation Law Journal, 73–105, 91.

[319] Diplock, 526.

[320] Ibid., 526.

[321] Peck, 91.

[322] *See United States v. Carroll Towing Co.,* 159 F.2d 169, 173, 1946 A.M.C. 35, (C.A.2 1947).

[323] Peck, 93.

4.4 Conclusion

The development of knowledge, experience and higher skills resulted in technological advances that have substantially improved the shipping industry. The introduction of steamships, now bigger and more sophisticated vessels, equipped with modern equipment for navigation and communications, and the implementation of new methods of stowage, carriage and inspections, have made the maritime adventure safer than ever. Even calling it an "adventure" nowadays may be an inappropriate label. With the scientific advances, it is now possible to foresee potential risks of the voyage, as well as the potential needs required by certain cargo during its carriage. Certainly, total accuracy has not been achieved yet, but it must be recognized that the level of knowledge regarding risk prevention is higher than when the HR Convention was adopted. The access to this knowledge and the availability of new technological tools allow carriers to address them more diligently. This is what has been recognized as an evolution of the seaworthiness condition. An evolution that not only affects the seaworthiness condition of the vessels, but also, the methods to provide better care of the cargo. Such evolution demands a higher level of diligence on the part of carriers with the passage of time. The existence of more technological resources for safer carriage creates the expectation that a diligent professional carrier will take advantage of those resources in the performance of his contractual and statutory duties.

But as this expectation is not always fulfilled,the international community has showed an increasing interest in promoting safety in navigation through the adoption of some minimum international standards. These standards include the application of new technological equipment or new methods of stowage. Most of these requirements do not address primarily the care of the cargo, but the safety of the vessel and the crew, although indirectly, it certainly benefits the cargo as well. Therefore, courts have clearly assigned liability when the lack of successful implementation of such technologies is the cause of the damage or loss.

Through technological advances, the shipping industry now offers as part of their service the carriage or better care of certain delicate or temperature sensitive cargoes previously not possible. It is up to the shipper to pay additional fees for that. If accepted, the correct implementation becomes a contractual obligation. Liability arises from the lack of application or defectiveness of a technology stated in the contract.

In the two aforementioned cases, liability is not under discussion. Once the causal link between the failure and the damage is proven, the carrier will be held liable. Aside from these cases, the case law shows the judges' prudence or hesitation in requiring the application of new technologies in two other situations. Technologies commonly used or those that are considered imperative precautions also raise liability when their absence on board or defectiveness causes loss or damage. To determine whether a technology fits within these two categories requires a higher evidentiary exercise. To determine whether it is in common usage, or that the special circumstances demanded a higher level of care will require in many cases further litigation. Another aspect that has been taking into account by judges when measuring the reasonability of requiring a precautionary measure is its cost efficiency. A technological precaution will be reasonable within the context of the H/H-VR obligations if the economic investment required is less than the damage or loss it intends to prevent. Judges have followed this approach when dealing with this issue, though calculations of its cost efficiency are not always accurate.

Summary of the Study

This work intended to present the past and present of the sea common carriers liability and the impact in the last decades of new technologies. For many centuries, carrier's liability was judged according to an absolute standard that held them as insurers of the goods. The origin of that standard is found in the Roman law. Such a standard responded to public policy measures taken by the legislators of that epoch and to the special needs of international trade during that time. It was necessary to hold carriers strictly liable in order to protect society against unethical practices on the part of carriers at a time when the commercial activity was gaining special relevance. It was also a pragmatic solution to assure the honesty and efficiency of the carriers as well as to prevent the difficulties for cargo owners in proving the cause of the damages. This reasoning guided the maritime law of Europe for centuries. It was evidenced in many codes of maritime laws adopted during the Middle Ages that were later incorporated into the English Black Book of Admiralty. English judges applied the same reasoning with particular clarity from the 18th through the beginning of the 19th century. It concurred with the expansion of the international trade of England resulting from the technological advances brought on by the industrial revolution. During the middle of the 19th century this standard experienced its decline. It was mostly caused by the ideological currents of the early 19th century. The liberalism theories in England prompted some conceptions of freedom of contract beyond the normal parameters of reasonability that allowed even some unethical practices. The increasing insertion of negligence clauses into the bill of ladings exonerated carriers from any liability, even from the damages or losses caused by their own negligence. With no liability at all, shipowners had no incentive to provide due care in the preparation of ships and in the care of the cargo. Such disinterest gave rise to the so called "coffin ships", badly maintained and equipped ships that sank at sea.

As a response, the international community moved for the adoption of statutory regulation against such practices, but without much success. The enactment of the Harter Act in 1893 was the response of the United States to the problem. But still, this regulation did not have much effect against the practices performed by the British shipowners. This required more international efforts to create a unified regulation capable of achieving wider acceptance.

V.H. Chacón, *The Due Diligence in Maritime Transportation in the Technological Era*,
Springer Series on Naval Architecture, Marine Engineering,
Shipbuilding and Shipping 5, DOI 10.1007/978-3-319-66002-8

Such a regulation came with the Hague Rules. Years of negotiation between the main parties of the business resulted in an international convention that responded to the very specific circumstances of that moment in history. Those circumstances created a situation of unbalanced allocation of risks. The shift of the liability rule from an absolute standard to a negligence-based standard brought on a necessity to re-distribute the risk allocation within the shipping contract. The carrier assumed the consequences of his negligence in the performance of two very specific and minimum duties: to practice due diligence in making the ship in all aspects seaworthy and the care of the goods. Shippers, on their part, must assume the consequences of 17 risks listed as exculpatory causes of liability for carriers, including the long criticized nautical fault. In addition, a limitation of liability was granted to carriers for the negligence in the performance of his duties. This unbalance, as well as the difficulties in determining what due diligence is has evidenced the need for updating this legal instrument. In consequence, there are two protocols amending this Convention, the Visby Amendment (1968) and the S.D.R. protocol (1979), but they do not substantially change the Hague Rules regime. Moreover, there are two newer and updated Conventions creating slightly different risk distribution regimes, the Hamburg Rules (1978) and the Rotterdam Rules (2009), but they are not yet into force. Therefore, the Hague Rules continue to govern the overwhelming majority of shipping contracts under bill of ladings worldwide.

The HR Convention was incorporated into the national legislations of the United Kingdom (1924), United States (1936) and also in Germany (1937). Though many countries have adopted the Hague Rules or its amended form known as the Hague-Visby Rules, England and the United States have developed perhaps the largest body of jurisprudence regarding the duties set in Article 3, Sections 1 and 2 of the rules. Apart from their large maritime activity, this is due to the fact that many carriers continued to include jurisdiction clauses in the bills of lading identifying the courts of these two countries as the exclusive forum to decide cargo claims arising under their contracts.

The courts of the United States and England have interpreted the main obligations set out in the H/H-VR as the diligence exercised by a prudent and competent shipowner to undertake a genuine and reasonable effort to make the ship seaworthy under the circumstances as well as to carefully and properly care for the cargo. For the first obligation, the courts have fragmented the concept of seaworthiness into the many parts of the ship's integrity. This was intended to assure the shipowner's attention to every specific part of the ship that are essential for her safe navigation and to provide appropriate conditions in order to carry the cargo on board. Courts have also recognized that this concept implies a permanent evolution according to the experience, development of knowledge, skills and scientific advances. This evolution requires the courts to analyze the carriers' obligation in relation to new technologies applied in the shipping industry over time.

To make the ship seaworthy and to care for the cargo, carriers have access to new technologies that produce a better and safer performance. The obligation of practicing due diligence requires the carrier to do everything that is reasonably possible according to the particular circumstances to assure the result of the contract. Prudent professional shipowners must take advantage of technologies available for the safety of the crew, the vessel and the cargo. This mandates the implementation of new technologies. Technologies offer the possibility to prevent many of the risks that have commonly threated the carriage of goods by sea.

To the question of what kind of technologies can be enforced as part of this obligation, the case law defined four groups: (1) Technologies ordered by statutes or international regulations; (2) Technologies contractually agreed upon; (3) Technologies commonly used in the industry; and, (4) Technologies required as imperative precautions. Once it is determined if a technology falls within one of these groups, liability can arise in three situations: (a) if it is not on board; (b) if on board but defective; and (c) if on board and in proper working condition, the master, officer or crew lack the necessary knowledge for its correct use, operation or interpretation of the information it provides. Of course, liability will be assign if there is a causal link between the failure and the damage or loss of cargo.

The reasonability of the application of any of these technologies has been subjected, apart from its reliability, to its cost efficiency. The relation between the cost of the technological precautionary measure and the damage or loss it intends to prevent plays an important role too. The precautionary measures should cost less than the possible damage it seeks to prevent. It is expected that such assessment is made by the international maritime organization when ordering the technological devices set in the standard established in the conventions. The list of safety equipment set in the SOLAS Convention for example, should have passed a test of cost efficiency before being ordered in the Convention. However, because this equipment is required to grant a safety of human life at sea, its cost efficiency does not play a fundamental role in the decision of ordering it statutorily. The same is expected when the technology is agreed upon between the parties. In these cases, it is essential for the carriage or is an extra precautionary measure stated in the contract. The shipper assumes the additional cost depending on its cost efficiency. For those technologies that are commonly used, its common use demonstrates in and of itself its profitability. No carrier would apply a technology that increases unreasonably its operative costs.

The determination of liability may present some more difficulties when it is demanded as imperative precautionary measures. In situations that represent a greater harm and higher possibility of damage, the reasonable care required must be proportionally higher. The problem is in determining when a situation demands higher care. While the other three situations may be solved by direct settlement between the parties, cases, where imperative measures are argued as part of the obligation, may require litigation.

However, this approach of limiting the application of precautionary measures to its cost efficiency has serious shortcomings when the carriage of goods by sea is analyzed in a wider social and economic context. Efficiency should not be reduced to the mere profitability for the parties, when the damage to or loss of cargo, in this time, have consequences beyond the mere loss of profits of the cargo owner. The carriage of goods by sea has been perhaps the most essential component of the growth and expansion of international trade. Such importance is not only relevant for today. The historic maritime law reveals the importance of the carriage activity in the past which demanded severe regulations. The steady development of every society is directly connected to its economic growth. Roman jurists and English judges of the 18th century applied a strict standard as a response to the needs of the international trade of their respective epochs. Similar needs are present in today's world economy. Every country's economic growth, in a global market economy, depends, perhaps more than ever, on the international exchange of products. The success of this exchange obviously requires the safe arrival of those products at destination. Safe carriage takes such a paramount importance in the current economy that it has pushed governments to demand newer international regulations with higher liabilities. Given the failure of those attempts, safe carriage can also be reached through the encouraging of and the implementation of new technologies as part of the due diligence obligation. Such encouragement may be reached through the adoption of more international instruments establishing specific standards for the protection of cargo, similar to those adopted by the IMO safety standards. The other possibility is through judicial rules.

Courts, when deciding cases under this regime, should take into account that the liability rules are already weak and insufficient for the current realities. These shortcomings must be dealt with until an updated liability regime comes into force. The judiciary in its function of adapting this legal instrument to the new issues of the maritime transportation must procure a fairer balance between the parties. It requires the enforcement of a better and more accurate level of care. Enforcing the application of new technologies as part of the ship's seaworthiness condition and as precautionary measures will foster the reduction not only of cargo damages or losses, but also the prevention of large maritime casualties. Giving the failure in the past to update the liability rules for the carriage of goods by sea under bill of lading, and the new challenges faced within international trade, the international community must take further steps in this regard.

List of Cases Cited and Reported

United States

Afran Transport Co. v. The Bergechief, 274 F.2d 469, 1960 A.M.C. 1380 (C.A.2 1960).

Albert v. R.P. Farnsworth & Co., 176 F.2d 198 (5th Cir. 1949).

Andean Trading Co. v. Pacific Steam Co. (1920), 263 Fed. 559 (2nd Cir. 1920).

Anglo-Saxon Petroleum Co., Ltd., of London, England v. United States (1950), 88 F. Supp 158 (D.C. Mass. 1950), 1950 A.M.C. 631 (D.C. Mass. 1950).

Anglo-Saxon Petroleum Co., v. United States, 222 F.2d 75, 1955 A.M.C. 985 (C.A. 1955).

Atlantic Transport Co. v. Rosenberg Bros. & Co., 34 F.2d 843 (C.A. 9 1929), 1929 A.M.C. 1539.

Atlas Assur. Co. v. Harper, Robinson Shipping Co., 508 F.2d 1381, 1975 A.M.C. 2358, 2689 (9th Cir. 1975).

B.M.A. Indus., Ltd. v. Nigerian Star Line, Ltd., 1986 AMC 1662, 786 F.2d 90 (2 Cir. 1986).

Bagwell v. Susman, 165 F.2d 412 (6th Cir. 1947).

Bank Line v. Porter, (The Poleric) 25 F.2d 843, 1928 A.M.C. 761 (4th Cir. 1928), *cert. denied,* 278 U.S. 623, 49 S.Ct. 25, 73 L.Ed. 544 (1928).

Bank v. Express Co., 93 U. S. 174 (1876).

Barge B.A. 1401 *(Hampton Roads Carriers, Inc. v. Allied Chemical Corp.)* 329 F2d. 387, 1964 A.M.C. 2458 (4th. Cir. 1964).

Bazin v. The Liverpool & Philadelphia Steam-Ship Co. (1857), 3 Wall. Jr. 229, 5 Am. Law. Reg. 459 (1857).

Binladen BSB Landscaping v. M.V. Nedlloyd Rotterdam 593 F. Supp. 546 (D.C. N.Y. 1984); 1985 A.M.C. 2113, 759 F. 2d 1006. 1012 (2nd Cir.).

Blommer Chocolate Co. v. *Nosira Sharon Ltd.* 776 F. Supp. 760, 1994 A.M.C. 1807 (S.D.N.Y. 1991), aff'd 963 F.2d 1522 (2nd Cir. 1992).

Boudoin v. Lykes Bros. S.S. Co., Inc., 348 U.S. 336, 75 S.Ct. 382, 99 L.Ed. 354 (1955).

*C. Itoh & Co. (America) Inc.*v. *M/V Hans Leonhardt,* 719 F. Supp. 479, 1990 A.M.C. 733, 743 (E.D.La. 1989).

V.H. Chacón, *The Due Diligence in Maritime Transportation in the Technological Era*,
Springer Series on Naval Architecture, Marine Engineering,
Shipbuilding and Shipping 5, DOI 10.1007/978-3-319-66002-8

Caemint Food v. Lloyd's Brasileiro 647 F. 2d 347, 1981 A.M.C. 1801 (2nd Cir. 1981).

Calmaquip Engineering West Hemisphere Corp., v. West Coast Carriers, Ltd. 650 F.2d 633, (5th Cir.1981).

Chicago, M & St. P. Ry v. Lindeman, 143 F. 946 (8th Cir. 1906).

China Union Lines, Ltd. v. A. O. Anderson & Co. (*The Union Reliance*), 364 F.2d 769 (5th Cir. 1966).

Cole v. Goodwin and Story (1838), 19 Wend. 251, 32 Am. Dec. 470 (1838).

Commonwealth Petrochemicals, Inc. V. S/S Puerto Rico, 455 F. Supp. 310, 1978 A.M.C. 2153 (D.C. Md. 1978), *rev'd on other grounds,* 607 F2d. 322, (4th Cir. 1979).

Compañia General de Tabacos de Filpinas v. United States (The Elkton), 49 F.2d 700, 1931 A.M.C. 1040 (C.A. 2 1931).

Complaint of Tecomar S.A., 765 F. Supp. 1178 (S.D.N.Y. 1991).

Consolidated & Barge Co. v. *Marcona Conveyor Corp.,* 716 F. 2d 1077, 1985 A.M.C. 117 (5th Cir. 1983).

Constructores Tecnicos S. de R.L. v. Sea–Land Serv., Inc., 945 F.2d 841, 1992 A.M.C. 1284 (5th Cir. 1991).

*Consumers Import Co., v. Kawasaki Kisen Kabushiki Kaisha, (The Venice Maru),*133 F.2d 781, 1943 A.M.C. 277 (C.A.2 1943).

Coryell v. Phipps (*The Seminole*), 1943, 317 U.S. 406, 409, 63 S.Ct. 291, 87 L. Ed. 363 (1943).

CTI Container Leasing Corporation v. Oceanic Operations Corp. 652 F.2d 377, 1982 A.M.C. 2541 (2nd cir. 1982).

Dallas W. Dietrich, AS Atlantic Seaboard Flour Mill v. United States Shipping Board Emergency Fleet Corporation, (The Panola) 1925 AMC 1173 (2nd. Cir.).

Dorr v. The New Jersey Steam Navigation Company (1850), 4 Sandford, 136 (1850).

Draper v. Commercial Ins. Co., 4 Duer (N.Y.) 234.

Du Pont de Nemours Inter, S.A. v. S.S. Mormacvega 493 F.2d 97, 1974 A.M.C. 67, (C.A.N.Y. 1974).

Dupond v. Vance (1856). 19 How. (60 U.S.) 162, 167, 15. L.Ed. 584 (1856).

Eastman Kodak Co. v. *S.S. Sealand Voyager,* 1991 A.M.C. 2356, 2361 (S.D.N. J. 1991), aff'd 958 F.2d 362, 1992 A.M.C. 1520 and 1816 (3rd Cir. 1992).

Electro-Tec Corp. v. *S/S Dart Atlantica,* 598 F. Supp. 929, 1985 A.M.C. 1606, (D.Md. 1984).

Elliot v. Russell 10 John 1 (1813).

Encyclopaedia Britannica v. S.S. Hong Kong Producer, 422 F.2d 7 (2nd Cir. 1969), cert. denied, 397 U.S. 964, 90 S.Ct. 998, 25 L.Ed.2d 255 (1970).

Erie & St. Lawrence Corpotaration v. Barnes Ames Co., 52 F.2d 217 (W.D.N.Y. 1931).

Express Co. v. Caldwell, 21 Wall. 264, 268 (1874).

Farrell Lines, Inc. v. S.S. Birkenstein, 207 F. Supp. 500, 1963 A.M.C. 1846 (D. C.N.Y. 1962).

Federazione Italiana Dei Consorzi Agrari v. Mandask Compania De Vapores, S.A., 284 F. Supp. 356, *aff'd in part, rev'd in part,* 388 F.2d 434 (2nd Cir.), *cert. denied,* 393 U.S. 828, 89 S.Ct. 92, 21 L.Ed.2d 99 (1968).

Federazione Italiana v. Mandask Compania 388 F.2d 434. 1968 A.M.C. 315 (2nd Cir. 1962).

Flexivan Leasing, Inc. v. M/V C.C. San Francisco, 1986 A.M.C. 333.

Francosteel Corp. v. Nederlandsch Amerikaansche Stoomvaart-Maatschppij, 249 Cal. App.2d 880, 1897 A.M.C. 2440, (Cal. App. 1967), cert. denied, 389 US 931 (1967).

General Foods Corp. v. The Troubador, 98 F. Supp. 207 (S.D.N.Y. 1951).

Gleadhill vs. Thompson, 56 N.Y. 194 (1874).

Greater New Orleans Expressway Commission v. Tug Claribel, (The Claribel) 222 F. Supp. 521, 1964 A.M.C. 967. (D.C.La. 1963), *aff'd sub nom Coleman v. Jahncke Service Inc.*, 341 F.2d 956, (C.A. La. 1965), 1965 A.M.C. 535.

Hart v. Pennsylvania Railroad Co., 112 U. S. 331, 5 Sup. Ct. Rep. 151 (1884).

Hearne v. Marine Ins. Co., 87 U.S. (20 Wall.) 488, 22 L.Ed. 395 (1874).

Hendry Corporation v. Aircraft Rescue Vessels, 113 F. Supp. 198, 201 (D.C. La.1953).

Herd & Co. Inc. V. Krawill Machinery Corp., 359 U.S. 297, 1959 A.M.C. 879.

Holland v. Seven Hundred & Twenty-Five Tons of Coal, 36 Fed. 784 (D.C.E.D. Wis).

Hostetter v. Park, 137 U.S. 30, 40; 11, S. Ct. 1, 34 L. Ed. 568 (1890).

Houlden & Co. v. *S.S. Red Jacket,* 1977 A.M.C. 1382, (S.D.N.Y. 1977), aff'd without opinion, 582 F.2d 1271 (2nd Cir. 1978).

Iligan Integrated Steel Mills, Inc. v. S.S. John Weyerhaeuser, 1975 AMC 33, 507 F.2d 68 (2nd Cir. 1974), cert. denied, 421 U.S. 965, 1975 AMC 2158 (1975).

In re Marine Sulphur Transport Corp., 312 F. Supp. 1081, 1105 (S.D.N. Y.1970), *aff'd in part, rev'd in part,* 460 F.2d 89 (2nd Cir.), *cert. denied,* 409 U.S. 982, 93 S.Ct. 318, 34 L.Ed.2d 246 (1972).

In re Pacific Mail S.S. Co. 130 Fed. 76 (9th Cir. 1904).

Ingersoll Milling Mach. Co. v. M/V BODENA, 829 F.2d 293 (2nd Cir.1987).

Inman v. South Carolina Ry. CoRailway Co., 129 U. S. 128, ante, 249 (1889).

Insurance Co. of North America v. Blue Star (North America), Ltd. Not reported in F. Supp., 1997 WL 345235 (S.D.N.Y. 1997), 1997 A.M.C. 2434.

Phoenix Ins. Co. v. Erie & Western Transp. Co., 117 U.S. 312, 6 Sup. Ct. Rep. 750, 29 L.Ed. 873 (U.S. Wis. 1886).

Intercontinental Trading Co. Inc. v- M/V Zenith Sun, 684 F. Supp. 861 (E.D.Pa. 1988).

International Navigation Company v. Farr & Bailey Manufacturing Company, 181 U.S. 218, 45 L.Ed.830, 21 S.Ct.591. (1901).

Irish Spruce (Irish Shipping Ltd. Lim. Procs.) 1975 A.M.C. 2259, 2568 (S.D.N. Y. 1975), 548 F.2d. 56, 1977 A.M.C. 780 (2nd Cir. 1977).

J. Gerber & Co. v. SS Sabine Howaldt, 437 F.2d 580, 596 (2nd Cir. 1971).

James Knott v. Botany Worsted Mills, 179 U.S. 69, 21 S.Ct. 30, 45 L.Ed. 90 (1900).

Jones v. The Flying Clipper, 116 F. Supp. 386 (S.D.N.Y.1953).

June T., Inc. v. King, 290 F.2d 404, 406 (5th Cir. 1961).

Kane v. Branch Motor Express Co., 290 F.2d 503, 507 (2nd Cir. 1961).

Konica Bus. Machines v. The Vessel Sea-Land Consumer, 47 F.3d 314, 315 (9th Cir. 1995), 1995 A.M.C. 1065 (9th Cir. 1995).

Konica Business Machines, Inc. v. Vessel "Sea-Land Consumer" 153 F3d. 1076. 1998 A.M.C. 2705, 98 Cal. Daily Op. Serv. 6955, 98 Daily Journal D.A.R. 9595.

Kulack v. The Pearl Jack, 79 F. Supp. 802, (W.D.Mich. 1948)

Kutger v. United States, 169 F. Supp. 104 (N.D.Fla., 1958)

Lagerloef Trading Co. v. United States, 43 F.2d 871, 872 (S.D.N.Y.1930).

Leather Best, Inc. v. S.S. Mormaclynx, 313 F. Supp., 1373, 1970 A.M.C. 1310 (E.D.N.Y. 1970).

Lekas & Drivas v. Goulandris, 306 F.2d 426, 1962 A.M.C. 2366 (2nd Cir. 1962).

Liverpool and Great Western Steam Company v. Phoenix Insurance Co., 129 U. S. 397, 9 S.Ct. 469, 32 L.Ed. 788. (1889).

Louis Dreyfus Corp. v. 27,946 Long Tons of Corn, 830 F.2d 1321, 1324, 1326. (5th Cir.1987).

Lyon v. *Mells*, 5 East, 428 (1804).

M/V *Tuxpan Lim. Procs.*, 765 F. Supp. 1150 (S.D.N.Y. 1991).

Mamiye v. Barber, 360 F.2d 774 (2nd Cir. 1966).

Martin v. Southwark,(The Southwark) 191 U.S. 1, 24 S.Ct. 1, 48 L.Ed. 65 (1903).

Matter of Ta Chi Navigation (Panama) Corp., S.A. 513 F. Supp. 148, 1981 A.M. C. 2350 (E.D.LaA.).

May v. Hamburg-Amerikanische Ppacketfahrt Aktiengesellschaft (The Isis), 290 U.S. 333, 54 S.Ct. 162, 78 L.Ed. 348 (1933).

McClellan v. Pennsylvania R. Co., 62 F.2d 61 (2nd Cir. 1932).

McGill v. Michigan Steamship Co., 144 F. 788 (9th Cir. 1906).

Metals & Mineral Corp. v. M/N Arktis Sky, 978 F 2d. 47, 1993 A.M.C. 509 (2nd Cir. 1992).

Michalic v. Cleveland Tankers, Inc., 364 U.S. 325, 81 S.Ct. 6, 5 L.Ed.2d 20 (1960).

Mitchell v. Trawler Racer, Inc., 362 U.S. 539, 80 S.Ct. 926, 4 L.Ed.2d 941 (1960).

Mitsui & Co., Ltd. v. American Export Lines, Inc. 636 F.2d 807, 1981 A.M.C. 331 (C.A.N.Y. 1981).

Mobil Shipping and Transp. Co. v. *Wonsild Liquid Carriers Ltd.,* 190 F.3d 64, 1999 A.M.C. 2710 (2nd Cir. 1999).

Navegacion Castro Riva, S.A. of Panama v. The Nordholm, 178 F. Supp. 736, (E.D.La.1959) *aff'd,* 287 F.2d 398 (5th Cir. 1961).

Navigazione Libera Triestina v. Garcia & Maggini Co., 30 F2d. 62 (C.A. 9 1929).

New Jersey Steam Navigation Co. v. Merchants' Bank of Boston, 47 U.S. 6 How. 344, 12 L.Ed. 465 (1848).

New Rotterdam Ins. Co. v. The S.S. Loppersum, 215 F. Supp. 563, 567 (S.D.N. Y.1963).

Nichimen Company v. M/V Farland, 462 F. 2d. 319, 1972 A.M.C. 1573 (2nd Cir. 1972).

Niles-Bement-Pond Co. v. Kampkiesa Kieselskabet Balto, 282 Fed. 235 (2nd Cir. 1922).

Northeast Marine Terminal Co., Inc. v. Caputo, 432 U.S. 249, 97 S. Ct. 2348, 53 L.Ed.2d 320 (1977).

Nuzzo v. *Rederi A/S Wallenco* 304 F.2d 506, 1962 A.M.C. 1871 (2nd Cir. 1962).

O'Connell Machinery Co. v. M/V Americana, 797 F.2d 1130, 1986 A.M.C. 2822 (2nd Cir. 1986).

Oliver v. Maryland Ins. Co., 11 U.S. (7 Cranch) 487, 3 L.Ed. 414 (1813).

Orient Ins. Co. v. United S.S. Co. 1961 A.M.C. 1228 (S.D.N.Y. 1961).

Orient Mid-East Lines v. A Shipment of Rice, 496 F.2d 1032 (5th Cir. 1974).

Orient Steam Nav. Co. v. U.S., 231 F. Supp. 469, 1964 A.M.C. 2163 (D.C.Cal. 1964).

Pacific Tow Boat Co. v. States Marine Corp. (The Cotton State), 276 F.2d 745 (9th Cir. 1960).

Palmer v. Ins. Co., 116 N.Y. 599, 23 N.E. 5.

Parnass International Trade & Oil Corp. v. Sea-Land Service, Inc., 595 F. Supp. 153, 1985 AMC 485 (S.D.N.Y. 1984).

Peter Paul Inc. v. M/S Christer Salen, 152 F. Supp. 410, 1957 A.M.C. 2141 (S. D.N.Y. 1957).

Petition of Long (The Smith Voyager), 293 F. Supp. 172(S.D.N.Y., 1968).

Petition of T.J. Howard, 53 F. Supp. 556, 1943 A.M.C. 1263 (E.D.N.Y.), 155 F.2d 780 (2nd Cir. 1946).

Philippine Sugar C. Agency v. Kokusai Kisen, Etc. (The Naples Maru) 106 F.2d 32, 1939 A.M.C. 1087 (2nd Cir.).

Pioko Fashions, Inc. v. American President Lines, Ltd., 1993 AMC 2615 (W.D. Wash. 1993).

Polarus Steamship Co. v. The T/S Sandejord, 236 F.2d 270 (C.A. 2 1956), 1956 A.M.C. 1779.

Propeller Niagara v. Cordes, 62 U.S. (21 How.) 7 (1858).

Pure Oil Co. v. Snipes, 293 F.2d 60 (5th Cir. 1961).

Quesoro v. Lykes Bros. Steamship Co., Inc., 1995 AMC 2054 (S.D.N.Y. 1995).

R.T. Jones Lumber Co. v. Roen Steamship Co. 270 F.2d 456, 1960 A.M.C. 46 (2nd Cir. 1959).

Railroad Co. v. Lockwood, 17 Wall. 357 (1873).

Railroad Co. v. Pratt, 22 Wall. 123, 134 (1874).

Cleveland, C., C. & ST. L. RY. CO., v. Ballentine, 84 Fed. 935, 28 C.C.A. 572 (C.A. 7 1898).

Railway Co. v. Stevens, 95 U. S. 655 (1877).

Richelieu & Nav. Co. v. Boston Marine Ins. Co. (The Spartan), 136 U.S. 408, 10 S.Ct. 934, 34 L.Ed. 398 (1890).

Robertson vs. National S.S.Co., 139 N.Y. 416(1891).

Rockwell International Corp. v. M.V. Incotrans Spirit, 1989 AMC 887 (S.D. Tex. 1989).

Rubens vs. Ludgate Hill S.S.Co. (1892), 1st Dept., 65 Hun 625, 20 N.Y. Supp. 481 at 185-86 (1892), affirmed without opinion (1894), 143 N.Y. 629 (1894).

S.S. Willdomino v. Citro Chemical Co., 272 U.S. 718 (1927).

Sanford & Brooks Co. v. Columbia Dredging Co., 177 Fed. 878, 101 C.C.A. 92 (4th Cir. 1910).

Santomarco v. United States, 277 F.2d 255, 257 (2nd Cir. 1960), cert. den'd, *American Stevedores Inc. v. United States*, 364 U.S. 823, 81 S.Ct. 59, 5 L.Ed.2d 52 (U.S. 1960).

Schiefflen v. Harvey, 6 John. 1709 (1810).

Schlichter v. Port Arthur Towing Company, 288 F.2d 801, (5th Cir. 1961).

Schnell & Co. v. S.S. Vallescura, (The Valescura) 293 U.S. 296, 1934 A.M.C. 1573 (1934).

Sedco, Inc. v. S.S. Strathewe, 1986 AMC 2801, 800 F.2d 27 (2nd Cir. 1986).

Senator Linie GmbH v. Sunway Line, Inc. 291 F.3d 145, 2002 A.M.C. 1217 (2nd Cir. 2002).

Sequros Banvenez. S.A. v. S/S Oliver Drescher, 761 F.2d 855, 1985 A.M.C. 2168 (2nd Cir.1985).

Sigri Carbon Corp. v. Lykes Bros. S.S. Co. Inc., 655 F. Supp.1435, 1988 A.M.C. 1787 (W.D.Ky.Y 1987).

Smith v. Clarke Hardware Co., 100 Ga. 163, 28 S. E. 73, 39 L. R. A. 607

Sogem-Afrimet, Inc. v. *M/V Ikan Selayang,* 951 F. Supp. 429, 1998 A.M.C. 1366 (S.D.N.Y. 1996).

Spang Chalfant & Co. v. Dimon S.S. Corporation (The *Pacific Fir),* 57 F.2d 965, 1932 A.M.C. 738, (C.A. 2, 1932).

Spang Chalfant & Co. v. Dimon S.S. Corporation 57 F.2d 965, 1932 A.M.C. 738, (C.A. 2, 1932).

Spartus Corp. v. S/S Yafo, 1979 AMC 2294, 590 F.2d 1310 (5th Cir. 1979).

St. Johns Corp. v. Companhia Geral Commercial Do Rio de Janeiro, 263 U.S. 119, 44 S.Ct. 30, 68 L.Ed. 201 (1923).

Standard Electrica, S.A. v. Hamburg Sudamerikanische Dampfschifffahrts-Gessellschaft, 373 F.2d 943(2nd Cir.), 1967 A.M.C. 881, cert. denied, 389 U.S. 831 (1967).

States S.S. Co. v. United States (The Pennsylvania), 259 F.2d 458 (9th Cir. 1958).

Steel v. State Line Steamship Co., 3 App. Cas. 72 (1877).

Sunkist Growers, Inc. v. Adelaide Shipping Lines, Ltd. 603 F.2d 1327, 1979 A. M.C. 2787 (9th Cir. 1979), cert denied, 444 US 1012, 1980 A.M.C. 2102 (1980).

T.J. Hopper, 53 F.2d 107, 1931 A.M.C. 1764 (D.C.N.Y. 1931), *aff'd*, 60 F.2d 737(2nd Cir. 1932), cert. den'd; *Eastern Transportation Company v. Northern Barge Corporation*, 287 U.S. 662, 53 S.Ct. 220, 77 L.Ed. 571.

Tenneco Resins, Inc. v. Davy Int'l, A.G., 881 F.2d 211, 1990 A.M.C. 402 (5th Cir. 1989).

Texaco North Dakota, Lim. Procs. 570 F. Supp. 1272, 1985 A.M.C. 1650 (E.D. La. 1983).

Texas & Pacific Railway Co. v. Behymer, 189 U.S. 468, 23 S.Ct. 622, 47 L.Ed. 905 (1903).

The Glenville, 1962 A.M.C. 2311 (S.D. Tex.).

The Aakre 122 F.2d. 469, 141 A.M.C. 1263 (2nd Cir. 1941).

The Abbazia, 127 F. 495 (D.C.N.Y. 1904).

The American Smelting v. Irish Spruce (1977), 548 F.2d. 56, 1977 A.M.C. 780 (2nd Cir. 1977).

The Anastasia, 11 F. Supp. 314, 1935 A.M.C. 579 (E.D.Va.)

The Archer, 29 F2d. 134, 1928 A.M.C. 1615 (E.D.N.Y.).

The Asturias, 40 F. Supp. 168, 1941 A.M.C. 761 at 762 (S.D.N.Y. 1941).

The Benjamin Noble, 244 F. 95, 97 (6th Cir. 1917) affirmed *sub nom. Capitol Transp. Co. v. Cambria Steel Co.* 249 U.S. 334, 39 S.Ct. 292 (1919).

The Black Gull, 250 F.2d 777, 1958 A.M.C. 277 (2nd Cir.*); second appeal*, 269 F.2d 68, 1960 A.M.C. 163 (2nd Cir.)

The Bolivia, E.D.N.Y., 1890, 43 F. 173, modified on other grounds, 49 F. 169 (2nd Cir. 1891).

The Caledonia, 157 U. S. 124, 15 Sup. Ct. 537. (1895).

The Carib Prince, 170 U. S. 655, 42 L. Ed. 1181, 18 Sup. Ct. Rep. 753. (1898).

The Chester Valley, 1940 AMC 555, 110, F.2d 592 (5th Cir. 1940).

The Chickasaw- (1966), 265 F. Supp. 595, 1966 A.M.C. 2219 (S.D. Cal. 1966), *upheld in* 414 F.2d 724, 1969 A.M.C. 1682, [1970] 1 Lloyd's Rep. 437 (9th Cir. 1969).

The Ciprya, 137 F.2d 326, 1946 A.M.C 947 (2nd Cir).

The Cleveco, 154 F.2d 605 (6th Cir. 1946).

The Colima (1897), 82 F. 665 (S.D.N.Y. 1897).

The Condor, The Nordpol, 84 F.2d 3, 1936 A.M.C. 1010 (2nd Cir.).

The Cygnet, 126 F. 742 (1st Cir. 1903).

The Delaware, 161 U.S. 459, 16 S.Ct. 516, 2010 A.M.C. 1803, 40 L.Ed. 771 (1896).

The E. Madison Hall, 140 F.2d 589, 1944 A.M.C. 202 (4th Cir.).

The Edith,10 F.2d 684, 1926 A.M.C. 281 (2nd Cir.).

The Edwin I. Morrison, 153 U. S. 199, 14 Sup. Ct. 823, 38 L.Ed. 688 (1894).

The Esso Providence, 112 F. Supp. 631, 1953 A.M.C. 1317 (S.D.N.Y.).

The Folmina, 212 U.S. 354, 29 S. Ct. 363, 53 L.Ed. 546 (1909).

The Friesland 104 F. 99 (D.C.N.Y. 1900).

The Georgian, 4 F. Supp. 718, 1933 A.M.C. 1540 (S.D. Fla.), *aff'd*, 76 F.2d 550 (5th Cir. 1935).

The Germanic, 196 U.S. 589, 25 S.Ct. 317, 49 L.Ed. 610 (1905).

The Gonzenheim, 36 F 2d. 869, 1930 A.M.C. 122 (5th Cir.).

The Heddernheim 39 F. Supp. 558, 1941 A.M.C. 730 (S.D.N.Y. 1941).

The Hindoo (1947), 74 F. Supp 145, (D.C.N.Y. 1947), 1947 A.M.C. 1630, affirmed on this point in *United States v. The Australian Star*, 172 F.2d 472 (2nd Cir. 1949), certiorari denied, 338 U.S. 823, 70 S.Ct. 69, 94 L.Ed. 499 (1949).

The Horaisan Maur, 73 F.2d 526, 1925 A.M.C. 96 (2nd Cir.).

The Indien, 5 F. Supp. 349, 1933 A.M.C. 1342 (S.D. Cal.), 71 F.2d 752, 1934 A. M.C. 1050 (9th Cir.).

*The Ionian Pione*er, 236 F.2d 78, 1956 A.M.C. 175 (5th Cir.).

The Iowa, D. Or., 1940, 34 F. Supp. 843(D. Or., 1940).

The Irrawaddy, 171 U. S. 192, 193, *sub nom. Flint* v. *Christall*, 43 L.Ed. 132, 18 Sup. Ct. Rep. 833. (1898).

The Jason, 162 F. 56 (D.C.), 178 F. 414 (C.C.A.), 225 U.S. 32, 32, S.Ct. 560, 56 L.Ed. 969 (1912).

The Jean Bart 197 Fed. 1002 (D. Cal. 1911).

The Kensington, 183 U.S. 263, 268 (1902).

The Manitoba 104 F. 145 (D.C.N.Y. 1900).

The Maria, 15 F. Supp. 745, 1936 A.M.C. 1314 (S.D.N.Y.); 91 F.2d 819, 1937 A.M.C. 934, (C.A. 4 1937)

The Martello, 1894, 153 U.S. 64, 74, 14 S. Ct. 723, 38 L.Ed. 637.

The Medford (1946). 65 F. Supp. 622 (D.C.N.Y. 1946).

The Murrell (1911), 200 F. 826 (D.C.); 195 F. 483 (C.C.A.).

The Naiwa, 3 F.2d 385, 1925 A.M.C., 85 (4th Cir.).

The Nancy Moran, 80 F. Supp. 623, 1948 A.M.C. 1609 (S.D.N.Y.).

The Negus, 298 Fed. 749, 1923 A.M.C. 625 (S.D.N.Y.).

The Neptune, 17 Fed.Cas. 1343, 1344 (No. 10,118) (C.C.S.D.N.Y.1868).

The Niagara, 84 F. 902 (2nd Cir. 1898).

The Northern Belle, 9 Wall. (U.S.) 526, 19 L.Ed. 746.

The Pennsylvania (1873), 86 U.S. 125, 1873 WL 16023 (U.S.N.Y.), 1998 A.M. C. 1506, 22 L.Ed. 148, 19 Wall. 125(1873),

The Pensilvania, (1956 A.M.C. 1810 (D. Ore), 259 F.2d 458, 1957 A.M.C. 1181 (9th Cir.), *on rehearing*, 259 F.2d 463, 1957 A.M.C. 2277 (9th Cir.), *on second rehearing*, 259 F.2d 470, (9th Cir.).

The Persiana, 185 Fed. 396 (2nd Cir. 1911).

The Phoenicia (D.C.) 90 Fed. 116, affirmed in 40 C.C.A. 221, 99 Fed. 1005

The Pinellas, 45 F.2d 174, 1930 A.M.C. 1875 (4th Cir.).

The Pocone, 159 F.2d 661, 665, 1947 A.M.C. 306 (2nd Cir.).

The Point Brava, 1 F. Supp. 1150, 1991 A.M.C. 2432 (S.D.N.Y.).

The President of India v. West Coast Steamship Co., (The Portland Trader) 213 F. Supp. 352, 1963 A.M.C. 649 (D.C.Or. 1962).

The President Polk, 43 F.(2d) 695 (C.C.A. 2).

The Prussia (1899), 93 F. 837, 35 C.C.A 625.

The Roseina, 1937 A.M.C. 359 (S.D.N.Y.).

The Sagamore (1924), 300 Fed. 701, 1924 A.M.C. 961 (2nd Cir.),

The Salvore, 60 F.2d 683, 685 (2 Cir.), cert. denied, 287 U.S. 653 (1932);

The Seeger, 104 F.2d 167, 1939 A.M.C. 792 (2nd Cir.);

The Severance, (4th Cir. 1945), 152 F.2d 916 (4th Cir. 1945), 922.

The Silverway, 15 F.2d 648, 1926 A.M.C. 1645 (5th Cir.).

*The Spart*an, 47 F.2d 189, 1931 A.M.C. 1 (2nd Cir.)

The Steel Navigator, 23 F.2d. 590 (2nd Cir. 1928).

The Steel Scientist, reported as *Daisy Philippine Underwear Co. v. United States Steel Products Co.,* 11 F. Supp. 175 (D.C.).

The Temple Bar, 45 F. Supp. 608 (D.Md.1942), affirmed 137 F. 2d 293 (4th Cir. 1943).

The Thames, 10 C.C.A. 232, 8 U.S.App. 580, 61 Fed. 1014 (C.A. 4 1894).

The Titania (1883) 19 F. 101 (D.C.N.Y 1883).

The Trave, 55 F. 117, 119-120 (S.D.N.Y., 1893), rev'd on other grounds, 68 F. 390 (2nd Cir. 1895), cert. denied, 163 U.S. 692, 16 S.Ct. 1203, 41 L.Ed. 309, (1895). *The Troubador* (1951), 98 F. Supp. 207, 210 (S.D.N.Y.1951).

The Turret Crown, 297 Fed. 766, 1924 A.M.C. 253 (2nd Cir.).

The Vestris, 60 F.2d 273, 1932 A.M.C. 863 (S.D.N.Y.).

The W.W. Bruce, 94 F.2d 834, 1938 A.M.C. 232 (2nd Cir.).

The *Waterville Victory* (1951), 1951 A.M.C. 320 (S.D.N.Y.), *aff'd* per curiam, *Continental Ins. Co. v. United States* 195 F.2d 527, 1952 A.M.C. 669 (2nd Cir.).

The William Crane, 50 Fed. 444 (D.Md.1889).

Tidmarsh v. Washington Fire & Marine Ins. Co., 4 Mason 439, 441; 23 F. Cas. 1197, 1198 (C.C. Mass.).

Trinidad Shipping Co. v. Frame Alston Co. 88 Fed. 528 (S.D.N.Y.1898).

Troupe v. Chicago D. & G. Bay Transit Co., 1956, 234 F.2d 253, 260 (2nd Cir.)

Tug Ocean Prince, Inc. v. United States, 584 F.2d 1151 (2nd Cir.1978), *cert. denied,* 440 U.S. 959, 99 S. Ct. 1499, 59 L.Ed. 2d 772 (1979).

U.S. v. Ultramar Shipping Co., Inc., 685 F. Supp. 887, 1988 A.M.C. 527 (S.D. N.Y., 1987).

Union Insurance Co. Philadelphia v. Smith, 124 U. S. 405, 8 Sup. Ct. Rep. 534 (1888).

United States Steel Products Co. v. American & Foreign Insurance Co. 82 F.2d. 752 (C.C.A.).

United States v. American Trading Co. (The Glymont), 66 F.2d. 617 (2nd Cir. 1933).

United States v. Carroll Towing Co., 159 F.2d 169, 1946 A.M.C. 35, (C.A.2 1947).

United States v. The Australian Star, 172 F.2d 472 (2nd Cir. 1949), certiorari denied, 1949, 338 U.S. 823, 70 S.Ct. 69, 94 L.Ed. 499.

Van Camp Sea Food Co. v. Di Leva, 171 F.2d 454(9th Cir. 1948).

W.T. Locket & Co. v. Cunard S.A. Co. (1927) 21 F.2d 191, 1927 A.M.C. 1057 (E.D.N.Y.).

Wellesly v. Hooper, 185 Fed. 733 (9th Cir. 1911).

Wirth Ltd. V. S/S Acadia Forest, 537 F.2d 1272, 1976 A.M.C. 2178 (C.A.LA. 1976).

Wood v. United States (*The Wilson Victory*), 125 F. Supp. 42 (S.D.N.Y., 1954).

Work v. Leathers, 97 U. S. 379 (1878).

United Kingdom

Actis Steamship Co. Ltd. v. The Sanko Steamship Co. Ltd., (The Aquacharm) [1982] 1WLR 119 (CA)

Albacora v. Westcott and Laurance Lin. [1966] 2 Lloyd's Rep. 53, 1966 S.L.T. 253

Amies v. Stevens, 1 Str. 128.

Athenian Tankers Management S.A. v. Pyrena Shipping (The Arianna). [1987] 2 Lloyd's Rep. 376.

Balli Trading Ltd v. Afalona Shipping Co. Ltd. (*The Coral*) [1993] 1 Lloyd's Rep 1.

Bodenham v. Bennet 4 Price Exch. 31.

Boucher v. Lawson (1815) Cas T H 194

Brass v. Maitland, (1856) 6 El & Bl 470; 119 ER 940.

Briddon v. Great Northern Ry. Co. (1958) 28 L.J. Ex. 51.

Brown v. Nitrate Producers S.S. Co. (1937) 58 Ll.L.R. 188.

Bukton v Tounesende (the Humber Ferryman) YB 22 Liber Assisarum No. 41, f.94 (1348)

Caltex Refining Co. Pty. V. BHP Transport Ltd. (The Iron Gippsland) [1994] 1 Lloyd's Rep. 335, 337

Cargo per Maori King v. Huges [1895] 2 QB 550.

Charles Brown & Co. v. Nitrate Producer Steamship Co. (1937) 58 Ll.L.R. 188.

Cheikh Boutros Selim El-Khoury v. Ceylon Shipping Lines Ltd. (The Madeleine), [1967] 2 Lloyd's Rep. 224, 241.

Coggs v. Bernard, (1703) 2 Lord Raymond 909; 92 E.R. 107; 90 E.R. 971 (1703).

Cranfield Bros. v. Tatem S.N.Co. (1939) 64 Ll.L.R. 264.

Dimitrios N. Rallias (1922) 13 Ll.L.R. 363.

Duncan Köster (The Teutonia) (1982) LR 4 PC 171.

Effort Shipping Co. Ltd. v. Linden Management, S.A. [1998] AC 605, [1998] 1 Lloyd's Rep. 337, 1998 A.M.C. 1050 (HL).

Empresa Cubana Importadora de Alimentos Alimport v. Iasmos Shipping Co. S. A. [1984] 2 Lloyd's Rep. 586.

Empresa Cubana Importadora de Alimentos Alimport v. Iasmos Shipping Co. S. A. (The Good Friend). [1984] 2 Lloyd's Rep 586.

FC Bradley & Sons v. Federal Steam Navigation Co. [1926] 24 Ll.L. Rep. 446 (CA), (1927) 27 Ll.L.Rep. 395 (HL).

Fiumana Societa di Navigazione v. Bunge & Co. Ltd. [1930] 2 K.B. 47.

Forward v. Pittard (1785), 99 E.R. 953, (1785) 1 Term Reports 27.

Freight Services Ltd. v BP Oil International Ltd (The Fiona) (1993). [1993] 1 Lloyd's Rep. 257, affirmed by the Court of Appeal (Civil Division) at [1994] 2 Lloyd's Rep. 506.

Garnett v. Willan, 5 B. & Ald. 53.

GH Renton & Co. Ltd. v. Palmyra Trading Corporation of Panama [1957] AC 149.

Gibbon v. Paynton 98 E.R. 199; (1769) 4 Burrow 2298.
Great China Metal Industries Co. Ltd. V. Malasyan International Shipping Corp. Berhard (The Bunga Seroja) [1999] 1 Lloyd's Rep 512 (HCA).
Guan Bee v. Palembang Shipping Co. [1969] 1 May L.J. 90.
Hang Feng Shipping & Trading Co. Ltd. v. Mullion & Co. Ltd., [1966] 1 Lloyd's Rep. 511.
Heaven v. Pender, (1883) 11 QBD 506.
Heinrich C. Horn v. Cia de Navegacion Fruco [1969] A.M.C. 1495.
Hong Kong Fir Shipping Co. v. Kawasaki Kisen Kaisha [1961] 1 Lloyd's Rep. 159, [1962] 2 QB 26.
Hyde v. Trent and Mersey Navigation Company, (1793) 5 T.R. 389
Islander Shipping Enterprises, S.A. v. Empresa Maritima del Estado, S.A. (The Khian Sea), [1979] 1 Lloyd's Rep 545.
Jindal Iron Steel Co Ltd and others v. Islamic Soldarity Shipping Co Jordan Inc. (The Jordan II) [2005] 1 Lloyds Rep 57.
Jones v. Oceanic S.N. Co. [1924] 2 K.B. 730.
Kopitoff v. Wilson, (1876) 1 QBD 377.
Lane v. Cotton (1706), 12 Modern 472 (1706); 88 ER 1458.
Leesh River Tea Co. Ltd. V. British India Steam Navigation Co. Ltd., [1967] 2 QB 250 (CA).
Luke v. Lyde-, (1759) 2 Burr. 882,
M.D.C. v. N.V.Z.M. Beursstraat [1962] 1 Lloyd's Rep. 180.
Maxine Footwear Co. Ltd. v. Can. Government Merchant Marine. [1959] AC 589 (PC).
McFadden v. Blue Star Line (1905) 1 K.B. 697
McIver v. Tate Steamer [1903] 1 K.B. 362; *Northumbrian Shipping Co. v. Trimm* [1939] AC 397.
MDC Ltd. v. NV Zeevaart Maatschapij Beursstraat (1962) [1962] 1 Lloyd's Rep. 180
Mitchell, Cotts v. Steel [1916] 2 KB 610
Morse v. Slew (1671), 86 E.R. 129.
Nugent v. Smith (1875) 1 C.P.D. 19.
Papera Traders Co. Limited & Others v Hyundai Merchant Marine Co. Limited, (The Keihin Co. Limited), [2002] 1 Lloyd's Rep. 719.
Parson Corp. v. CV Scheepvaatonderneing Happy Ranger (The Happy Ranger), (2006). [2006] EWHC 122 (Comm); [2006] 1 Lloyd's Rep. 649.
*Paterson Steamship Ltd. v. Robin Hood Mill*s (1937) 58 LI.L.R. 33.
Phillips Petroleum Co. v. Cabaneli Naviera S.A. (The Theodegmon), [1990] 1 Lloyd's Rep. 52.
Pyrene Co. Ltd. v. Scindia Navigation Co. Ltd. [1954] 2 QB 402.
RE Missouri S.S. Co. (1889) 42 Ch. D. 321.
Renton v. Palmyra Trading Corporation of Panama [1957] A.C. 149.

Rey Banano del Pacifico CA v. Transportes Navieros Ecuatorianos (The Isla Fernandina) [2000] 2 Lloyd's Rep. 15.

Riley v. Horne 130 E.R. 1044; (1828) 5 Bingham 217.

Rio Tinto Co. Ltd. v. The Seed Shipping Co. Ltd. (1926) 24 Ll. L.R. 316.

Riverstone Meat Co. Pty. v. Lancashire Shipping Co. (The Muncaster Castle), [1961] A.C. 807, [1961] 1 Lloyd's Rep. 57.

Robin Hood Flour Mills Ltd. v. NM Paterson & Sons Ltd. (The Farrandoc), [1967] 2 Lloyd's Rep. 276.

Rowson v. *Atlantic Transp. Co.*, [1903] 1 K.B. 114.

Scaramanga & Co. v. Stamp, (1880) 5 CPD 295 (CA).

Smith v. Horne, 8 Taunt. 144.

Southcott v. Bennett, (1601) 4 Coke. Rep. 83b; 76 E.R. 1061.

Stag Line LTD v. Foscolo, Mango and Co. Ltd., [1932] AC 328.

Standard Oil Company of New York v. Clan Lines Steamers, Ltd. [1924] A.C. 100, (1923) 17 Ll. L. Rep. 120.

Stanton v. Richardson (1874) 9 CP 390.Stanton v. Richardson, L. R. 7 C.P., 421.

State Trading Corp. of India v. Doyle Carrier Inc. (The Jute Express) [1991] 2 Lloyd's Rep. 55.

Tattersall v. National Steamship Co. (1984) 12 QBD 297.

The Apostolis, [1997] 2 Lloyd's Rep. 241, 257 (C.A.).

The Arawa, [1977] 2 Lloyd's Rep. 416.

The Assunzione, [1956] 2 Lloyd's Rep. 468.

The Australia Star, (1940) 67 Ll.L.R. 110.

The Chusan, [1956] 1 W.L.R. 72.

The Easter City, [1958] 2 Lloyd's Rep 127.

The England, [1973] 1 Lloyd's Rep. 373.

The Esso Plymouth, [1955] 1 Lloyd's Rep. 429.

The Eurasian Dream, [2002] 1 Lloyd's Rep. 719.

The Farrandoc, [1967] 2 Lloyd´s Rep. 276.

The Fjord [1999] 1 Lloyd´s Rep. 307.

The Fogo [1967] 2 Lloyds Rep. 208.

The Friso [1980] 1 Lloyd´s Rep. 469.

The Gang Cheng (1998) 6 MLJ 488.

The Glenfruin, 10 Prob. Div. 103.

The Hellenic Dolphin [1978] 2 Lloyd's Rep. 336.

The Hill Harmony [1999] 2 Lloyd Rep. 209.

The Hollandia [1983] 1 A.C. 565.

The Kapitan Sakharov [2000] 2 Lloyd's Rep. 255.

The Lady Gwendolen [1965] 1 Lloyd's Rep. 335.

The Makedonia [1962] 1 Lloyd´s Rep 316.

The Maori King v. *Hughes* [1895] 2 QB 550.

The Marion [1984] 1 AC 563.

The Mekhanik Evgrafov and The Ivan Derbenec, [1987] 2 Lloyd's Rep. 634.

The Mountoswald, 15 Lloyd's List L. Rep. 144 (C.A.1923).

The Nora [1956] 1 Lloyd's Rep. 617.

The President Monroe [1972] 1 Lloyd's Rep. 385.

The Rafaela S, [2003] 2 Lloyd's Rep. 113 (CA).

The River Rima [1987] 3 All ER 1 [1987] 2 Lloyd's Rep 106: [1988] 1WLR 758 [1988] 2 Lloyd's Rep 193, HL.

The Schwan, L.R. (1909) App. Cas. 450.

The Sea Rambler, 83 Lloyd's List L. Rep. 84 (K.B. Div. 1949).

The Sedgepool, [1956] 2 Lloyd's Rep. 668, 679.

The Star Sea, [1997] 1 Lloyd's Rep. 360.

The Toledo, [1995] 1 Lloyd's Rep. 40.

The Torenia, [1983] 2 Lloyd's Rep. 210.

The Torepo, (2002), [2002] EWHC 1481 (Admlty).

The Tres Flores, [1973] 2 Lloyd´s Rep. 247.

The Vortigen, [1899] Prob. 140 (C.A.).

Thin v. Richards, [1892] 2 QB 141 (C.A.).

Union of India v. N.V. Reederij Amsterdam (*The Amstelslot),* [1963] 2 Lloyd's Rep 223.

Vardinoyannis v. The Egyptian General Petroleum Corp. (The Evaggelos Th) [1971] 2 Lloyd's Rep. 200.

Woodlife's Case (1596) Moore 462.

Canada

De Carvalho v. Kent Line (1951): 32 M.P.R. 282.

Grain Growers Export Co. v. *Canada Steamship Lines Ltd.,* 43 O.L.R. 330, 344-45 (Ont. S.C. App. Div.), upheld (1919) 59 S.C.R. 643 (Supr. C. of Can.).

Health Steel Mines Ltd. v. The Erwin Schroeder. [1970] Ex.C.R. 426

The Niagara, 1944 A.M.C. 1307 (B.C.S.C.).

Netherlands

The *NDS Provider*, 1st of February, 2008, nr C06/082HR, published in RvdW 2008, 177.

References

Books and Articles

Abraham, H. J. (1978). *Das Seerecht in der Bundesrepublik Deutschland: Kommentar und Materialsammlung*, Großkommentare der Praxis (4th ed.). Berlin [u.a.]: de Gruyter.
Abraham, H. J. (1974). *Das Seerecht: Ein Grundriss mit Hinweisen auf d. Sonderrechte anderer Verkehrsmittel, vornehml. d. Binnenschiffahrts- u. Luftrecht* (4th ed.). Berlin, New York: de Gruyter.
Alcantara, J. M. (2010). *The Grand Liability Unified Theory*, Scritti in Onore di Francesco Berlingieri, Il Diritto Marittimo (2 Vols.) Genova.
Anderson, H. E., III. (2009). 'Risk, Shipping, and Roman Law. *Tulane Maritime Law Journal, 34*, 183–210.
Anderson, O. E., JR. (1953). *Refrigeration in America: A history of a new technology and its impact*. United States of America: Princeton University Press.
Anderson, P. (2005). *ISM code: A practical guide to the legal and insurance implications, Lloyd's practical shipping guides* (2nd ed.). London: LLP.
Ashburner, W. (1976). *Homos Rodion Nautikos*, The Rhodian Sea-Law, Repr. of the ed. Oxford 1909. Aalen: Scintia.
Baatz, Y. (2009). *The Rotterdam Rules: A practical annotation*. London: Informa Maritime Law.
Baer, H. R. (1979). *Admiralty Law of the Supreme Court* (3rd ed.). Charlottesville, Va: Michie Co.
Basedow, J. (1983). Common Carriers Continuity and Disintegration in U.S transportation law. *Transportation Law Journal, 13*, 1–42.
Beale, J. H., JR. (1897). 'The Carrier's Liability: Its History. *Harvard Law Review, 11*, 158–68.
Benedict, E. C., & Knauth, A. W. (1969). *Benedict on admiralty* (Vol. 1, 7th ed.). New York, NY: Matthew Bender, 1969, i.e. 1958]-; v. 1.
Benedict, R. D. (1909). The Historical Position of the Rhodian Law. *The Yale Law Journal, 18*, 223–242.
Billah, M. M. (2014). *Effects of insurance on maritime liability law: A legal and economic analysis*. Cham: Springer.
Black, H. C. M. (1910). *A law dictionary* (2nd ed.). St. Paul, Minn.: West Publishing Co..
Booker, M. D. (1987). *Containers: Conditions, law and practice of carriage and use*. London: Derek Beattie.
Bordahandy, P.-J. (2005). Containers: a Conundrum or a concept?. *The Journal of International Maritime Law, 11*, 342–371.
Bouvier, J. (1914). *Bouvier's law dictionary and concise encyclopedia*, 3 rev. (being the eigth Edition) by Francis Rawle, 2 vols. Kansas City, Mo.: Vernon Law Book Company, II.
Bruce, G. a. J. C. F. (1902). *A treatise on the jurisdiction and practice of the English Courts in Admiralty actions and appeals* (3rd ed.). London: Maxwell.
Buckland, W. W. (1925). *A manual of Roman private law*. Cambrige: University Press.

V.H. Chacón, *The Due Diligence in Maritime Transportation in the Technological Era*,
Springer Series on Naval Architecture, Marine Engineering,
Shipbuilding and Shipping 5, DOI 10.1007/978-3-319-66002-8

Burdick, W. L. (1989). *The principles of Roman law and their relation to modern law*, (Repr.). Holmes Beach, Florida: Gaunt & Sons, Inc..

Campbell, J. C. B. (1851). *The lives of the chief justices of England: from the Norman conquest till the death of Mansfield* (2 Vols). Philadelphia, 1851.

Clarke, M. A. (1976). *Aspects of the Hague Rules: A comparative study in English and French law*. The Hague: Nijhoff.

Cleirac, E. (1981). *The ancient sea-laws of Oleron, Wisby, and the Hanse-towns, still in force. Taken out of a French book, intitled, Les us & coutumes de la mer* (31686th ed.). Abingdon, Oxon: Professional Books, 1981.

Colinvaux, R. P. (1954). *The carriage of goods by Sea Act, 1924*. London: Stevens & Sons Limited.

Colinvaux, R. (1982). *Carver's carriage by sea*. British shipping laws (13th ed., 2 Vols.). London: Stevens, 1982.

Cooter, R., & Ulen, T. (2008). *Law and economics* (5th ed.). Pearson Addison-Wesley: Harlow.

Critchell, J. T. a. R. J. (1912). *A history of the frozen meat trade: An accound of the development and present day methods of preparation, transport, and marketing offrozen and chilled meats*. London: Constable & Company Ltd.

Diamond, A. Q. (1978). The Hague-Visby Rules. *Lloyd's Maritime and Commercial Law Quarterly, 2*, 225–266.

Diplock, K. (1970). Conventions and morals—Limitation clauses in International Maritime Conventions. *Journal of Maritime Law and Commerce, 1*, 525–536.

Dönges, T. E. (1928). *The liablility for safe carriage of good in Roman-Dutch Law*. Cape Town: Juta & Co., Ltd.

Driver, G. R., & Miles, J. C. (1952). *The Babylonian Laws* (2 Vols.). Oxford: Clarendon Press, 1952.

Editorial. (2006). Seaworthiness - the ilussion of the Hague compromise. *The Journal of International Maritime Law, 12*, 87–88.

Edwards, C. (1904). *The Hammurabi Code: and the Sinaitic Legislation*. London: Watts & Co.

Fanning, A. E. (1986). *Steady as she goes: A history of the compass department of the admiralty*. London: H.M.S.O; National Maritime Museum [distributor].

Farthing, B. (1993). *International shipping: An introduction to the policies politics and institutions of the maritime world* (2nd ed.). London [etc.]: Lloyd's of London Press.

Fitzgerald, R. C. (1948). Admiralty and prize jurisdiction in the British commonwealth of nations (1948). *The Juridical Review, 60*, 106–122.

Fletcher, E. G. M. (1932). *The Carrier's liability*. London: Stevens.

Force, R. (1996). A comparison of the Hague, Hague-Visby and the Hamburg Rules: Much Ado About (?). *Tulane Law Review, 70*, 2051–2089.

Frederick, D. C. (1991). The Political participation and legal reform in the international maritime rulemaking process: from the HR to the Hamburg rules. *Journal of Maritime Law and Commerce, 22*, 81–117.

Ganado, M. a. K. H. M. (1990). *Marine Cargo delays*, The law of delay in the carriage of general cargoes by sea. London.

García Garrido, M. J. (2004). *Derecho privado romano: Casos, acciones, instituciones, Derecho/ediciones académicas* (13th ed.). Madrid: Ediciones académicas, DL.

Garner, B. A., & Black, H. C. (2009). *Black's law dictionary* (9th ed.). St. Paul, Minn: West.

Gaskell, N. J. J., Chorley, R. S. T. C., Debattista, C., & Swatton, R. J. (1987). *Chorley and Giles' shipping law* (8th ed.). London: Pitman Publishing.

Gilmore, G., Black, C. L., JR. (1975). *The law of admiralty* (2nd ed.). Mineola N.Y: The Foundation Press, Inc.

Girvin, S. (2011). *Carriage of goods by sea* (2nd ed.). Oxford, New York: Oxford University Press.

Gold, E. (1981). *Maritime transport: The evolution of international marine policy and shipping law*. Lexington, MA: Lexington Books.

Gordley, J. (1991). *The philosophical origins of modern contract doctrine, Clarendon law series (Oxford [England].* New York: Clarendon Press; Oxford University Press.
Gorton, L. (1971). *The concept of the common carrier in Anglo-American law*, Scandinavian university books (Gothenburg: Läromedelafërl. (Akad.-förl.) (Vol. 43).
Graveson, R. H. (1948). The Spirit of English Law. *The Juridical Review, 60*, 83–105.
Hall, K., Ely, J. W., & Grossman, J. B. (2005). *The Oxford companion to the Supreme Court of the United States* (2nd ed.). Oxford, New York: Oxford University Press.
Hancock, H. E. (1950). *Wireless at Sea: The first fifty years*. London: Percy Lund, Humphries & Company Limited.
Hausmaninger, H. (1985). Diligentia Quam In Suis: A Standard of Contractual Liability from Ancient Roman to Modern Soviet Law. *Cornell International Law Journal, 18*, 179–202.
Herber, R. (1999). *Seehandelsrecht: Systematische Darstellung*. Berlin, New York: de Gruyter.
Hochhaus, K.-H., & Glandien, H. (2006). *Cool: Reefer-Technik mit Zukunft: Kühlschiffe; Markt, Transport und Perspektive*. Hamburg: Seehafen-Verl.
Holdsworth, W. (1936). *A history of English law* (4th. reprint., Vol. 2). London: Mathuen & Co. Ltd.
Holmes, O. W. (1968). *The common law*. London: MacMillan., In Mark De Wolfe Howe (Ed.).
Honnold, J. O. (1993). Ocean Carriers and Cargo; Clarity and Fairness—Hague or Hamburg? *Journal of Maritime Law and Commerce, 24*, 75–109.
Hughes, A. D. (1999). *Casebook on carriage of goods by sea* (2nd ed.). London: Blackstone.
Hummels, D. (2007, Summer). Transportation cost and International Trade in the second era of globalization. *The Journal of Economic Perspectives, 21*, 131–154.
Ibbetson, D. (2013). The challenges of comparative legal history. *Comparative Legal History, 1*, 1–11.
Jados, S. S. (1975). *Consulate of the sea, and related documents*. University, Ala: University of Alabama Press.
Jowitt, W. A., & Burke, J. (1977). *Jowitt's dictionary of English law* (2nd ed.). London: Sweet & Maxwell.
Karan, H. (2004). *The carrier's liability under international maritime conventions: The Hague, Hague-Visby, and Hamburg rules*. Lewiston, NY [u.a.]: E. Mellen Press.
Kent, J. (1896). *Commentaries on American Law* (Vol. 14). In J.M. Gould (Ed.) Boston: Little, Brown, and Company.
Knauth, A. W. (1953). *The American Law of ocean bills of lading* (4th ed.). American Maritime Cases: Baltimore.
Koh, P. S. K. (1986). *Carriage of goods by sea*. Singapore: Butterworths.
Krieger, K. F. (1970). *Ursprung und Wurzeln der Rôles d'Oléron*, Quellen und Darstellungen zur Hansischen Geschichte, n. F (Vol. 15). Köln: Böhlau.
Longley, H. N. (1967). *Common carriage of Cargo*. Mathew Bender: Albany, N. Y.
Lucas, J. D. (1978). *Cases and materials on admiralty, University casebook series* (2nd ed.). Mineola N.Y: Foundation Press.
Mackenzie, T. M., & Kirkpatrick, J. (1991). *Studies in Roman law: With comparative views of the laws of France, England and Scotland* (7th ed.). Wm. W. Gaunt: Holmes Beach, Fl.
Mangone, G. J. (1997). *United States admiralty law*. Kluwer Law International: The Hague; Boston.
Margetson, N. J., 'Liability of the carrier under the Hague (Visby) Rules for cargo damage caused by unseaworthiness of its containers' (2008) 14, *The Journal of International Maritime Law*, 153–61.
McLintock, Hare A., Wilson J. O., C D. S., COM M., A A. L., Librarian C., Library G. A., Wellington and New Zealand Ministry for Culture, & Heritage Te Manatu Taonga. (2013). *DUNEDIN—Te Ara Encyclopedia of New Zealand*. Retrieved November 18, 2013, from http://www.teara.govt.nz/en/1966/ships-famous/page-6.
Meyer-Termeer, A. J. M. Die Haftung der Schiffer im griechischen und römischen Recht.

Miller, G. (1996). *The legal and economic basis of international trade*. Westport, Conn: Quorum Books.
Nikaki, T. (2010). The Carrier's Duties Under the Rotterdam Rules: Better the Devil You Know? *Tulane Maritime Law Journal, 35*, 1–44.
Payne, W. a. I. E. R. H. (1989). *Carriage of Good by Sea* (13th ed.) E.R. Hardy Ivamy. London: Butterworths.
Peck, D. S. (1998–1999). Economic analysis of the allocation of liability for cargo damage: The case for the carrier, or is it? *Transportation Law Journal, 26*, 73–105.
Rabe, D. (2000). *Seehandelsrecht: Fünftes Buch des Handelsgesetzbuches : mit Nebenvorschriften und Internationalen Übereinkommen* (4th ed.). München: C. H. Beck.
Sanborn, F. R. (1930). *Origins of the early English maritime and commercial law*. New York: The Century Co.
Sánchez Calero, F. (2010).*El contrato de transporte marítimo de mercancías: (Reglas de La Haya-Visby, Hamburgo y Rotterdam)*, Colección Grandes tratados Aranzadi (2nd ed., Vol. 170). Cizur Menor (Navarra): Thomson Reuters-Aranzadi.
Schoenbaum, T. J., & Yiannopoulos, A. N. (1984). *Admiralty and maritime law: Cases and materials, Contemporary legal education series*. Charlottesville, Va: Michie Co.
Schoenbaum, T. J. (2011). *Admiralty and maritime law*, Practitioner treatise series (5th ed.). [St. Paul, MN]: Thomson/West, 2011.
Schomberg, A. C. (1786). *A treatise on the maritime laws of Rhodes : By Alexander C. Schomberg* (Oxford: sold by D. Prince and J. Cooke. J. F. and C. Rivington, P. Elmsly, and T. Payne and Son, London, 1786).
Scrutton, T. E., & Eder, B. (2011). *Scrutton on charterparties and bills of lading* (22nd ed.). London: Sweet & Maxwell.
Sherman, C. P. (1924). *Roman Law in the modern world* (2nd ed., 3 Vols., Vol. 1). New York: Baker, Voorhis & Co.
Simon, S. (1973). Latest Development in the Law of Shipping Containers. *Journal of Maritime Law and Commerce, 4*, 441–454.
Simon, S. (1974). The law of shipping containers: Part I: A look at the factual and legislative background, and a critique of a new judicial formula. *Journal of Maritime Law and Commerce, 5*, 507–532.
Story, J. (1878). *Commentaries on the law of Bailments: with illustrations* (9th ed.). Boston: Little, Brown.
Stott, J. R. (1978). *Refrigeration at sea* (2nd ed.). London: Applied science.
Street, T. A. (1906). *The foundation of legal liability*. A presentation of the theory and development of the common law (3 Vols., Vol. 2). Northport, Long Island, N.Y.: Thompson.
Sturley, M. F. (1993). changing liability rules and marine insurance: conflicting empirical arguments about Hague, Visby, and Hamburg in a Vacuum of Empirical Evidence. *Journal of Maritime Law and Commerce, 24*, 119–149.
Sturley, M. F. (1991). The History of COGSA and the Hague Rules. *Journal of Maritime Law and Commerce, 22*, 1–57.
Sturley, M. F., Fujita, T., & van Ziel, G. J. D. (2010). *The Rotterdam rules: The UN convention on contracts for the international carriage of goods wholly or partly by Sea*. London: Sweet & Maxwell.
Sturley, M. F. (1990). *The legislative history of the carriage of good by Sea Act and the Travaux préparatoires of the Hague Rules* (3 Vols., Vol. 2). Littleton, Colorado.: Rothman.
Sweeney, J. C. (1975–1976). The UNCITRAL Draft Convention on Carriage of Good by Sea (Part I). *Journal of Maritime Law and Commerce, 7*, 69–125.
Tetley, W. (2004). Interpretation and construction of the Hague, Hague/Visby and Hamburg rules. *The Journal of International Maritime Law, 10*, 30–70.
Tetley, W. (2008). *Marine Cargo Claims* (4th ed., 2 Vols., Vol. 1). Cowansville, [Toronto]: Blais; Thomson Carswell.

Tetley, W., McDonough, B. G., & Nixon, E. B. (1988). *Marine cargo claims* (3rd ed.). Montreal, QC, Canada: International Shipping Publications, BLAIS.
The Notes. (1873). Three degrees of negligence, The. *American Law Review, 8*, 649–668.
Torrent, A. (2002). *Manual de derecho privado romano*. Madrid: Edisofer.
United Nations Convention on Contracts for the International Carriage of Goods Wholly or Partly by Sea. Vienna: United Nations, 2009.
Vvan Den Bergh, G. C. J. J. (1975). Custodiam Praestare: Custodia-Liability or Liability for Failing Custodia? *Tijdschrift voor rechtsgeschiedenis—The Legal History Review, 43*, 61–72.
Warren, C. (1999). *History of the Harvard Law School and of early legal conditions in America*. Lawbook Exchange: Union, NJ.
Wheeler, E. P. (1890). *The modern law of carriers: The limitation of the common-law liability of common carriers under the law merchant, statutes and special contracts*. New York: Baker, Voorhis & Co., 1890.
Wilson, J. F. (2010). *Carriage of goods by sea* (7th ed.). New York: Pearson/Longman.
Ziegler, A. v., *Haftungsgrundlage im internationalen Seefrachtrecht: Evolution der Haftungsgrundlage von den Haager Regeln zu den Hamburger Regeln unter Berücksichtigung des nationalen und internationalen Seerechts von Grossbritannien, den Vereinigten Staaten von Amerika, Kanada, Frankreich, Deutschland und der Schweiz* (Zürich, Baden-Baden: Schulthess; Nomos, 2002).
Zimmermann, R. (2002). Europa und das römische Recht. *Archiv für die civilistische Praxis, 202*, 243–316.
Zimmermann, R. (1990). *The law of obligations: Roman foundations of the civilian tradition* (1st ed.). Cape Town: Juta & Co., Ltd.

Reports

Comité Maritime Internatinel 'CMI Yearbook 2013'.
UNCTAD, *Bills of Lading: Report by the Secretariat of UNCTAD* (1971).
UNCTAD, '*Review of maritime transport 2011*'.
UNCTAD, '*Review of maritime transport 2012*'.
World Shipping Council. Retrieved November 04, 2013, from http://www.worldshipping.org/.

IMO Documents

CSS code: Code of safe practice for cargo stowage and securing (3rd ed., 2011 ed.). London: International Maritime Organization, 2011.
General requirements and performance standards for shipborne radiocommunications and navigational equipment: Including related operational recommendations and guidance; incorporating the 2008 Amendments and all amendments adopted up to December 2010, IMO publication (6th. ed., 2011 ed.). London: IMO, 2011.
International Convention for Safe Containers. (2012). *1972: CSC* (5th ed.). London: IMO.
International Maritime Dangerous Goods Code. (2008). *IMDG-Code; incorporating amendment 34-08*. London: International Maritime Organization.
SOLAS: Consolidated text of the International Convention for the Safety of Life at Sea, 1974, and its Protocol of 1988: Articles, annexes, and certificates (5th ed.). London: International Maritime Organization, 2009.
IMO | FAQs. Retrieved May 08, 2014, http://www.imo.org/About/Pages/FAQs.aspx.

MIX
Papier aus verantwortungsvollen Quellen
Paper from responsible sources
FSC® C105338

If you have any concerns about our products,
you can contact us on
ProductSafety@springernature.com

In case Publisher is established outside the EU,
the EU authorized representative is:
Springer Nature Customer Service Center GmbH
Europaplatz 3, 69115 Heidelberg, Germany

Printed by Libri Plureos GmbH
in Hamburg, Germany